CAT

P9-DDQ-449

FTC Orlando Campus

000000620

RELEASE 5

Mechanical Desktop
A Process-Based Approach

Thomas Short
CommandTrain Inc.
Troy, Michigan

Anthony Dudek
A.F. Dudek & Associates
Chicago, Illinois

Florida Technical College Library
12900 Challenger Prkwy,
Orlando, FL 32826

 Autodesk.

Certified Instructor

Autodesk.

Registered Author/Publisher

Publisher

The Goodheart-Willcox Company, Inc.

Tinley Park, Illinois

FLORIDA TECHNICAL COLLEGE LIBRARY
12689 CHALLENGER PARKWAY
SUITE 130
ORLANDO, FL 32826

Learning Mechanical Desktop is not an Autodesk product. Autodesk takes no responsibility with regard to the selection, performance, or use on non-Autodesk products. All understandings, agreements, or warranties must take place directly between Goodheart-Willcox and the prospective user. AUTODESK SPECIFICALLY DISCLAIMS ALL WARRANTIES, EXPRESSED OR IMPLIED, INCLUDING, BUT NOT LIMITED TO, THE IMPLIED WARRANTIES OR MERCHANTABILITY AND FITNESS FOR A PARTICULAR PURPOSE.

The Autodesk logo is registered in the U.S. Patent and Trademark Office by Autodesk, Inc.

Copyright 2001

by

THE GOODHEART-WILLCOX COMPANY, INC.

All rights reserved. No part of this book may be reproduced, stored in a retrieval system, or transmitted in any form or by any means, electronic, mechanical, photocopying, recording, or otherwise, without the prior written permission of The Goodheart-Willcox Company, Inc. Manufactured in the United States of America.

Library of Congress Catalog Card Number 00-067701
International Standard Book Number 1-56637-818-4

1 2 3 4 5 6 7 8 9 10   01   05 04 03 02 01

"The Goodheart-Willcox Company, Inc., and the authors make no warranty or representation whatsoever, either expressed or implied, with respect to any of the software or applications described or referred to herein, their quality, performance, mechantability, or fitness, for a particular purpose. Further, The Goodheart-Willcox Company, Inc., and the authors specifically disclaim any liability whatsoever for direct, indirect, special, incidental, or consequential damages arising out of the use or inability to use the software or applications described or referred herein.
The contents of the Mechanical Desktop software is subject to change between maintenance releases. The material in this book is based on the most recent Mechanical Desktop software release available at the time of publication."

Library of Congress Cataloging-in-Publication Data

Short, Thomas.
  Learning Mechanical Desktop R5 : a process-based approach / by Thomas Short, Anthony Dudek.
    p. cm.
  ISBN 1-56637-818-4
  1. Engineering graphics. 2. Mechanical desktop.
3. Engineering design--Data processing.
  I. Dudek, Anthony. II. Title.
T353.S467 2001
620'.0042'02855369--dc21   00-067701
                                             CIP

Introduction

Learning Mechanical Desktop is a text designed with the student as well as the engineering professional in mind. You will find it is presented in a manner designed to facilitate learning—using practical examples and clear instructions. If you are looking for loads of theory, you will not find it in this text. The intention of this text is to recreate the actual workflow experienced by professionals as they use the software. After all, this text was written by those very same professionals. By the time you have finished this text, you will have a keen understanding of the methods used to produce a viable solid model part or assembly in Mechanical Desktop (MDT).

Parametric design is very important in MDT. You will encounter this a great deal. Throughout this text you will find example after example of parametric design principles. This text cannot stress this enough. The fact that your solid model should not be a static part but a dynamic part, able to withstand revision after revision is of great importance. You will hear this principle echoed throughout this text. The software was designed for this capability, so take advantage of it.

The goal of *Learning Mechanical Desktop* is to present a process-based approach to the MDT commands, options, and techniques. Each topic is presented in a logical sequence where they naturally fit in the design process of real-world products. In addition, this text offers the following features:

- MDT commands are introduced in a step-by-step manner.
- Easily understandable explanations of how and why the commands function as they do.
- Numerous examples and illustrations to reinforce concepts.
- Professional tips explaining how to use MDT effectively and efficiently.
- Practices involving tasks to reinforce chapter topics.
- Chapter tests for reviewing commands and key MDT concepts.
- Chapter exercises to supplement each chapter.

WHAT YOU NEED TO KNOW

This text was written for people with some experience in AutoCAD—either 2D or 3D and either a formal introductory course or practical experience. Since MDT runs "on top" of AutoCAD, the more you know about AutoCAD the easier it will be to concentrate on the ideas and techniques of MDT. When more advanced AutoCAD techniques or commands are used, they will be explained. However, basic techniques, such as drawing a circle, are expected to be understood. It is also expected that the user is familiar with the use of the mouse, keyboard, and basic Windows operations.

FONTS USED IN THIS TEXT

Different type faces are used throughout the chapters to define terms and identify MDT and AutoCAD commands. Important terms always appear in ***bold-italic face, serif*** type. MDT menus, commands, variables, dialog box names, and button names are printed in **bold-face, sans serif** type. Filenames, directory names, paths, and keyboard-entry items appear in the text in Roman, sans serif type. Keyboard keys are shown inside brackets [ ] and appear in Roman, sans serif type. For example, [Enter] means to press the Enter (Return) key.

Prompt sequences are set apart from the body text with space above and below, and appear in Roman, sans serif type. Keyboard entry items in prompts appear in **bold-face, sans serif** type. In prompts, the [Enter] key is represented by the enter symbol (↵).

Introducing Commands

There are several ways to select MDT drawing and editing commands. The format is slightly different in typing commands from the keyboard when compared to selecting commands from the toolbar or pull-down menus. All commands and related options in this text are introduced by providing all the command entry methods.

In many cases, the MDT command entries are shown as if they were typed at the keyboard. This allows you to see the full command name and the prompts that appear on screen. Since you are encouraged to enter commands in the most convenient manner, shortened command aliases are also presented. Commands, options, and values you must enter are given in **bold** text as shown in the following example. Pressing the [Enter] key is indicated with the enter symbol (↵). (Also, refer to the earlier section *Fonts Used in This Text*.)

> Command: **PP** *or* **AMPROFILE**↵
> Select objects for sketch: **END**↵

General input tasks such as picking a point or selecting an object are presented in *italics*.

> Command: **PP** *or* **AMPROFILE**↵
> From point: *(pick circle A)*

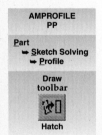

The command line, pull-down menu, and toolbar button entry methods are presented throughout the text. When a command is introduced, these methods are illustrated in the margin next to the text reference. The toolbar in which the button is located is also identified. The example in the margin next to this paragraph illustrates the various methods of initiating the **AMPROFILE** command.

Flexibility in Design

Flexibility is the key word when using ***Learning Mechanical Desktop***. This text is an excellent training aid for individual as well as classroom instruction. ***Learning Mechanical Desktop*** teaches you MDT and its applications to real-world problems. It is also a useful resource for professionals using MDT in the work environment.

Notices

There are a variety of notices you will see throughout the text. These notices consist of technical information, hints, and cautions that will help you develop your Mechanical Desktop skills. The notices that appear in the text are identified by icons and rules around the text. The notices are as follows:

PROFESSIONAL TIP	These are ideas and suggestions aimed at increasing your productivity and enhancing your use of MDT commands and techniques.

NOTE	A note alerts you to important aspects of the command or activity that is being discussed.

CAUTION	A caution alerts you to potential problems if instructions or commands are used incorrectly, or if an action could corrupt or alter files, folders, or disks. If you are in doubt after reading a caution, always consult your instructor or supervisor.

Reinforcement and Evaluation

The chapter examples, practices, tests, and exercises are set up to allow you to select individual or group learning goals. Thus, the structure of *Learning Mechanical Desktop* lends itself to the development of a course devoted entirely to MDT training. *Learning Mechanical Desktop* offers several ways for you to evaluate your performance. Included are:

- **Examples.** The chapters include mini-tutorial examples that offer step-by-step instructions for producing MDT drawings, parts, and/or assemblies. The examples not only introduce topics, but also serve to help reinforce and illustrate principles, concepts, techniques, and commands.
- **Practices.** Each chapter is divided into short sections covering various aspects of MDT. A practice composed of several instructions is found at the end of most sections. These practices help you become acquainted with the commands and techniques just introduced. They emphasize a specific point.
- **Chapter Tests.** Each chapter also includes a written test. Questions may require you to provide the proper command, option, or response to perform a certain task.
- **Chapter Exercises.** A variety of drawing exercises follow each chapter. The exercises are designed to make you think and solve problems. They are used to reinforce the chapter concepts and develop your skills.

The User's Drawing CD

Each chapter consists of examples, practices, and exercises. Most of these activities require a drawing file that has been created and supplied to you on the *User's Drawing* CD. This CD is packaged with the text. At various points throughout this text, you will be instructed to open a file. The file will be used to develop, emphasize, and reinforce MDT concepts and techniques.

The CD is set up so each chapter has a file folder with specific chapter-related files. You may install these files on your hard drive or run them from the CD. Before installing any file to your hard drive, be sure to check with your instructor or network administrator.

About the Authors

The authors have been using the software since its infancy. They can safely say that you have the best of all releases in *Mechanical Desktop Release 5*.

Thomas Short

Thomas Short is a nationally recognized expert in Mechanical Desktop and 3D solid/surface modeling. He is a registered mechanical engineer in Michigan and has his B.S. and M.S. in Mechanical Engineering. He was a faculty member in the Mechanical Engineering department at Kettering University (formerly General Motors Institute). He is also a member of the Society of Manufacturing Engineers and the Society of Automotive Engineers.

Thomas has been using, teaching, and consulting in CAD since 1975 and AutoCAD since 1983. In 1984, Tom founded CommandTrain, Inc., which is a Premier Authorized AutoCAD Training Center. He is a Certified AutoCAD Instructor and Certified Technical Instructor. He has written for and taught many courses in Autodesk software including AutoCAD, AutoCAD 3D, AutoLISP, and Mechanical Desktop. He has taught at every Autodesk University. He has also taught AutoCAD classes in the United States, Mexico, Brazil, Canada, and England.

Thomas is a CAD consultant and trainer for many companies, including Ford Motor Company, General Motors, Visteon, 3M, and McDonnell Douglas. He has helped several tooling and manufacturing companies in the Detroit area implement successful strategies for using Mechanical Desktop as their solid modeling system.

Anthony Dudek

Anthony Dudek has been working with AutoCAD and Mechanical Desktop in the mechanical engineering industry since 1985. He is now a nationally recognized expert in Mechanical Desktop and 3D solid/surface modeling. Anthony is a Certified Autodesk Instructor and Autodesk Training Specialist. For the past several years, he has been a speaker at Autodesk University.

Anthony has been a Mechanical Designer, 3D Modeler, CAD Manager, programmer, and management/networking consultant. For the past 8 years, he has run A.F. Dudek & Associates, which is a CAD consulting firm in the Chicago area. He also is a Mechanical Desktop instructor at Moraine Valley Community College's Premier Authorized AutoCAD Training Center, as well as other venues. He is the author of numerous articles and training materials.

A Note from the Authors

We hope you find this text helpful as you expand your skills into the world of 3D design. It was written with much thought and effort to be as effective a text and guide as we could make it.

NOTICE TO THE USER

This text is designed as a complete entry-level MDT teaching tool. The authors present a typical point of view. Users are encouraged to explore alternative techniques for using and mastering MDT. The authors and publisher accept no responsibility for any loss or damage resulting from the contents of information presented in this text. This text contains the most complete and accurate information that could be obtained from various authoritative sources at the time of production. The publisher cannot assume responsibility for any changes, errors, or omissions.

ACKNOWLEDGMENTS

The authors and publisher would like to thank the following individuals and companies for their assistance and contributions.

Jerry McNaughton, Allied Reaction Systems

Bruce Edmonds; Lamina, Wheeler Boyce, and Brauer

Justin Shimp, Piedmont Virginia Community College

Table of Contents

Fully-Constrained Sketches

Part Mode, Drawing Mode, and Views

Constructing Holes

CHAPTER 6

Dimensions, Equations, and Design Variables

The Sketch Plane and More Features

Work Planes

CHAPTER 12

More Features

CHAPTER 13

Threads, Springs, and Sweeps

CHAPTER 14

Multiple Parts in One File

CHAPTER 15

Building Assemblies

CHAPTER 16

Assembly and Part Analysis

CHAPTER 17

Local and External Parts

CHAPTER 18

Assembly Drawings, Balloons, Parts Lists, and Bill of Materials

CHAPTER 19

Scenes and Assembly Drawings

CHAPTER 1

Getting Started with Mechanical Desktop

Objectives

After completing this chapter, you will be able to:

* Describe the principles and methodology used by Mechanical Desktop to create dynamic models.
* Describe the Mechanical Desktop user interface.
* Understand the layout, function, and components of the different modes.
* Select and use the various command entry methods.
* Use keyboard aliases to change the viewports and viewpoints.

Drawings Needed

The following is a list of files that you will need to work through this chapter. These files can be found on the **User's Drawings** *CD included with this text.*

Examples	Practices	Exercises
Example-01-01.dwg	Practice-01-01.dwg	Exercise-01-01.dwg

INTRODUCTION TO MECHANICAL DESKTOP

Mechanical Desktop (MDT) actually started with AutoCAD R12 as a product called *AutoCAD Designer*. Designer was the first parametric solid modeler to be released by Autodesk. It was great for what it did—piece parts. There was no assembly modeling at all, or surface modeling for that matter. If you wanted to create surface models, you had to turn to another Autodesk product called AutoSurf, which provided NURBS surfaces. If you wanted to do an assembly model in Designer, you just stuck your parts together and called it an assembly. There was no actual relationship between the parts as there is in MDT. Right around AutoCAD R13, Autodesk married Designer and AutoSurf into one product called Mechanical Desktop Release 1. It was a hit and has not slowed down since—garnering a top spot for itself in the market as of this writing.

Although MDT can be used to create surface models, this will not be covered in this book. Surface modeling is on the wane in this industry because solid modeling has been traditionally easier. In addition, solid modeling now is able to model most of the topology found in surface modeling.

THE BASIC UNDERLYING PRINCIPLES

The basic principles that were learned in Designer are applicable today in MDT. Those principles are based on the fact that all 3D parts are made up of *features*. Features are, in turn, made up of *sketch profiles*. Sketch profiles are 2D shapes, closed or not, that are constrained and dimensioned to be considered fully described, or fully *constrained*. Think of 2D detailing—everything has to be dimensioned so that the part can be fully described to the machinist.

Parametric control can be applied at this point by relating dimensions to one another instead of static numeric values. An even higher level of control can be applied by using *design variables*. These are words like *length, width, radius,* etc. They can be referred to in equations, such as LENGTH/2.0. This method provides the highest level of control in MDT. If you want to drive your design with a spreadsheet, you will need to use design variables.

Always remember that your final solid model should be a dynamic model capable of changing shape, in some cases radically, without any errors or failures of features. Applying parametric control to your model does take some additional time. Can the software be used without it? Yes, it can, but if you want *control* over your revisions (all design projects have revisions), then you have to think about relationships between features. For example, if you have a plate with an array of holes in the corners and the length and width of the plate needs to be changed, the holes need to stay in the corners. Parametrics will be useful to ensure the holes stay the proper distance from the edges of the plate. You will learn to apply all of these principles in this text.

There is a wide range of feature modeling commands in the software—fillets, chamfers, holes, arrays, shelling, etc. It includes almost everything you need to model any shape. If you need to model an organic free-curved shape, such as the MS Mouse, then you will need to look into the surface modeling capabilities of the software. This is beyond the scope of this text.

ASSEMBLY MODELING

There are assemblies all around you in the real world. Very rarely will you need to just model a part. In the beginning you will start out by modeling single parts. This is the easiest way to approach learning the software. This text progresses logically from piece parts to assemblies. Oftentimes, you will need to model other parts that will mate with the first in a parametric assembly.

A parametric assembly is one that will update position dynamically when one or more of its parts change shape or position. AutoCAD 3D Solids can model parts, but it has no assembly capability. Often, in AutoCAD 3D Solids you make your parts, move them into position and that is it—you have a static assembly. Change the shape of one part, and the others could care less. The assembly will not update dynamically. Contrast that with MDT's assembly modeler. The shape of a part in an assembly is changed and all others that are affected will adjust position so that they are still assembled in the manner in which you intended.

2D DRAWING LAYOUTS

As you know, 2D drafted views of our 3D parts are still produced in engineering today. If you have ever drafted a difficult view, or even an isometric view, you will appreciate what this software can do. Most of the work involved in creating 2D views is done by the software. Consider it your own personal drafter. You simply tell it which views you want—top, front, side, section, isometric, etc.—and it creates them. The views are tied to the 3D model so that if it changes shape the views update accordingly. There is a large amount of annotation that can be applied to an engineering document: dimensions, surface finish, geometric tolerances, datums, weld symbols, etc. The software manufacturer has been very good over the years of adding this functionality into the product.

BEFORE GETTING STARTED

Before starting MDT, it is a good idea to set the default directory. This is where MDT looks for drawing (dwg) files when you open a file. Put your mouse on the MDT icon on your Windows desktop, right-click, and select **Properties**. Click on the **Shortcut** tab in the dialog box that is displayed. See **Figure 1-1.** The **Start in:** field sets the default drive and folder. This figure shows it set to drive C. The quotes are only required if there are spaces in the folder names. The example, practice, and exercise files can be run directly from the *User's Drawings* CD supplied with the text or can be copied to your hard drive or server.

The **Shortcut key:** field sets a keyboard combination of the [Ctrl] key and the [Alt] key and a single letter to start MDT. Set this combination by clicking on the field, holding down the [Ctrl] and [Alt] keys simultaneously and entering a single letter, in this case M. Click on **OK** and then start MDT.

Starting Mechanical Desktop

In the startup process, MDT displays the **Mechanical Desktop Today** dialog box shown in **Figure 1-2.** From this dialog box you can create new drawings, open a recently used drawing (an existing file), browse for lists of existing drawings, or open the symbol libraries. Click on **Browse** and a list of the dwg files in the default folder is displayed. See **Figure 1-3.**

Find and click on the file Example-01-01.dwg and open it. MDT completes loading and displays this assembly on the screen.

Figure 1-1.
The **Properties** dialog box is used to set the default directory for MDT.

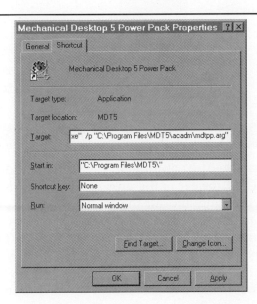

Figure 1-2.
The **Mechanical Desktop Today** is used to start new drawings, open existing drawings, and work with symbols libraries.

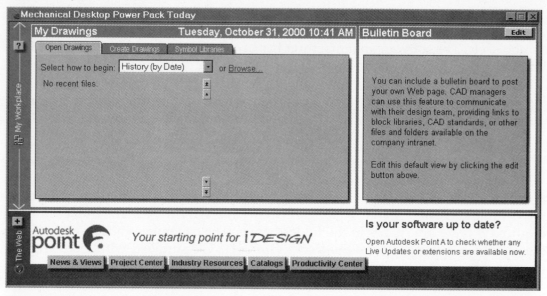

Figure 1-3.
The **Select File** dialog box is used to start new drawings, open existing drawings, and work with symbols libraries.

Enter file to open

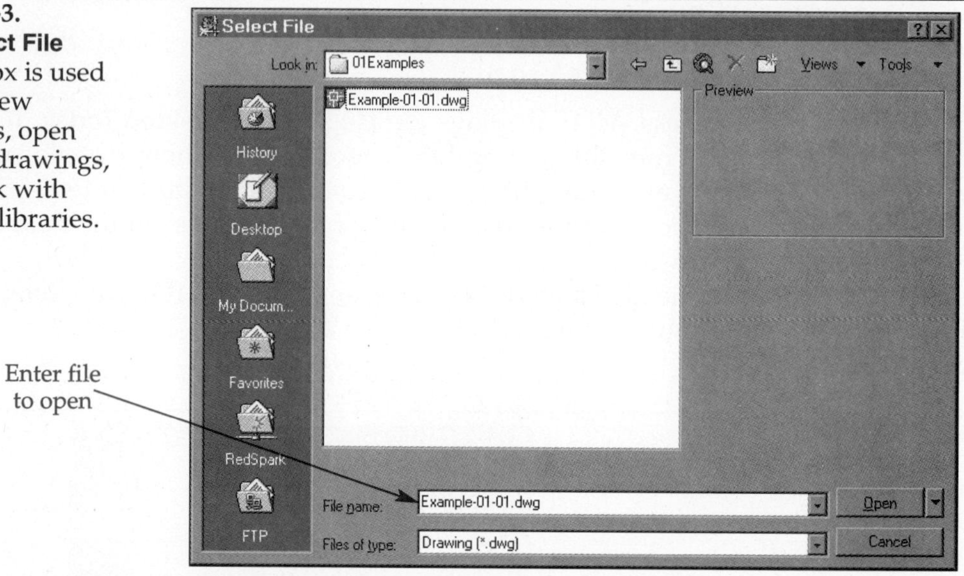

THE MDT USER INTERFACE

With Example-01-01.dwg open, the default user interface can be reviewed. See **Figure-1-4.** The **Browser** is the area on the left side of the screen and it lists all the parts in the drawing and all the features used to create each part. It is extremely useful for changing the geometry and the properties of parts and features. The **Browser** can be expanded by clicking on the plus sign in front of the part name—much as folders are expanded in Windows Explorer. Clicking on the minus sign will collapse the list.

The **Browser** can be made smaller and made to float by doubling-clicking in the gray area above the **Scene** tab. It can then be enlarged and docked by double-clicking in the blue area next to the works **Desktop Browser**. It can also be shut off, but it is so useful and important that this is not recommended.

Figure 1-4.
The basic MDT user interface.

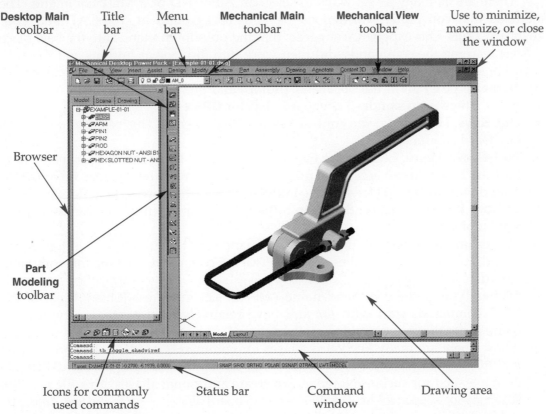

Desktop Main toolbar · Title bar · Menu bar · Mechanical Main toolbar · Mechanical View toolbar · Use to minimize, maximize, or close the window · Browser · Part Modeling toolbar · Icons for commonly used commands · Status bar · Command window · Drawing area

Toolbars and Modes

By default, there are four toolbars that open automatically. These are the **Mechanical Main**, **Mechanical View**, **Desktop Main**, and **Part Modeling**. By default, these toolbars are docked around the drawing area.

MDT has three different modes or screens, which are **Model**, **Scene**, and **Drawing**. There are four different toolbars that open when the mode is changed. These are the **Part Modeling**, **Assembly Modeling**, **Scene**, and **Drawing Layout** toolbars. Selecting a different mode with the *mode selector tabs* changes the mode, as well as displays a different toolbar. Selecting a different toolbar with a button on the **Desktop Main** toolbar displays the toolbar, as well as changes the displayed mode.

The exception is the **Model** mode that uses both the **Part Modeling** and the **Assembly Modeling** toolbars. The **Model** mode is used to construct parts and build assemblies, which is why it has two selections of toolbars. You will spend most of your time in this screen.

Scene mode

Go to the **Scene** mode by clicking on the **Scene** selector tab at the top of the **Browser**. Scenes are used to create exploded views and views with different sets of parts visible and invisible. Note that the display changed to the **Scene** toolbar.

Drawing mode

Go to the **Drawing** mode by clicking on the **Drawing** tab or the **Drawing Layout** button on the **Desktop Main** toolbar. The screen displays 2D orthographic drawings of one of the parts in the assembly and the **Drawing Layout** toolbar is displayed. Much, but not all, of the work required to create these drawings is done automatically. Return to the **Model** mode by selecting the **Model** tab or the **Part Modeling** button on the **Desktop Main** toolbar.

Selecting MDT Commands

There are as many as six ways to select an AutoCAD or a MDT command. The methods that you prefer will depend on your background in AutoCAD and your keyboard skills. Our experience suggests that the pull-down menus are the best place to start learning the commands and then work into the toolbar buttons and the alias keys. For example, to issue the AutoCAD command to open a new drawing you could:

- **Name.** Type OPEN at the Command: prompt and press [Enter]
- **Alias.** There is no standard keyboard alias for **OPEN**.
- **Hot Keys.** Hold down the control key and the letter O. This referred a [Ctrl]+[O] keyboard combination.
- **Pull-Down Menu.** Click on the **File** menu and select **Open**. The underlined letters indicate how you can access the pull-down menus from the keyboard. For **OPEN**, hold down the [Alt]+[F] keyboard combination and then press the letter O.
- **Toolbar Button.** Click on the **Open** button, the second button on the **Mechanical Main** toolbar.

An example of one of the MDT commands that you will use often is the command used to profile sketches—the **AMPROFILE** command. To issue this command you could:

- **Name.** Type in the command name AMPROFILE and press [Enter]. Most of the MDT commands start with AM and have relatively long names, so typing the name is not the recommended method.
- **Alias.** There are many keyboard aliases for commands in MDT. The one for the **AMPROFILE** command is PP. Type in PP and press [Enter]. Editing the MCAD.PGP file with a text editor such as Notepad can create additional aliases. This file is in the \MDT\desktop\support folder. For details on keyboard aliases, refer to the text *AutoCAD and its Applications—Basics, AutoCAD 2000* published by Goodheart-Willcox. A list of some of the default aliases sorted by command name is shown in **Figure 1-5.**
- **Pull-Down Menu.** Click on the **Part** menu and select **Sketch Solving** and select **Profile** from the cascading menu.
- **Right-Click Pop-Up Menu.** Put the cursor in the drawing area, right-click, and a *context-sensitive* pop-up menu is displayed. Pick **Sketch Solving** and then Profile from the cascading submenu. *Context sensitive* means that the menu varies depending on where the cursor is on the screen, what MDT mode you are in, and what command is active.
- **Toolbar Button.** Click on the **Profile a Sketch** button—the fifth one down in the **Part Modeling** toolbar.

Figure 1-5.
The MDT commands and their keyboard aliases.

Command Name	Alias	Command Name	Alias
3DORBIT	OO	AMNEW	N
AMADDCON	JJ	AMPARDIM	II
AMCATALOG	NN	AMPROFILE	PP
AMDELTRAIL	YY	AMREVOLVE	GG
AMDELTWEAKS	KK	AMSKPLN	SS
AMDWGVIEW	Q	AMTRAIL	Y
AMEDITFEAT	LL	AMTWEAK	K
AMEDITVIEW	QQ	AMUPDATE	CC
AMEXTRUDE	G	AMVARS	WW
AMFILLET	EE	AMVISIBLE	VV
AMHOLE	HH	mnu_assm_update	AA

Viewing the MDT Part

There are also many keyboard aliases in the MCAD.PGP file that call up AutoLISP routines. Many of these change the viewports and viewpoints. Some of these are extremely useful and will increase your productivity.

Changing the number of viewports

With Example-01-01.dwg still open, enter the number 4 at the Command: prompt and the screen is divided into four viewports. Displayed are the top, front, right side, and isometric views. A zoom extents is done on each viewport. The aliases that change viewports and their effects are as follows:

- **1.** One view showing the viewpoint in the current viewport.
- **2.** Two views showing the top and isometric.
- **3.** Three views showing the top, front, and isometric.
- **4.** Four views showing the top, front, right side, and isometric.

These same routines are the last four toolbar buttons on the flyout on the **Mechanical View** toolbar as shown in **Figure 1-6.**

Changing the viewpoint

Again with Example-01-01.dwg still open, enter 1 at the Command: prompt to get one viewport. There are nine aliases that change the viewpoint, which is the direction that you are looking at the part. Enter 5 to get the top view and a zoom extents. Enter 8 to get the right front isometric and then enter 88 to get the left front isometric. Aliases and their views are as follows:

- **5.** Top.
- **55.** Bottom.
- **6.** Front.
- **66.** Back.
- **7.** Right side.
- **77.** Left side.
- **8.** Right front isometric.
- **88.** Left front isometric.
- **9.** The sketch view (the same as the plan view in AutoCAD).

These routines are also on toolbar buttons on the flyout on the **Mechanical View** toolbar as shown in **Figure 1-7.** A list of the other aliases for AutoLISP routines and their effects are shown in **Figure 1-8.**

NOTE A complete list of command aliases can be found in the *Appendix.*

Figure 1-6.
Changing the
number of
viewports can also
be done from the
Mechanical View
toolbar.

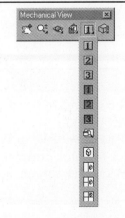

Figure 1-7.
Changing the
viewpoint can also
be done from the
Mechanical View
toolbar.

Figure 1-8.
Additional aliases for AutoLISP routines and their effects.

Command Name	Alias	Effect
MCADBB	BB	Activate an assembly
MCADD	D	Starts the **DVIEW** command
MCADDD	DD	Sets UCS to the view and locates the view depth
MCADF	F	Saves the current view as view 1
MCADFF	FF	Saves the current view as view 2
MCADJ	J	Draws screen position vertical Xlines
MCADUU	UU	Sets the UCS to the view position
MCADW	W	Changes to the drawing mode
MCADX	X	Does a Zoom 0.5X
MCADXX	XX	Does a Zoom 2.0X
MCADZZ	ZZ	Starts real time Zoom
HIDE	0	Hides hidden lines on the part
MCAD[	[	Rotates part 15° CCW about vertical screen axis
MCAD]	]	Rotates part 15° CW about vertical screen axis
MCAD=	=	Rotates part 15° CCW about horizontal axis
MCAD-	-	Rotates part 15° CW about horizontal axis
MCAD5U	5U	Top view and UCS to view
MCAD6U	6U	Front view and UCS to view
MCAD7U	7U	Side view and UCS to view
MCAD5D	5D	Top view, UCS to view, and set Z depth
MCAD6D	6D	Front view, UCS to view, and set Z depth
MCAD7D	7D	Side view, UCS to view, and set Z depth

PRACTICE 1-1

❑ Open Practice-01-01.dwg and expand the part listing in the **Browser** to see how the part was created.
❑ Right-click on MODULEHOUSING in the **Browser** and select **Properties** and **Color** and change the color to color number 40.
❑ Get four viewports on the screen and change the isometric to the left front isometric view.

CREATING A NEW DRAWING FILE

To create a new drawing file, you have two choices. If you click on the **File** pull-down menu, you can select **New...** or **New Part File**. The file created when selecting **New Part File** can only contain one part. Therefore, the screen does not have the **Scene** tab. Its benefit is the simpler menu structure. The file created by **New...** can contain many parts both internal and external. Unless otherwise stated, *create a new file* means use the **New...** option.

Chapter Test

Answer the following questions on a separate sheet of paper.

1. With AutoCAD Release 12 the Mechanical Desktop product was actually called _____.
2. Describe MDT's basic principle dealing with the relationship between features, sketch profiles, and 2D shapes.
3. What does it mean when it is stated that a final solid model should be a dynamic model capable of change without any errors or failures of features?
4. What is a *parametric assembly*?
5. What are the differences between an assembly of AutoCAD 3D Solids and a parametric assembly created by MDT's assembly modeler?
6. Describe how to dock a floating **Browser.**
7. What are the three different modes or screens available in MDT?
8. Describe how to get four views of a part on screen.
9. Describe a quick way to get a right front isometric view of a part.
10. Describe the difference between creating a new drawing file with the **New...** option and with the **New Part File** option.

Chapter Exercises

Exercise 1-1.   Open the drawing Exercise-01-01.dwg and expand the **Browser** to see the steps used in its creation.

Exercise 1-2.   Start a new file using the **New...** option in the **File** pull-down menu. Start a new file by choosing **Start from Scratch** in **Select how to begin:** drop-down box and select **Metric.** Notice the differences in the MDT user interface.

This 4 bar clamp was created and rendered in MDT.

CHAPTER 2

Sketching, Profiling, Extruding, and Revolving

Objectives

After completing this chapter, you will be able to:

* Sketch, profile, and extrude a simple sketch for the base feature.
* Use the **AMREPLAY** command to understand how a part was constructed.
* Access and use the **AMPROFILE**, **AMEXTRUDE**, and **AMREVOLVE** command.
* Apply the six **Termination** selections for the **AMEXTRUDE** command.
* Apply the **Join** and **Cut** operations.
* Use a part edge to close an open profiled sketch.
* Create an offset, parallel work plane as a reference plane for extrusion.
* Use the **Browser** to edit a feature of the part.
* Dynamically rotate the shaded model of the part.
* Construct parts using the **AMEXTRUDE** and **AMREVOLVE** commands.

Drawings Needed

*The following is a list of files that you will need to work through this chapter. These files can be found on the **User's Drawings** CD included with this text.*

Examples	Practices	Exercises
Example-2-01.dwg	Practice-2-01.dwg	Exercise-2-01.dwg
Example-2-01-2D.dwg	Practice-2-01-2D.dwg	Exercise-2-02.dwg
Example-2-01-solid.dwg		Exercise-2-03.dwg
Example-2-02-2D.dwg		Exercise-2-04.dwg
Example-2-02-E.dwg		Exercise-2-05.dwg
Example-2-02-R.dwg		Exercise-2.06.dwg

EXTRUDING SOLIDS

The fundamental process for creating a solid model of a part follows a consistent, straightforward procedure. The first steps are as follows:

1. *Create* a closed 2D shape, called a sketch, using standard AutoCAD drawing commands.
2. *Profile* the shape with the **AMPROFILE** command; MDT attaches some information (called *constraints*) to the sketch.

3. *Extrude* the sketch into a solid (called a *part*).
4. *Repeat* the process to add features to the part.

PROFESSIONAL TIP

The *base sketch,* the very first sketch, should be as simple as possible and yet reflect a basic feature of the part.

For the part shown in **Figure 2-1A**, a good choice for the base feature is the large cylinder on the left end. The sketch is an AutoCAD circle. We are going to build this part from the existing 2D top view in **Figure 2-1B**. Mechanical Desktop (MDT) remembers each step in the building process and can replay the steps one by one. Open the drawing Example-2-01.dwg. To see each step and get a preview of how we are going to build the part, use the **AMREPLAY** command. To access this command, enter AMREPLAY at the Command: prompt, select **Replay** from the **Part** cascading menu in the **Part** pull-down menu, or pick the **Feature Replay** button on the **Part Modeling** toolbar.

All construction starts with a base sketch that is extruded or revolved into a solid part. What can you use for a base sketch? Any closed shape consisting of any combination of 2D AutoCAD objects—lines, arcs, splines, or polylines. One or more circles would also qualify. The shape must close and cannot cross itself as shown in **Figure 2-2**. Any islands in the sketch will be extruded as voids in the part.

AMREPLAY

Part
➥ Part
  ➥ Replay

Part Modeling
toolbar

Feature Replay

Figure 2-1.
A—Shade model of part to be created.
B—The 2D views of the part.

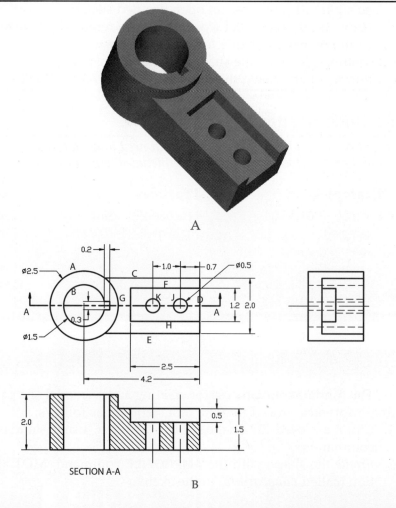

Learning Mechanical Desktop R5

Figure 2-2.
A base sketch can consist of any closed shape that does not cross itself. Any islands in the sketch will be extruded as voids in the part.

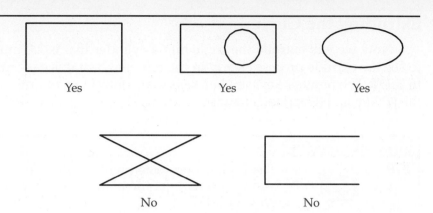

Yes   Yes   Yes

No   No

Profiling the Sketch

MDT requires you to *profile* a sketch with the **AMPROFILE** command before extruding or revolving it into a solid. Profiling does several things:

- Tests the sketch to make sure it closes and has no overlapping lines.
- "Cleans up" the sketch using a set of rules that apply "2D Constraints" to the objects such as making nearly vertical lines vertical. More about these rules in *Chapter(s) 3* and *10*.
- Analyzes the sketch to determine how many additional constraints and dimensions are required to fully define the shape. MDT uses the term "constrain" for define.
- Prepares the profile for extruding or revolving by exploding polylines into individual lines and arcs and yet combining the objects into one profile that can be selected with one pick.

Open the drawing Example-2-01-2D.dwg. Make the 1-top-view layer current and freeze all others, except 1-letters and Am-vis. Now, zoom in on the top view as shown in **Figure 2-3.** Use the large circle at the left end for the base sketch. The first step is to profile this sketch. To access the **AMPROFILE** command, enter AMPROFILE or PP at the Command: prompt, select **Profile** from the **Sketch Solving** cascading menu in the **Part** pull-down menu, or pick the **Profile** button on the **Part Modeling** toolbar.

Once the **AMPROFILE** command is initiated, you are asked to select the object. Select the circle A. The command sequence is as follows:

AMPROFILE
PP

Part
➥ Sketch Solving
   ➥ Profile

Part Modeling
toolbar

Profile

    Select objects for sketch: *(select circle A)*
    Select objects: 1 found
    Select objects: ↵
    Computing…
    Solved under constrained sketch requiring 1 dimensions or constraints.
    Computing…
    Command:

We will cover the problem of Solved under constrained sketch requiring 1 dimensions or constraints. prompt in *Chapter 3*.

Figure 2-3.
The top view of the part with the 1-top-view layer current and all other layers frozen, except 1-letters and Am-vis.

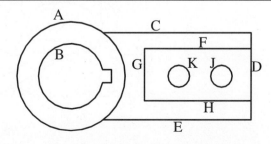

Extruding the Circle

Now we will extrude the circle into a cylinder that is 2.0 units high. Its difficult to see the extrude process clearly in the top view, but it is easy to change viewpoints in MDT. The number keys are *hot keys* that control viewpoints and viewports. Press the [8] key and [Enter] to get an isometric view.

PROFESSIONAL TIP

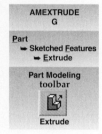

The [5] key and [Enter] will get the top view and the [9] key and [Enter] will get the plan view, which is called the *sketch view* in MDT.

Since there is only one sketch in the drawing, the **AMEXTRUDE** command will automatically work with this *current sketch*. To access this command, enter AMEXTRUDE or G at the Command: prompt, select **Extrude...** from the **Sketched Features** cascading menu in the **Part** pull-down menu, or click the **Extrude** button in the **Part Modeling** toolbar.

The **Extrusion** dialog box does not allow many choices since the circle is the first or *base feature*. Input the 2.0 value as shown in **Figure 2-4A**.

On screen, the isometric view shows an arrow (the default color is blue) indicating the direction of extrusion. The length of the arrow is proportional to the value in the distance field. The **Flip** box will switch the arrow to extrude the sketch up or down. Set the arrow so it points up and press [Enter]. The resulting solid is shown in **Figure 2-4B**.

Note that the **Browser** on the left side of the screen keeps track of every operation. **Figure 2-5** shows the **Browser** and the construction done so far. We will use the **Browser** later to edit the distance of the extrusion.

Figure 2-4.
A—The **Extrusion** dialog box showing settings.
B—The result of the extrusion.

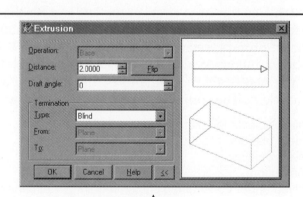

A

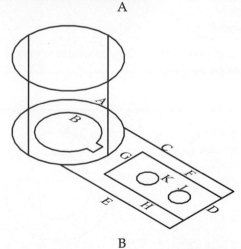

B

Figure 2-5.
The **Browser**
showing the
extrusion.

PROFESSIONAL TIP

If the **Browser** is not on, select **Desktop Browser** from the **Display** cascading menu in the **View** pull-down menu. This toggles the **Browser** on and off.

USING SOLID EDGES TO CLOSE AN OPEN SKETCH

The next step demonstrates the technique of using the edge of the solid model to close an open sketch. Issue the **AMPROFILE** command and select the lines C, D, and E, which are labeled in **Figure 2-4B.**

CAUTION

Line D is actually 3 lines and this will make the sketch much more difficult to change when you want to edit the size of the part. Existing 2D drawings often contain multiple lines or lines on top of lines. In this case, erase the three lines and redraw a single line.

Since they do not form a closed shape and this is not the base sketch, MDT will prompt you to select an edge to close the profile. Select circle A. To perform this profile, the command sequence is as follows:

```
Command: AMPROFILE or PP↵
Select objects for sketch: (select line C)
Select objects for sketch: 1 found (select line D)
Select objects for sketch: 1 found, 2 total (select line E)
Select objects for sketch: 1 found, 3 total
Select objects for sketch: ↵
Select part edge to close the profile: (select circle A)
Select part edge to close the profile: ↵
Solved under constrained sketch requiring 3 dimensions or constraints.
Computing...
```

Now, use the **Amextrude** command to extrude the sketch blind a distance of 1.5 units. Do not forget to change the operation to **Join**. In the **Extrusion** dialog box the previous operation setting, which was **Cut**, is the current until it is changed. In this case, if it is not changed the operation will cut "air" and you will wonder where your solid went. The **Extrusion** dialog box and the settings are shown in **Figure 2-6A.** Make sure the blue arrow is pointing up and then click **OK**. The results of the extrusion are shown in **Figure 2-6B.**

Figure 2-6.
A—The **Extrusion** dialog box showing settings.
B—The result of the extrusion.

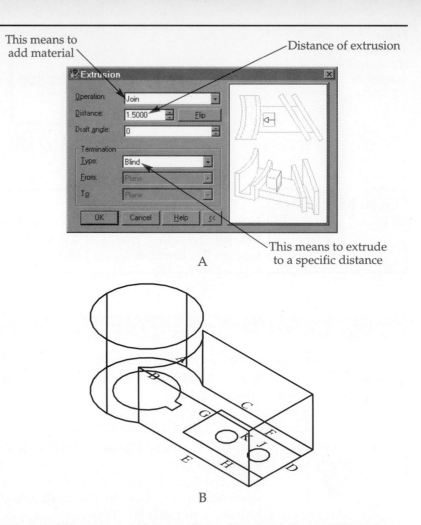

This means to add material

Distance of extrusion

This means to extrude to a specific distance

A

B

CREATING THE INNER CIRCLE AND KEYWAY

The next step is to profile the closed 2D sketch of the inner arc B and the three lines that form the keyway. The four objects can be picked one at a time, by window, or window-crossing. The command sequence is as follows:

Command: **AMPROFILE** or **PP**↵
Select objects for sketch: *(pick arc B)*
Select objects: 1 found
Select objects: *(select the three lines that form the keyway by windowing)*
Other corner: 3 found, 4 total
Select objects: ↵
Solved under constrained sketch requiring 6 dimensions or constraints.
Computing...

Issue the **Amextrude** command. There are now more choices in the **Extrusion** dialog box since this is a feature on the base part. To remove material, select **Cut** in the **Operation** field and select **Through** in the **Termination** field. See **Figure 2-7A**. Make sure the blue arrow is pointing up and then click **OK**. The results are shown in **Figure 2-7B.**

Figure 2-7.
A—The **Extrusion**
dialog box showing
settings.
B—The result of the
extrusion.

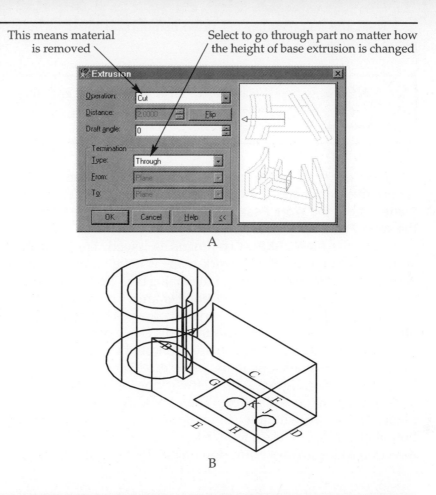

This means material
is removed

Select to go through part no matter how
the height of base extrusion is changed

A

B

EXTRUDING TO A WORK PLANE AND TO A FACE

The sketch formed by lines DHGF needs to be cut away from the surface LMN, which Mechanical Desktop calls a *face*. See **Figure 2-8**. One of the most important choices MDT users must consider is: *"What is the design intent?"* Do you want the bottom of the cut to be a specific distance from face CDE or from face LMN? For this part, assume the cut needs to be 0.5 units deep regardless of the extrusion height of CDE. A reference plane will be placed 0.5 units below LMN. This reference plane is called a *work plane,* which is a tool that allows you to construct features that must reside on a part where there is no face to sketch on or no place to put a feature. It can be used to create an artificial part face for feature construction. If a work plane is moved because of changes in dimensions or constraints, then it and any features attached to it are also moved.

Figure 2-8.
Part showing the
surface LMN.

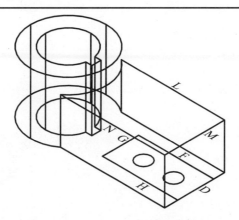

Then, the shape will be extruded using the **From-To** termination section, the work plane, and the face.

> **NOTE**  Work planes have many uses and these will be covered in more detail in *Chapter 8*.

AMWORKPLN

Part
➥ Work Feature
  ➥ Work Plane...

Part Modeling
toolbar

Work Plane

First the work plane must be created using the **AMWORKPLN** command. To access this command, enter **AMWORKPLN** at the Command: prompt, select **Work Plane...** from the **Work Feature** cascading menu in the **Part** pull-down menu, or click the **Work Plane** button on the **Part Modeling** toolbar.

Once the **AMWORKPLN** command is issued, the **Work Plane Feature** dialog box appears. See **Figure 2-9**. The work plane is to be parallel to face LMN, so select **Planar Parallel** in the **1st Modifier** area of the dialog box. The work plane is to be offset from face LMN, so select **Offset** in the **2nd Modifier** area. The offset distance is 0.5, so insert 0.5 in the **Offset:** text box in the lower-right corner of the dialog box. Make sure the **Create Sketch Plan** check box is *not* selected. This check box will be covered in *Chapter 3*.

When you click on **OK**, you are prompted to Select work plane or planar face. Select the plane LMN by picking in the plane or select an edge. MDT will then show an icon of a mouse. See **Figure 2-10**. Left-clicking on the mouse will alternate the plane between the two possibilities. When LMN is highlighted, right-click to accept that plane. The screen will now show the work plane with arrows pointing in the direction of offset. Again, use a left-click to change the direction of the arrows. Change the direction so the arrows point down and right-click to accept.

Figure 2-9.
The **Work Plane Feature** dialog box is used to insert a work plane.

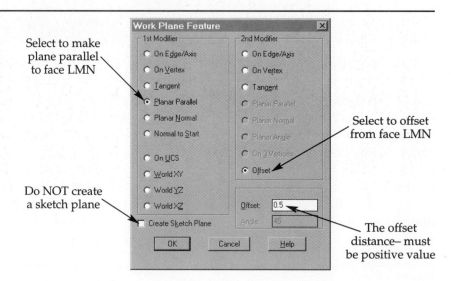

Select to make plane parallel to face LMN

Select to offset from face LMN

Do NOT create a sketch plane

The offset distance— must be positive value

Figure 2-10.
The mouse icon is shown when there are choices to be made on the screen. A left-click cycles through possible choices and a right-click accepts the current choice.

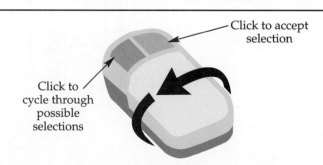

Click to accept selection

Click to cycle through possible selections

PROFESSIONAL TIP

To make work planes look different than part edges, change the color of the AM_WORK layer and set the linetype to PHANTOM2. This is shown in **Figure 2-11**.

To access the **LAYER** command, enter LAYER at the Command: prompt, or select **Layer...** from the **Format** cascading menu in the **Assist** pull-down menu.

This accesses the **Layer Properties Manager** dialog box. See **Figure 2-12**. Click on the color name for AM_WORK and select different color, such as green, in the **Select Color** dialog box. Then click on the linetype name and select PHANTOM2.

Figure 2-11.
The result of inserting a work plane.

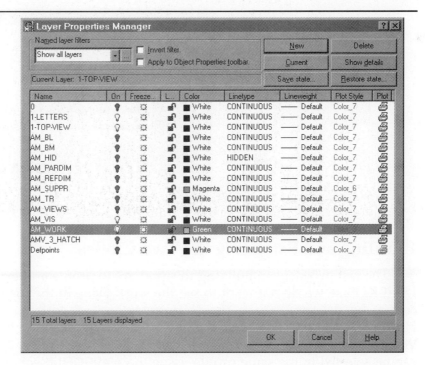

Figure 2-12.
The **Layer Properties Manager** is used to make changes to layers.

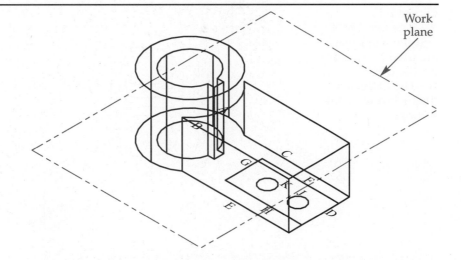

Profiling Lines F, G, and H

Use the **AMPROFILE** command to profile lines F, G, and H. Use edge D to close the profile. See **Figure 2-13A.** Use the **AMEXTRUDE** command to extrude the profile. Use the **Cut** operation and the **From-To** termination type and from **Plane** and **to Face**. See **Figure 2-13B**. Pick the **OK** button and you receive the following command sequence:

Select face or work plane (first termination): *(select the work plane)*
Select face (second termination): *(select the face LMN)*
Enter an option [Next/Accept] <Accept>: ↵
Computing...
Command:

Figure 2-13.
A—The face LMN and the work plane are used to extrude profile DFGH.
B—The **Extrusion** dialog box showing settings.

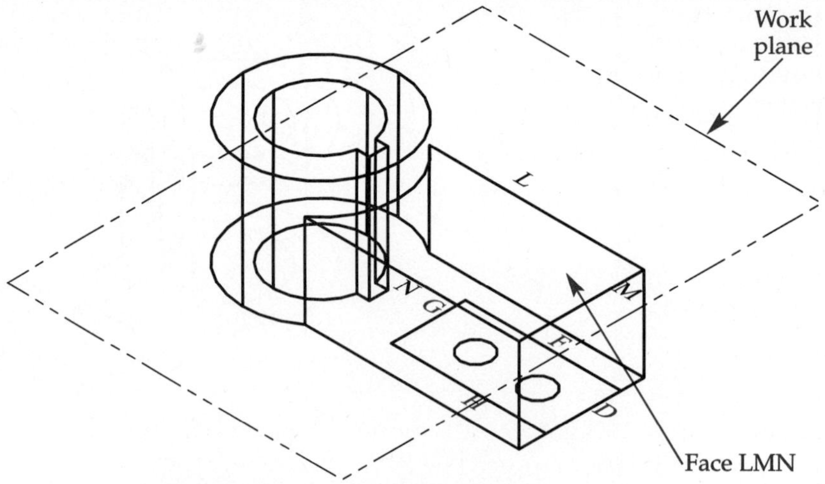

Profiling and Extruding the Two Circles

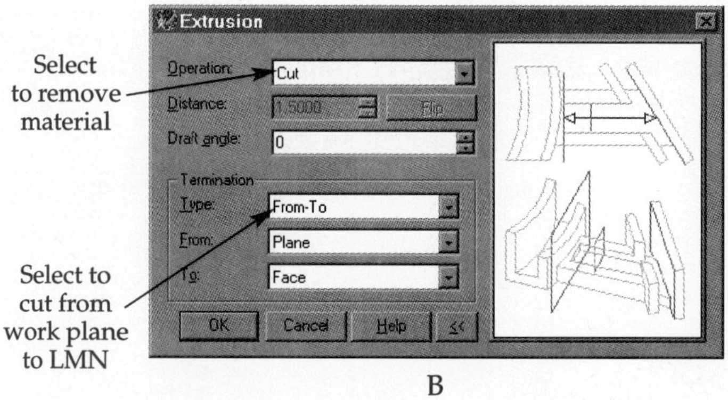

The last two steps in building this part are to profile and extrude the two circles J and K. First, we do not want to see the work plane in the display. The work plane cannot be deleted because the features associated with it will also be deleted. However, it can be made *invisible*. Right-click on WorkPlane1 in the **Browser** and select **Visible**. This toggles between visible and invisible. See **Figure 2-14A.** Circles J and K can be selected in the same profile. Issue the **AMPROFILE** command and select both circles. Issue the **AMEXTRUDE** command, select **Cut** and **Through**, and click on **OK**.

Now view the part with hidden lines removed. First set the **FACETRES** system variable, which controls the quality or number of facets in circles and curves, to 2.

Figure 2-14.
A—The result of the extrusion with the work plane invisible. B—The part with hidden lines removed.

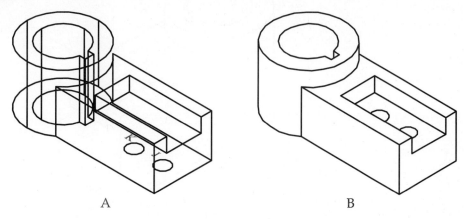

A   B

This value can be set from .01 to 10. Now, set the **DISPSILH** (DISPlay SILHouette) variable, which that controls the silhouette lines on the edges of cylinders and rounds, to 1. For the hide, enter 0 (zero) or enter HIDE at the Command: prompt. See **Figure 2-14B.**

EDITING SIZES OF THE PART

One powerful feature of MDT is the ability to easily change the size and shape of the part. To do so, right-click on the feature you wish to change in the **Browser**. Once the pop-up menu appears, select **Edit**. Note that when you pick on the name of feature in the **Browser**, that feature is highlighted on the screen.

First, change the height of the large cylinder. Put the cursor on ExtrusionBlind1 in the **Browser**, right-click, and select **Edit**. This brings up the **Extrusion** dialog box. Change the value for the **Distance:** to 3.0 and pick **OK** and then press [Enter]. Notice that the change was not made on screen and in the **Browser** ExtrusionBlind1 is highlighted. This means the part needs to be updated. Remember, several features can be edited before doing an update. See **Figure 2-15.**

Figure 2-15.
The change was not made on screen and in ExtrusionBlind1 is highlighted, which means the part needs to be updated.

Indicates that the part needs to be updated

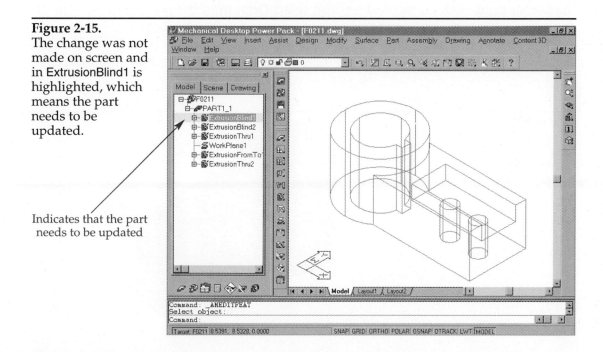

Figure 2-16.
The part after the edit and after it has been updated.

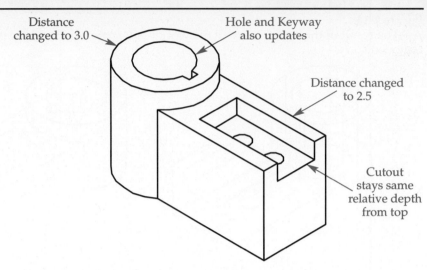

Distance changed to 3.0

Hole and Keyway also updates

Distance changed to 2.5

Cutout stays same relative depth from top

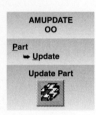

AMUPDATE
OO

Part
→ Update

Update Part

To update the part after editing use the **AMUPDATE** command. To access this command, enter AMUPDATE or OO at the Command: prompt, select **Update** in the **Part** pull-down menu, or select the **Update Part** button—this button is located several places on the screen. Notice that the hole with the keyway updates as well. This is because it was extruded using the **Through** termination option.

Now change the height of the rectangular portion of the part. Right-click on ExtrusionBlind2 in the **Browser**, pick **Edit**, change **Distance:** in the **Extrusion** dialog box to 2.5 and update the part. Note that the rectangular cutout stays the same depth relative to the top surface. See **Figure 2-16.**

VIEWING THE PART AS A SHADED MODEL

Desktop View
toolbar

Toggle Shading/
Wireframe

Desktop View
toolbar

3D Orbit

Now is the time for a little *appreciation.* One of the dramatic features of MDT is the dynamic rotation of a shaded model. MDT lets you view the part as a shaded model and dynamically rotate that shaded model. To shade the model, select the **Toggle Shading/Wireframe** button on the **Desktop View** toolbar. See **Figure 2-17.**

To dynamically rotate the model use the **3D Orbit** button on the **Desktop View** toolbar, or right-click on the desktop and select **Orbit** from the pop-up menu. The **3D Orbit** choice displays five circles and a new cursor. See **Figure 2-18.** The cursor indicates the type of rotation. Move the cursor on either the top or bottom small circle, click and drag, and the part rotates about a horizontal axis. Move it on either the right

Figure 2-17.
The part after using the **Toggle Shading/Wireframe** button.

Figure 2-18.
Screen after **3D Orbit** has been invoked.

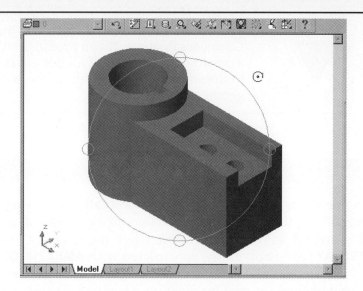

and left small circle, click and drag, and the part rotates about a vertical axis. Move it to any point outside the large circle (shown), click and drag, and the part rotates about its center. Move it to any point inside the large circle, click and drag, and it results in free rotation.

PRACTICE 2-1

❑ Open Practice-2-01.dwg and using **AMREPLAY** review how the part was constructed.

❑ Open Practice-2-01-2D.dwg as shown here. Construct the part from the 2D top view.

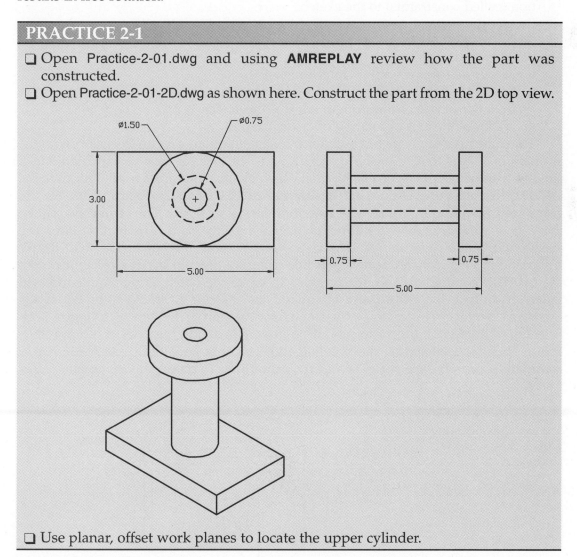

❑ Use planar, offset work planes to locate the upper cylinder.

<table>
<tr><td>
CAUTION</td><td>MDT will only extrude objects that are drawn with a Continuous linetype. If you profile circle A, it will be considered a *construction circle* and will not extrude. Change the linetype to Continuous before profiling. We will work on construction geometry in *Chapter 10*.</td></tr>
</table>

REVOLVING SOLIDS

Some solids are naturals for revolution rather than extrusion. The cap, shown in **Figure 2-19**, with the O-ring groove could be modeled with three circles and one work plane for a **From-To** extrusion. Open the drawing Example-2-02-E.dwg and use **AMREPLAY** to see how it was constructed with **AMEXTRUDE**.

It is easier to build using revolution. This can be reviewed in Example-2-02-R.dwg. The fundamental process for creating a rotational solid model of a part follows the same consistent, straightforward procedure. Remember, the steps are as follows:

1. *Create* a closed 2D shape, called a sketch, using standard AutoCAD drawing commands.
2. *Profile* the shape with the **AMPROFILE** command; MDT attaches some information (called constraints) to the sketch.
3. *Revolve* the sketch into a solid (called a part) with **AMREVOLVE**.
4. *Repeat* the process to add features to the part.

The base sketch must:
- be a closed shape with no interior islands.
- have an axis of rotation; this can be a construction line with a linetype other than Continuous or the straight edge of the sketch.
- not cross the axis of rotation or cross itself.

The same rules apply to the base sketch for a revolution as for an extrusion, with the addition of an axis to revolve the sketch around. The axis can be a construction line, a straight edge of the sketch, or a *work axis*. We will look at examples of all three of these.

The first example is the part in Example-2-02-2D.dwg. See **Figure 2-20**. We are going to revolve the right half of the section view into a part. The centerline will be the axis of rotation and will be included in the sketch as a construction line. The sketch will be the right half of the section view without the hatch. The **BOUNDARY** command is a fast way to create a polyline from existing objects. It works much like the **BHATCH** command. It's not necessary to erase the hatch as the command will ignore it. Initiate the **BOUNDARY** command by entering BOUNDARY or BO at the Command: prompt. This accesses the **Boundary Creation** dialog box. See **Figure 2-21**. Select the **Pick Point** button and pick at point A. The highlighted object is a polyline. Profile the object, including the polyline and the centerline. Now, use **AMREVOLVE** command to revolve the sketch about the centerline. To access this command, enter

Figure 2-19.
The cap with an O-ring groove shown in 3/4 section.

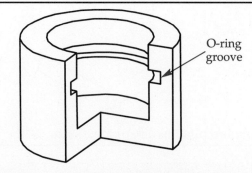

O-ring groove

Figure 2-20.
2D views of the cap. The right half of the section view and the centerline will be used to revolve into a part.

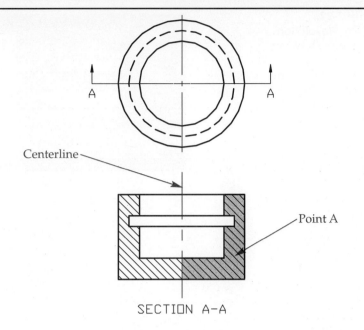

Centerline

Point A

SECTION A-A

Figure 2-21.
The **BOUNDARY** command and **Boundary Creation** dialog box are used to define the right half of the section view as a polyline.

Click and pick point A

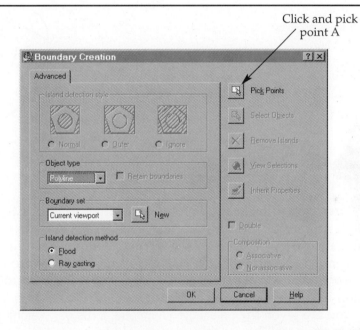

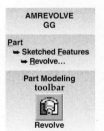

AMREVOLVE or GG at the Command: prompt, select **Revolve...** from the **Sketched Features** cascading menu in the **Part** pull-down menu, or pick the **Revolve** button on the **Part Modeling** toolbar.

When prompted to select a revolution axis, select the centerline. This accesses the Revolution dialog box. See **Figure 2-22A.** Input the information shown in the dialog box. The results of the revolution are shown in **Figure 2-22B.**

Figure 2-22.
A—The **Revolution** dialog box showing settings.
B—The result of the revolution after it has been moved.

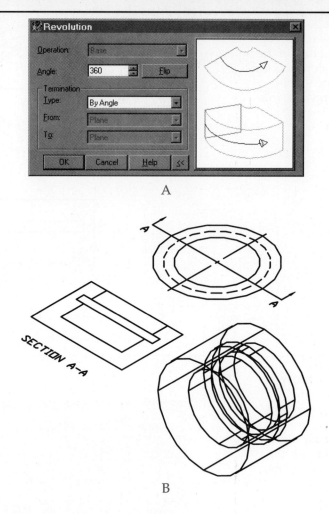

A

B

Adding Features to Revolved Parts

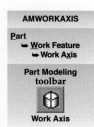

AMWORKAXIS

Part
→ Work Feature
→ Work Axis

Part Modeling
toolbar

Work Axis

First, zoom in on the revolution. Now, to add a feature such as a V-shaped groove on the outside of the part, it is necessary to again create an axis of revolution. In this case we will use the **AMWORKAXIS** command to insert a work axis. To access this command, enter AMWORKAXIS at the Command: prompt, select **Work Axis** from the **Work Feature** cascading menu in the **Part** pull-down menu, or pick the **Work Axis** button in the **Part Modeling** toolbar.

You are given the Select cylinder, cone, torus or [Sketch]: prompt. Pick a cylindrical edge in the part. The outside edge would be a good choice. See **Figure 2-23.**

Go to the view of the sketch plane by entering 9. Sketch a V-shaped polyline near the edge of the part. See **Figure 2-24A.** Now, profile the sketch of the "V" and close with the edge of the cylinder. To use the edge of cylinder to close a sketch, the Dispsilh system variable must be set to 1. Revolve this profile around the work axis with a Full termination and Cut operation. The results are shown in **Figure 2-24B.**

Figure 2-23.
Inserting a work axis.

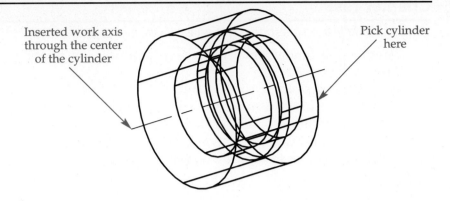

Inserted work axis through the center of the cylinder

Pick cylinder here

Figure 2-24.
A—Sketch a V-shaped polyline, profile the sketch using the cylinder edge to close the polyline, and revolve the profile around the work axis. B—The result of the revolution.

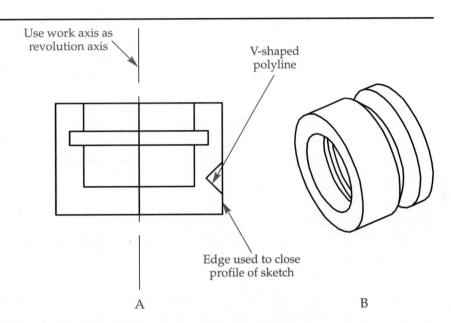

Use work axis as revolution axis

V-shaped polyline

Edge used to close profile of sketch

A

B

PRACTICE 2-2

❏ Start a new drawing and call it Practice-2-02. Use English or Metric units, whichever you prefer. The numbers are easy integers in both systems. Construct the part shown here.

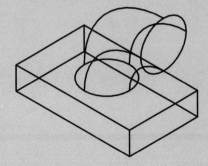

❏ Draw a rectangle 5″ by 3″ (125 by 75 mm).
❏ Profile the rectangle and extrude it 1″ (25 mm) down.
❏ Draw a circle at the center of the rectangle of diameter 2″ (50 mm).
❏ Profile the circle.
❏ Revolve it 90° about the short edge on the right side.

Chapter Test

Answer the following questions on a separate sheet of paper.

1. Describe the fundamental process (procedure) for creating a solid model of a part.
2. What is a *base sketch*?
3. What can be used to make up a base sketch? If any, what are the parameters and/or restrictions for a base sketch?
4. What command is used to review the steps used to build the part?
5. Match the *hot keys* to the viewpoints and viewports.
   a. An isometric view.   [9] key
   b. The top view.   [5] key
   c. The sketch view.   [8] key
6. In an on-screen isometric view during an extrusion, what does the blue arrow indicate?
7. What is a *work plane*?
8. After a work plane is used and no longer needs to be seen, should it be deleted? Why or why not?
9. Describe the process of changing the size and shape of the part.
10. Describe the process of changing the part to a shaded model and then dynamically rotating that shaded model.
11. The same rules apply to the base sketch for a revolution as for an extrusion, with the addition of what?

Chapter Exercises

There is a solid part for all exercises with the name Part-Ex-2-XX.dwg that you can open and AMREPLAY to review one way of constructing the part. These drawings are found on the User's Drawings CD.

Exercise 2-1.   Open the drawing Exercise-2-01.dwg and extrude the part from the top view.

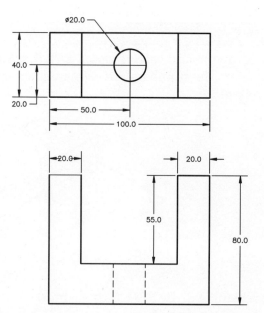

Exercise 2-2. Open the drawing Exercise-2-02.dwg and extrude the part from the top view.

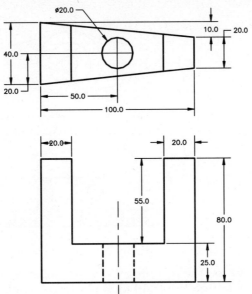

Exercise 2-3. Open the drawing Exercise-2-03.dwg and construct the part from the top view. Use **On Edge** and **Planar Angle** of 45° to define the work plane.

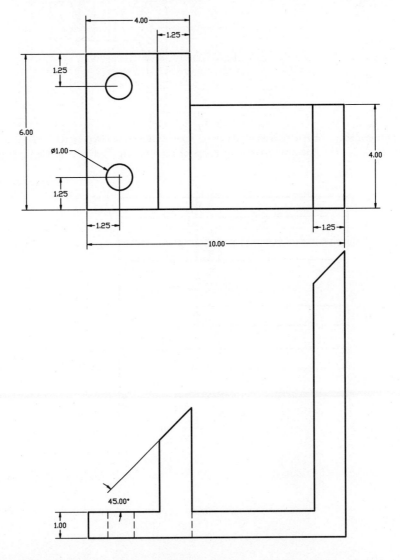

Exercise 2-4. Open the drawing Exercise-2-04.dwg and construct the part from the top view. Use a draft angle of 15° on the tapered hole.

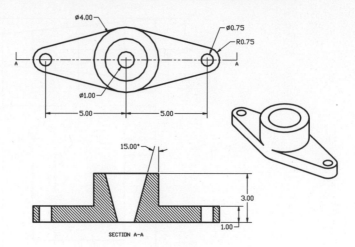

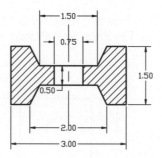

Exercise 2-5. Open the drawing Exercise-2-05.dwg and construct the steel wheel by rotating the right side about the centerline.

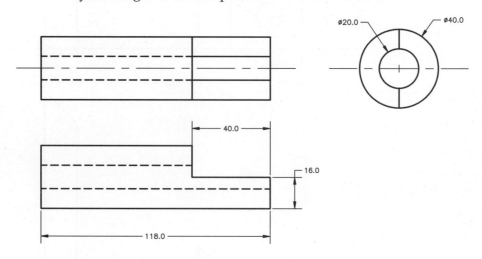

Exercise 2-6. Open the drawing Exercise-2-06.dwg and construct the stepped shaft by rotating the two shapes, one 360° and the other 180°.

CHAPTER 3

Fully-Constrained Sketches

Objectives

After completing this chapter, you will be able to:

* Understand the need for and definition of a fully-constrained sketch.
* Display the constraints on a profiled sketch.
* Add parametric dimensions to a simple sketch to fully constrain it.
* Change the value of a parametric dimension on a sketch.
* Understand the function of the fixed point.
* Add dimensions to the base sketch for an extruded part that was not fully constrained.
* Estimate the dimensions required to fully constrain a sketch.
* Apply the concentric constraint to circular features.
* Apply the tangent constraint to circular and linear features.
* Remove existing constraints on a profiled sketch.

Drawings Needed

*The following is a list of files that you will need to work through this chapter. These files can be found on the **User's Drawings** CD included with this text.*

Examples	Practices	Exercises
Example-3-01.dwg	Practice-3-01.dwg	Exercise-3-01.dwg
Example-3-02.dwg	Practice-3-02.dwg	Exercise-3-03.dwg
Example-3.03.dwg	Practice-3-03.dwg	Exercise-3-04.dwg
Example-3-04.dwg		Exercise-3-05.dwg
Example-3-05.dwg		Exercise-3-06.dwg
		Exercise-3-07.dwg
		Exercise-3-08.dwg

THE FULLY-CONSTRAINED SKETCH

To create a solid model of a part, an additional step is added to the fundamental process you learned in *Chapter 2*. That step (step 3) is to fully constrain the sketch. The process is now as follows:

1. *Create* a closed 2D shape, called a *sketch*, using standard AutoCAD drawing commands.

2. *Profile* the shape with the **AMPROFILE** command; MDT attaches some information to the sketch. This information is called geometric or *2D constraints*.
3. *Fully constrain* the sketch with parametric dimensions and additional 2D constraints.
4. *Extrude* the sketch into a solid, called a *part*.
5. *Repeat* the process to add features to the part.

Taking the time and effort to fully constrain a sketch is worthwhile, because any dimension in the extruded solid part can be changed at any time in the design process. Also, the parametric dimensions you add in the sketch will automatically appear in the 2D drawings that you generated from the solid model. The dimensions can be modified either in the model or the 2D drawings and both will be updated (changed).

Since the required number of dimensions depends upon the complexity of the sketch, the very first sketch (the base sketch) should be as simple as possible and yet reflect a basic feature of the part. If you draw the rectangle shown in **Figure 3-1A** and then profile it, you receive the following prompt:

> Solved underconstrained sketch requiring 2 dimensions or constraints.

If we add two dimensions, the sketch will be *fully constrained*. Add the length AD and the height AB as shown in **Figure 3-1B** and you will receive the following prompt:

> Solved fully constrained sketch.

This means that there is enough information, dimensions, and 2D constraints about this sketch to fully define its geometry. If you add an additional dimension, such as side CD, it will over constrain the sketch and the dimension will be automatically erased. MDT is acting as your checker. Think of the number of dimensions required to define a two dimensional drawing and this will generally be the same as the dimensions required to fully constrain a sketch.

Refer to the triangle in **Figure 3-2A**. This triangle requires two dimensions to fully define its geometry. The dimensions either are the height and length, or one angle and the height or the length. A circle, such as **Figure 3-2B**, requires the diameter only. How many dimensions are required for the shape in **Figure 3-2C**? The answer is six—three vertical dimensions and three horizontal dimensions.

Figure 3-1.
A—This rectangle requires two dimensions to be fully constrained.
B—Shown are the dimensions that make the rectangle a fully-constrained sketch.

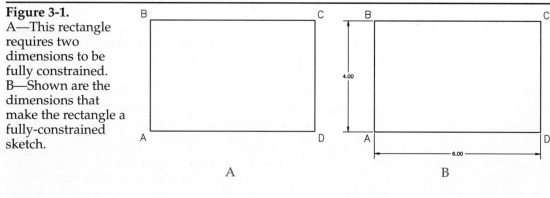

Figure 3-2.
A and B show sketches that are fully constrained. How many dimensions are required for the shape in C?

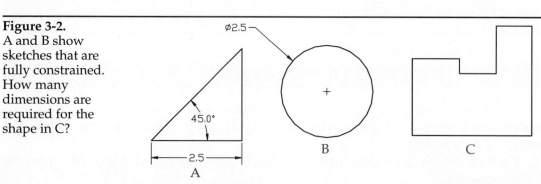

❏ Open drawing Practice-3-01.dwg.
❏ Profile each sketch to check the required number of dimensions. These are the same shapes as in **Figure 3-2**.

2D constraints

MDT automatically applies *2D constraints* to a sketch when it is profiled. *Constraints* are geometric properties of the objects in the sketch. The most fundamental constraints are the orientation of lines with respect to the current User Coordinate System (UCS). MDT will search for lines in the sketch that are parallel to the X axis and mark them as **Horizontal**, and lines that are parallel to the Y axis and mark them **Vertical**.

Open the drawing Example-3-01.dwg. This drawing was shown in **Figure 3-1**. Two dimensions and some information about its *squareness* are required to completely describe the rectangle. In a standard 2D drawing, this may be specified by geometric dimension and tolerance (GD&T) symbols. MDT controls the squareness by applying a horizontal constraint to BC and AD and a vertical constraint to AB and CD. Each edge of the sketch is numbered starting with zero and the symbols **H** (horizontal) and **V** (vertical) are put next to the number. MDT also fixes one corner, called a vertex, of the sketch in the coordinate system. The letter **F** indicates the fixed vertex and if a dimension is changed, the edges move relative to the fixed point.

Setting the constraint size

To set the constraint size, select **Part Options** as the last option in the **Part** pull-down menu. This displays the **Mechanical Options** dialog box and the **Part** tab. In this dialog box select the **Constraint Size...** button to access the **Constraint Display Size** dialog box. See **Figure 3-3**. Use the scroll bar to increase or decrease the size.

Figure 3-3.
The constraint display size can be changed using **Constraint Display Size** dialog box.

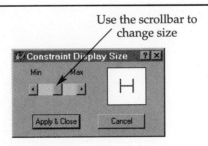

Use the scrollbar to change size

Displaying the constraints

To display the constraint symbols temporarily on the sketch, use the **AMSHOWCON** command. To access this command, enter AMSHOWCON at the Command: prompt, select **Show Constraints** from the **2D Constraints** cascading menu in the **Part** pull-down menu, or pick the **Show Constraints** button on the **Part Modeling** toolbar. If you enter this command at the command line, you will receive the following prompt:

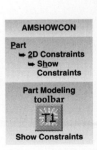

Enter an option [All/Select/Next/eXit] <eXit>: *(enter A to show all constraints)*
Enter an option [All/Select/Next/eXit] <eXit>: *(press [Enter] to exit command)*

The *edge numbers* are displayed in the circles. See **Figure 3-4**. They always start with zero on every sketch, whether it is the base sketch or a feature sketch. The **H** and **V** symbols are shown next to and in the same orientation as the edge numbers and the **F** symbol is in the lower-left corner.

Figure 3-4.
A rectangle with the
constraints showing.

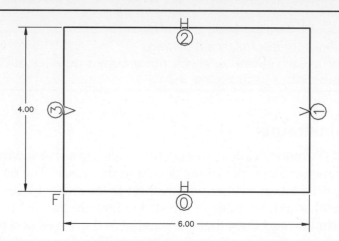

ADDING DIMENSIONS TO THE SKETCH

DIMSTYLE

Assist
→ Format
  → Dimension
    Style...

Use the standard AutoCAD dimension style command **DIMSTYLE** to set the dimension parameters, such as size and precision. To access this command, enter DIMSTYLE at the Command: prompt or select **Dimension Style...** from the **Format** cascading menu in the **Assist** pull-down menu. This accesses the **Dimension Style Manager** dialog box. See **Figure 3-5**.

Figure 3-5.
The **Dimension Style Manager** is used to set the dimension parameters, such as size and precision.

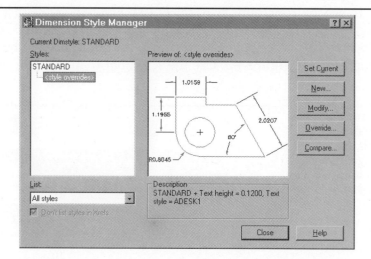

> **CAUTION**
>
> Do not use the AutoCAD dimension command to place dimensions on a sketch after it has been profiled. MDT will not recognize these dimensions as parametric dimensions and they will have no effect on the required constraints.

There are three methods of placing parametric dimensions on a profiled sketch. You can use new dimensioning, power dimensioning, or automatic dimensioning.

New Dimensioning

New dimensioning is done with the **AMPARDIM** command. To access this command, enter AMPARDIM or II at the Command: prompt, select **New Dimension** from the **Dimensioning** cascading menu in the **Part** pull-down menu, or pick the **New Dimension** button on the **Part Modeling** toolbar. The **AMPARDIM** command also displays the constraints and edge numbers on the sketch.

Now open the drawing Example-3-02.dwg. **Figure 3-6** shows the constraints automatically applied by MDT. The profiled sketch has 10 edges and it requires an additional 10 dimensions or constraints to be fully constrained.

If there is a continuous edge to be dimensioned, such as edge number 7 labeled AB, the technique is to pick anywhere on the edge. Then pick a point where you want the dimension placed. The command sequence for adding a vertical dimension to edge AB is as follows:

> Command: **AMPARDIM** *or* **II**↵
> Select first object: *(pick line AB)*
> Select second object or place dimension: *(pick the placement of dimension)*
> Enter dimension value or [Undo/Hor/Ver/Align/Par/aNgle/Ord/Diameter/pLace]
>   <1.5000>: **1.5**↵
> Solved underconstrained sketch requiring 9 dimensions or constraints.
> Select first object: ↵
> Command:

If there is no continuous edge, such as distance DS, make two picks. The first pick should be at D and the second at S. Then pick a point where you want the dimension placed. When 2D views are automatically created from the model, the parametric dimensions are included exactly as you created and placed them. Some care and forethought will make cleanup and manipulation of the 2D faster. If you can think that far ahead, put the dimensions where you will want them in the drawings. If not, do not worry. Changing dimension locations and styles in the drawings is easy to do.

Do *not* use osnaps to pick the dimension points. MDT automatically uses the end points of lines and arcs, and the centers of arcs and circles. When you place the dimension, the value is displayed both on the drawing and on the command line. You can then enter a different value for the dimension and both the dimension and the part will change. This is parametric dimensioning. The parametric dimensions are automatically placed on layer AM_PARDIM and the default color is green.

Figure 3-6.
The sketch with the constraints shown.

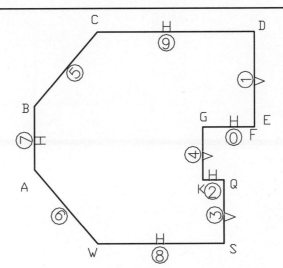

PROFESSIONAL TIP

The following are tips for dimensioning:

- Construct the base sketch accurately or at least close to accurate. Large changes in dimensions may distort the model so that further work is impossible.
- Input the overall dimension first. This will help to *lock* the model into shape and prevent major distortions.
- Keep the base sketch as simple as possible; a maximum of 6 or 7 constraints. **Figure 3-6** is too complex with 10 constraints. Some of the geometry would be better put in later as *features* such as chamfers.
- After inputting a dimension, press [Enter] to leave the **AMPARDIM** command. Then, press [Enter] again to reissue it. If you make a drastic mistake, you can use the **UNDO** command to undo one dimension and not lose all the previous dimensions.
- If you get the wrong type of dimension, for example you want Aligned and you get Parallel, you can change it by selecting the correct letter from the choices at the command line.

PRACTICE 3-2

- ❏ Open the drawing Practice-3-02.dwg.
- ❏ Profile the sketch and display the constraints.
- ❏ Modify the dimension style to display two places after the decimal.
- ❏ Input the 10 dimensions and set their values as shown here. This should result in a fully-constrained sketch.

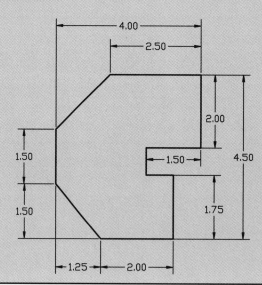

Power Dimensioning

AMPOWERDIM

Part
→ Dimensioning
  → Power
    Dimensioning

Part Modeling
toolbar

Power Dimensioning

Open Example-3-02.dwg if it is not already opened. Now insert dimensions using power dimensioning. This is done with the **AMPOWERDIM** command. To access this command, enter AMPOWERDIM at the Command: prompt, select **Power Dimensioning** from the **Dimensioning** cascading menu in the **Part** pull-down menu, or pick the **Power Dimensioning** button on the **Part Modeling** toolbar.

The object selection process requires picking endpoints with osnaps or pressing [Enter] to pick an existing edge. Pick the endpoints of line CD and place the dimension. When the dimension turns red, it is at the proper standard distance from the part. Click at that location to place the dimension. See **Figure 3-7**.

The **Power Dimensioning** dialog box is now displayed. See **Figure 3-8**. With this box you can do the following:
- Change the value of the dimension.
- Change the number of places after the decimal for this one dimension.
- Add tolerances and fits to this one dimension.
- Place text and symbols before or after the dimension.

The first of these is most useful in the sketching process. The others are useful for creating the detail drawings. The dimension value can be entered or edited in the **Expression** field.

Figure 3-7.
The fully-constrained sketch with 10 dimensions.

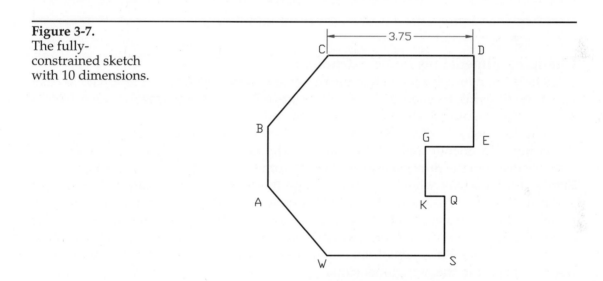

Figure 3-8.
With the **Power Dimension** dialog box, you can change the value of the dimension, change the number of places after the decimal for this one dimension, add tolerances and fits to this one dimension, and place text and symbols before or after the dimension.

Use this area to place text with dimension

Use to change value of dimension

Pick to add tolerances

Pick to add fits

Pick to change the number of decimal places

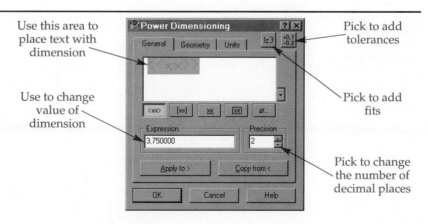

AMMODDIM

Part
→ Dimensioning
→ Edit
   Dimension

Part Modeling
toolbar

Edit Dimension

Changing dimensions on the sketch

When dimensioning the sketch, either the base or a feature, any parametric dimension can be erased and then reentered. The value can also be changed by selecting the modify dimension or **AMMODDIM** command. To access this command, enter AMMODDIM at the Command: prompt, select **Edit Dimension** from the **Dimensioning** cascading menu in the **Part** pull-down menu, or pick the **Edit Dimension** button on the **Part Modeling** toolbar. The command sequence for changing the line CD dimension is as follows:

> Command: **AMMODDIM**⏎
> Select dimension to change: *(pick the line CD dimension)*
> New value for dimension <3.750000>: **3.5**⏎
> Solved under constrained sketch requiring 9 dimensions or constraints.
> Select dimension to change: ⏎

The dimension is changed and the part is updated automatically.

PROFESSIONAL TIP When editing dimensions, the fastest way to change a value of a dimension is to double-click on the dimension that brings up the **Power Dimensioning** dialog box.

Changing dimensions on an extruded or revolved sketch

There are two ways to change the dimension of an extruded or revolved sketch. Open the drawing Example-3-03.dwg. See **Figure 3-9**. This part was extruded from a base sketch (the outside) and one feature (the hole).

The first way to change the dimension will only let you change an existing dimension. To change one of the outside dimensions, right-click the mouse on ExtrusionBlind1 in the **Browser** and select **Edit** from the pop-up menu. This displays the **Extrusion** dialog box where you can change the termination, distance, draft angle, and direction of extrusion. Click the **OK** button to show the existing dimensions of the sketch and the extrusion on the 3D part. See **Figure 3-10**. You are prompted to select an object. Select any dimension and enter in a new value. Change as many dimensions as you wish. Press [Enter] when you are done. Use the **AMUPDATE** command to update the part to the new dimensions.

Figure 3-9.
This part was extruded from a base sketch (the outside) and one feature (the hole).

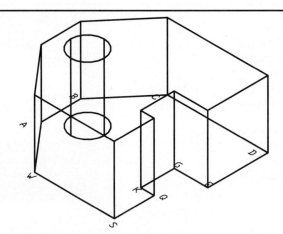

Figure 3-10.
Click the **OK** button in **Extrusion** dialog box to show the existing dimensions of the sketch and the extrusion on the 3D part.

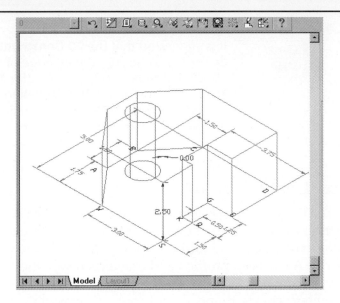

The second way is to select **Edit Sketch** from the pop-up menu. This will display only the sketch on the screen. Dimensions may be edited or erased and reentered. Power dimensioning can be used. For sketches that were not fully constrained, additional dimensions and constraints can be added.

The fix point

Every base profiled sketch has one fix point that is either located at the end of a straight edge or the center of an arc. The *fix point* locates the sketch in XY plane of the UCS coordinate system and is a 2D constraint temporarily displayed by a letter "F." When you change the value of a parametric dimension of the sketch, the fixed point does not move. It remains *fixed* in the coordinate system. The rest of the edges move relative to its position. Normally the fixed point does not have to be moved. But sometimes adding a dimension to the sketch will distort the sketch in a way you did not intend. Moving the fixed point may reduce distortion. Adding additional fixed points with the fix point constraint will be discussed in Chapter 10.

USING CONSTRAINTS TO CONSTRAIN GEOMETRY

There are 15 constraints that can be applied to geometry of the sketch. In many cases, they can be used instead of parametric dimensions. For example, if the centers of two concentric circles were dimensioned with respect to one another, the resulting dimension would be zero. A zero dimension can be used in the sketch. However, when drawings are automatically generated, you have to hide the zero dimensions to keep the drawings to standards. One of the 15 constraints will make two circles *concentric*, and is very easy to use. Concentric and tangent constraints are covered in this chapter. The rest are covered in Chapter 10.

To access the concentric and tangent constraints, select constraints, enter AMADDCON or JJ at the Command: prompt, and then enter CN (concentric) or T (tangent) when prompted for an option. You can also access these constraints by selecting **Concentric** or **Tangent** from the **2D Constraints** cascading menu in the **Part** pull-down menu, or pick the **Concentric** or **Tangent** button from the **2D Constraints** toolbar.

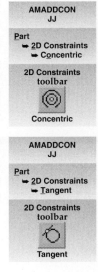

AMADDCON
JJ

Part
→ 2D Constraints
→ Concentric

2D Constraints
toolbar

Concentric

AMADDCON
JJ

Part
→ 2D Constraints
→ Tangent

2D Constraints
toolbar

Tangent

PROFESSIONAL TIP

It is suggested that the **2D Constraints** toolbar be launched and docked at some convenient edge of your screen. To launch the **2D Constraints** toolbar, pick the **Launches 2D Constraints Toolbar** button on the **Part Modeling** toolbar. This will also help in constructing sketches, since the last button toggles this toolbar to the **2D Sketch** toolbar.

Concentric Constraints

Open the part Example-3-04.dwg as shown in **Figure 3-11.** This will demonstrate the *concentric constraint.* The part is to be extruded from three circles—the base from circle A, the hub from circle B, and the hole from circle C. All three circles are in the same plane.

Circle A has been profiled and one dimension, the 4.0 diameter, has been added to fully constrain the sketch. The fix point constraint is at the center of the circle. At the moment, circles B and C and their dimensions are simply AutoCAD objects. Before we can profile circle B and relate it to circle A, circle A must be extruded using the **AMEXTRUDE** command. Extrude circle A downward 2.0 units. See **Figure 3-12.**

Once circle A has been extruded, profile circle B by issuing the **AMPROFILE** command and selecting both circle B and the 3.0 dimension. The dimension is selected because regular AutoCAD dimensions placed on the sketch before it is profiled can be used as parametric dimensions. This will result in a sketch requiring 2 more constraints—the X and Y coordinates of the circle with respect to the circular part. Even though A and B were originally drawn concentric, this relationship is not recognized. You must input the constraint.

To add the concentric constraint, use the **Concentric** option of the **AMADDCON** command. You are prompted to select the object to be reoriented—select circle B. Now you are prompted to select the object to be made concentric to—think of this as the reference object. Select either of the circular edges of the part (either the top or bottom). The order of this selection is important—circle B must be selected first.

Figure 3-11.
This sketch will be used to demonstrate concentric constraints.

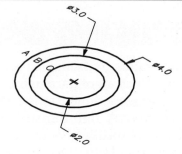

Figure 3-12.
Circle A has been profiled and extruded.

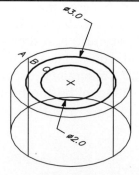

Figure 3-13.
Circle B has been
made concentric to
circle A, profiled,
and extruded.

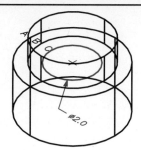

Figure 3-14.
Circle C has been
made concentric to
circle A, profiled,
and extruded.

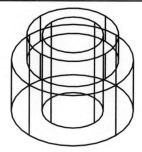

Circle B will be fully constrained. Now use **AMEXTRUDE** to extrude profile B **Join**, **Blind**, and up 1.0 unit. See **Figure 3-13.**

Repeat this process for circle C. It is good practice to constrain C to the base feature—the outside edge A of the cylinder. Now use **AMEXTRUDE** to extrude profile C **Cut** and **Mid-Through**. See **Figure 3-14.**

Tangent Constraints

Open the drawing Example-3-05.dwg as shown in **Figure 3-15A.** The completed part shown in **Figure 3-15B** demonstrates the tangent constraint. The 3.0 unit cylinder is tangent to three of the edges of the rectangular prism. To construct this part, the rectangle was drawn, profiled, and extruded. Then the circle for the cylinder was drawn. This circle can now be fully constrained with no dimensions by making it tangent to the three sides.

First profile the circle. To add the tangent constraint, use the **Tangent** option of the **AMADDCON** command. Select the circle and then one of the sides of the rectangle. Order is important—the circle must be selected first. Repeat this procedure for the other two sides. Use **AMEXTRUDE** to extrude the circle **Join**, **Blind**, and up 4.0 units. See **Figure 3-16.**

Figure 3-15.
Using the sketch in A, tangent constraint will be demonstrated and used to make the part shown in B.

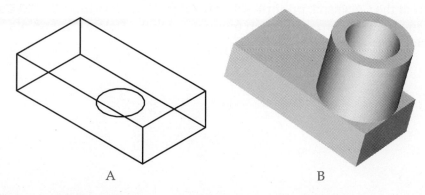

A   B

Figure 3-16.
The circle has been profiled and tangent constraint applied.

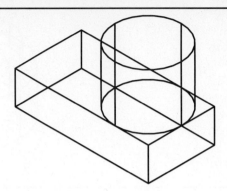

Figure 3-17.
A—The second circle is drawn, profiled, and dimensioned. B—The second circle is constrained concentric to the first circle and then profiled.

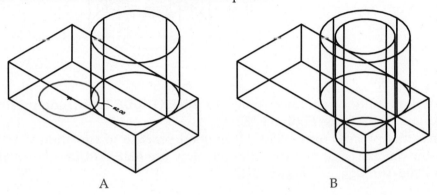

A   B

Next, draw and dimension a 2.0 diameter circle for the through hole. See **Figure 3-17A.** Remember, you are still working on the bottom of the block. Constrain the circle concentric to either edge of the cylinder and then extrude it **Cut**, **Through**, and up. The part on your screen should look like **Figure 3-17B.**

PROFESSIONAL TIP   If you wish to view the constraints, they can be temporarily displayed with the **AMSHOWCON** command. The letter "T" shows tangent constraints on circles.

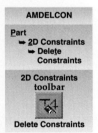

AMDELCON

Part
➥ **2D Constraints**
   ➥ **Delete**
     Constraints

2D Constraints toolbar

Delete Constraints

Deleting Constraints

If you make a mistake applying a constraint or MDT applies one that you do not want, you can delete it with the **AMDELCON** command. To access this command, enter AMDELCON at the Command: prompt, select **Delete Constraints** from the **2D Constraints** cascading menu in the **Part** pull down menu, or pick the **Delete Constraints** button on the **2D Constraints** toolbar.

❑ Open the drawing Practice-3-03.dwg.

❑ Construct the part from the 2D top view. This is the same example that was used in Chapter 2, but this time construct it using fully constrained sketches. One way of using dimensions and constraints to fully constrain is to use the following procedure.

❑ **Step 1.** Profile circle A and the 2.5 dimension and extrude 2.0 units.

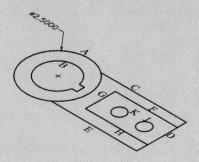

❑ **Step 2.** Profile lines C, D, and E and use circle A to close the profile. Add three dimensions and extrude 1.5 units and **Join**.

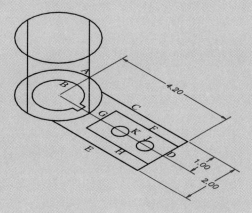

❑ **Step 3.** Profile circle B and the three lines of the keyway. Add a concentric constraint between A and B and then add four dimensions. Extrude through and cut.

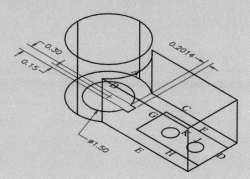

(Continued)

❑ Step 4. Put in a work plane, parallel and offset 0.5 units down from the top of the last extrusion. Do not create the sketch plane.

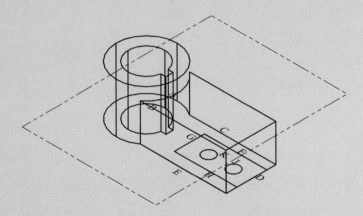

❑ Step 5. Profile lines F, G, and H and use line D to close the profile. Add three dimensions. Extrude from the work plane to the top of plane and cut.

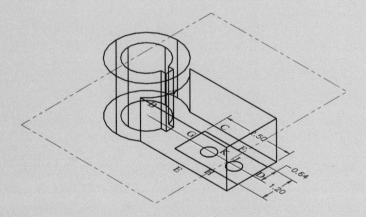

❑ Step 6. Profile circle J and input the three dimensions. Extrude through and cut.

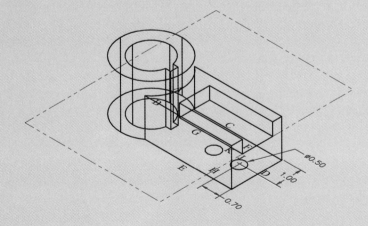

(Continued)

❑ Step 7. Profile circle K and input three dimensions. Extrude through and cut.

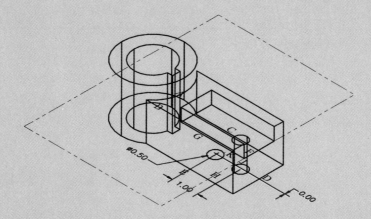

❑ Step 8. Change the length of ExtrusionBlind_2 to 5.5 units and the height to 1.0 unit. Update the part.

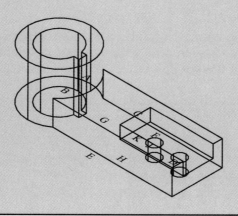

Chapter Test

Answer the following questions on a separate sheet of paper.

1. To create a solid model of a part, what additional step is added to the fundamental process learned in *Chapter 2?*
2. MDT automatically applies *2D constraints* to a sketch when it is _____.
3. What is the difference between a *constraint* and a *dimension* on a profiled sketch?
4. How many constraints and dimensions can be added to a profiled sketch?
5. What command is used to display the constraint symbols temporarily on the sketch?
6. When the constraints are displayed, what do the **H**, **V**, and **F** indicate?
7. What command is used to do new dimensioning?
8. Describe the four things that can be done using the **Power Dimensioning** dialog box.
9. Describe two ways to change the dimension of an extruded or revolved sketch.
10. What command is used to add concentric and tangent constraints?

Chapter Exercises

Exercise 3-1. Start a new drawing and construct the part shown here using fully-constrained sketches. Then change the overall length to 14 units. The drawing Exercise-3-01.dwg can be reviewed with **AMREPLAY** to see how this part was constructed.

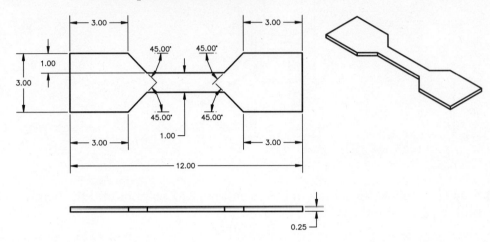

Exercise 3-2. Start a new drawing and construct the part shown here using fully-constrained sketches. Extrude both sketches downward to avoid changing the sketch plane. On the base sketch, change the fixed point to the upper-left corner. When the part is complete, change the overall length to 120 mm and note that the sizes move away from the fixed point.

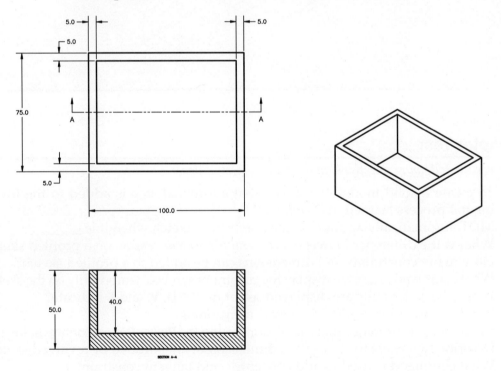

Exercise 3-3. Start a new drawing and construct the part using fully-constrained sketches. Use the concentric constraint to locate the hole with the boss. Then change the overall length to 8.0 units. The drawing Exercise-3-03.dwg can be reviewed with **AMREPLAY** to see how this part was constructed.

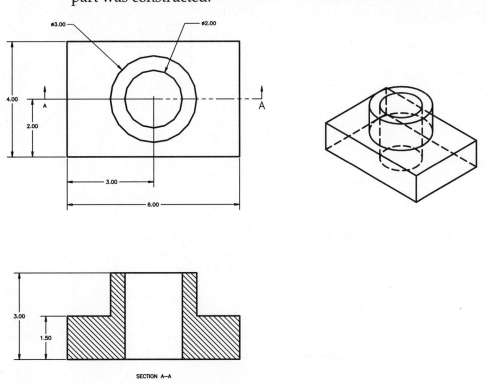

SECTION A—A

Exercise 3-4. Start a new drawing and construct the part using fully-constrained sketches. Use the concentric constraint to locate the hole with the boss. Then change the overall length to 130 units. The drawing Exercise-3-04.dwg can be reviewed with **AMREPLAY** to see how this part was constructed.

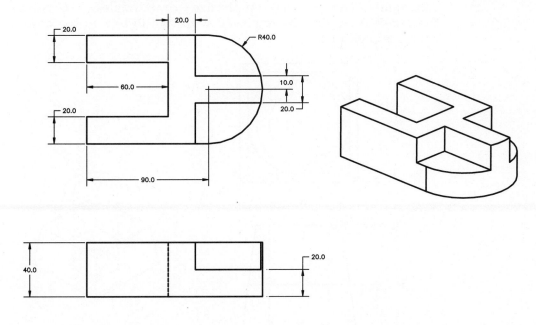

Exercise 3-5. Start a new drawing and construct the part shown below using fully-constrained sketches. Use the concentric constraint to locate the hole with the boss. Then change the overall length to 8.0 units. The drawing Exercise-3-05.dwg can be reviewed with **AMREPLAY** to see how this part was constructed.

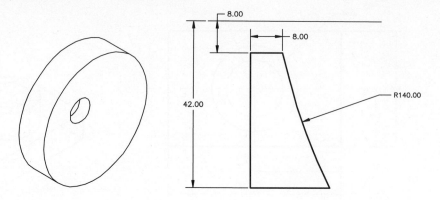

Exercise 3-6. Start a new drawing and construct the part using fully-constrained sketches. The drawing Exercise-3-06.dwg can be reviewed with **AMREPLAY** to see how this part was constructed.

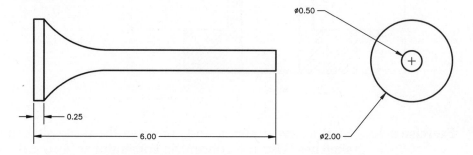

Exercise 3-7. Start a new drawing and construct the part. Fully constrain the sketches using three dimensions and the tangent and concentric constraints. This exercise emphasizes draft angles. The drawing Exercise-3-07.dwg can be reviewed with **AMREPLAY** to see how this part was constructed.

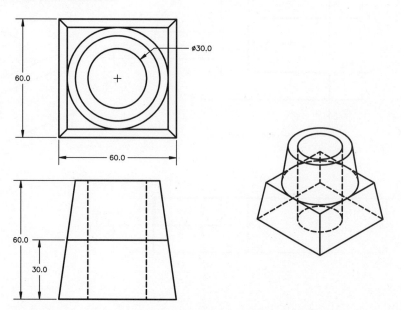

Exercise 3-8. Open the drawing Exercise-3-08.dwg. Profile the rough sketch at the top so that circle B is tangent to lines A and C and circles B and F are concentric and the sketch has the dimensions shown on the bottom sketch.

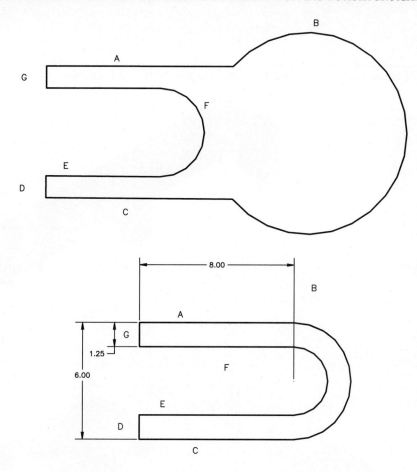

MDT is used to create assemblies such as this button assembly.

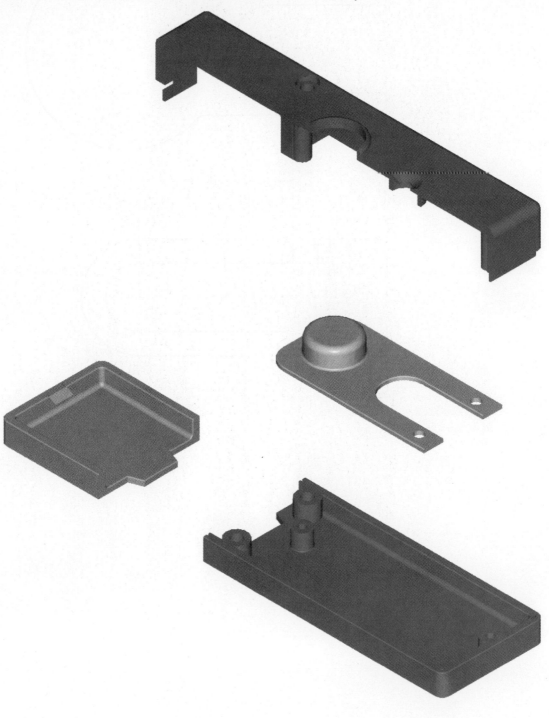

CHAPTER 4

Part Mode, Drawing Mode, and Views

Objectives

After completing this chapter, you will be able to:

* Move between the **Part Mode** and the **Drawing Mode**.
* Change parametric dimensions in the drawing view and update both part and views.
* Work with and create layouts in the **Drawing Mode.**
* Create a base view drawing from the model.
* Add orthographic and isometric views to the base view.
* Know the difference between parametric and reference dimensions.
* Create full and offset section views.
* Create and profile cut lines.
* Create detail views at different scale factors.
* Create an auxiliary view showing a true view of an inclined plane.

Drawings Needed

*The following is a list of files that you will need to work through this chapter. These files can be found on the **User's Drawings** CD included with this text.*

Examples	Practice	Exercises
Example-4-01.dwg	Practice-4-02.dwg	Exercise-4-01.dwg
Example-4-02.dwg		Exercise-4-02.dwg
Example-4-03.dwg		Exercise-4-03.dwg
Example-4-04.dwg		Exercise-4-04.dwg
Example-4-05.dwg		
Example-4-06.dwg		

MOVING BETWEEN THE MODEL MODE AND THE DRAWING MODE

Open the part Example-4-01.dwg. The solid model, called the *part*, was created in the *model mode*. (In previous releases and in AutoCAD 14, this was called *model space*. We will not use this old model space/paper space terminology.) The model mode screen shows the part, the **Browser** lists the operations used to create the part, and the toolbar shows the **Part Modeling** toolbar. See **Figure 4-1.**

Figure 4-1.
Shown is the basic MDT configuration when in **Model Mode.**

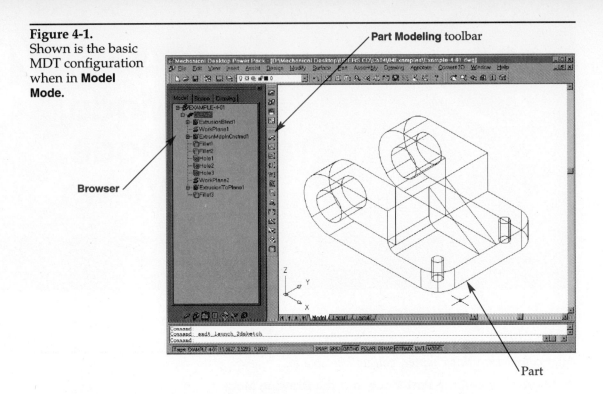

Part Modeling toolbar

Browser

Part

When the 2D drawing views are automatically created from the model, they go into a layout in the *drawing mode*. Moving between the part in **Model Mode** and the drawings in **Drawing Mode** is very easy in MDT. To move to the drawing mode, enter AMMODE at the Command: prompt. The command sequence is as follows:

Command: **AMMODE**
Restore mode [Drawing/Model] <Drawing>: **D**↵

The default is the opposite of the mode that is showing on the screen.

> **CAUTION**
> It is suggested you do *not* try to use the Release 14 commands and aliases such as **TILEMODE** and **PS**.

AMMODE
Desktop Main toolbar
Drawing Layout

Selecting Layout1 from the bottom of the drawing area also allows you to move to the **Drawing Mode**. You can also pick the **Drawing** tab at the top of the **Browser** or pick the **Drawing Layout** button on the **Desktop Main** toolbar.

When you switch to the **Drawing Mode** with the file Example-4-01.dwg open, the screen will show a title block, several views of the part with dimensions, the **Browser** listing the views created, and the **Drawing Layout** toolbar. See **Figure 4-2.**

Figure 4-2.
Shown is the basic
MDT configuration
when in **Drawing
Mode.**

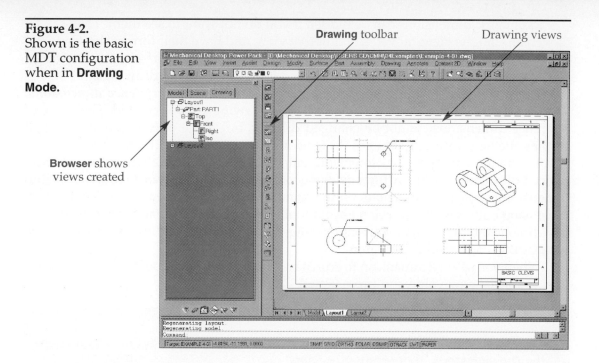

Browser shows
views created

Drawing toolbar   Drawing views

CHANGING PARAMETRIC DIMENSIONS

When looking at Example-4-01.dwg in the drawing mode, you will notice that some of the dimensions are green. See **Figure 4-3.** These are the parametric dimensions that were placed on the sketches when the part was created. They automatically appear on the drawing view when it is created. They are on the AM_PARDIM layer and are green by default. Do *not* erase any of these parametric dimensions. The ability to change that parameter of the part will be lost if they are erased.

Figure 4-3.
In **Drawing Mode,** the parametric dimensions are displayed in color (green).

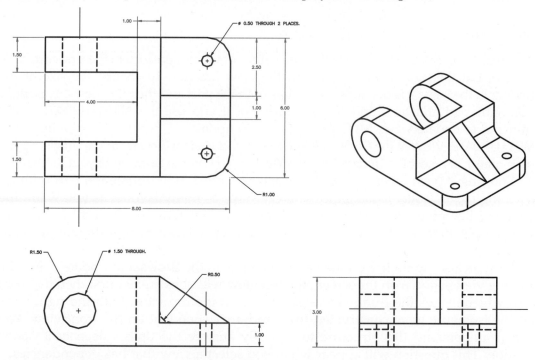

Parametric dimensions can be hidden if you do not want to see them. They can be moved around the view, moved to a different view, and have their format changed. You can also change their values. This will change the size of the part and the drawings. Conversely, editing the part in the model mode will change the drawings.

There are also some dimensions that are cyan. These are *reference dimensions.* They are dimensions manually added after the drawing view is created. The value of reference dimensions cannot be changed directly, but if the part is edited they will display the new values. Reference dimensions are on the AM_REFDIM layer.

AMMODDIM

Part
→ Di**m**ensioning
   → **E**dit
       Dimension

Drawing Layout
toolbar

Edit Dimension

To change a parametric dimension, enter AMMODDIM at the Command: prompt, select **E**dit Dimension from **Di**mensioning on the **P**art pull-down menu, select **E**dit Dimension from **E**dit Dimensions on the **A**nnotate pull-down menu, or select the **Edit Dimension** button on the **Drawing Layout** toolbar. Once the command is accessed, pick the dimension to change. You can change more than one dimension while the **AMMODDIM** command is activated.

When you select a dimension to edit, its current value will appear on the command line. Enter the new value and press [Enter].

PRACTICE 4-1

❑ Example-4-01.dwg should still be open. If it is not, please open it now and access the **Drawing Mode**. Refer to the top view.
❑ Change the value of the 8.00 horizontal dimension to 10.00 and change the 6.00 vertical dimension to 5.00 using **AMMODDIM**.
❑ Press [Enter] to accept the changes.

PROFESSIONAL TIP  The fastest way to change a dimension is to double-click on it bringing up the **Power Dimensioning** dialog box. With it you can change the value and several other characteristics including the number of decimals displayed. This was discussed in *Chapter 2.* If you double-click on a reference dimension, you can change characteristics, but not its value.

Updating the Part and Drawing

After doing *Practice 4-1*, notice that the values of the dimensions in the drawing have changed, but the size of the drawing in the view has not. To change the size of the drawing in the view, it is necessary to update both the part and then the views. You could update the part and then update the views, but if you use the **AMDT_UPDATE_VIEW** command to update views MDT prompts you to update the part first. To do this, enter AMDT_UPDATE_VIEW at the Command: prompt, select **Update Vie**w from the **D**rawing pull-down menu, or select the **Update Drawing View** button on the **Drawing Layout** toolbar.

AMDt_UPDATE_VIEW

Drawing
   → Update Vie**w**

Drawing Layout
toolbar

Update Drawing View

Entering Y (yes) when prompted Update part now?, will update the part without switching to the **Model Mode.** Since you can have as many layouts as you want, the command line gives the opportunity to update the current layout or all layouts. In this example, there are two layouts, Layout1 and Layout2, showing different views of the same part.

You can also right-click on the name of a view in the **Browser** to update it. For the drawing views shown on the screen, the top view was constructed first then the front view, which means the front view is *dependent* on the top view. Both the right view and the isometric were constructed from and are dependent on the front view. You need to understand this structure to logically answer the Update dependent views? question. This question will appear when you select a view that has dependencies.

WORKING WITH LAYOUTS IN THE DRAWING MODE

In the **Drawing Mode** of Example-4-01.dwg, you can see that there are two layouts, Layout1 and Layout2. This is displayed in two places, in the **Browser** and in the tabs at the bottom of the drawing window. You can have as many layouts as you wish. However, only one can be the active layout at any time. Inactive layouts are grayed out both in the **Browser** and in the tabs. To make Layout2 active, click on the **Layout2** tab or double-click on Layout2 in the **Browser**, or right-click on Layout2 in the **Browser** and select **Activate Layout**.

Renaming the Layout

Again, right-click on Layout2 in the **Browser**. Notice that there are now eleven choices in the menu. See **Figure 4-4.** The first available one is **Rename**; select it and change the name to FIRST ANGLE, because this drawing is done in the first angle projection system. This system is used in parts of Europe. Note that the name changes in both the **Browser** and the tabs.

Figure 4-4.
This pop-up menu used to work with views in **Drawing Mode.**

The Page Setup

Each layout has been set up with a specific plotter and paper size using the **Page Setup...** selection. Selecting **Page Setup** brings up the **Page Setup** dialog box. See **Figure 4-5A.** The **Layout Settings** tab shows that the paper size is ANSI D in inches and the orientation is Landscape. Click on the **Plot Device** tab. See **Figure 4-5B.** The plotter for this layout is DWF ePlot.pce, which is the *Drawing Web Format* and it sends the plot to a file. This is used because it will work on all computers regardless of what plotter you are using, if any. We will use this dialog box to create a new layout.

Figure 4-5.
A—The **Plot Device** tab is used to configure a plotter. B—The **Layout Setting** tab is used to set
paper size, as well as other plotting parameters.

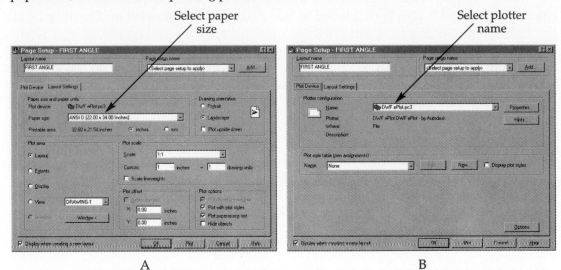

A   B

AMDT_NEW_LAYOUT

Drawing
➥ New layout

Drawing Layout
toolbar

New Layout

Creating a New Layout

To create a new layout, enter AMDT_NEW_LAYOUT at the Command: prompt,
select **New Layout** from the **Drawing** pull-down menu, or select the **New Layout** button
on the **Drawing Layout** toolbar.

When prompted, name the new layout SECTION VIEWS. Make this new layout
the active layout and then access the **Page Setup** dialog box. From the **Plot** tab, select
the DWF ePlot.pc3 plotter. From the **Layout Settings** tab, select ANSI C (17.00 × 22.00
Inches) for the paper size.

PROFESSIONAL TIP You can also create a new layout by right-clicking in the
Browser or drawing window and selecting **New Layout** from
the pop-up menu.

Right-click on SECTION VIEWS in the **Browser**. Pick **Insert Title Block** from the
pop-up menu. This accesses a text window. When prompted, select ANSI-C Size(in) by
entering 9. Press [Enter] to exit the command. The border is automatically placed near
the lower-left corner of the view. To move it exactly to the corner, use the **MOVE**
command and select the entire border with a window and press [Enter]. Pick the
lower-left corner of the border as the reference point and the lower-left corner of the
paper outline as the destination.

Creating layouts is an AutoCAD 2000 process and a full description of all the
details is beyond the scope of this book on MDT. For more information please refer
to *AutoCAD and its Applications—Basics, AutoCAD 2000*. What we have learned here
is enough to get you started.

PRACTICE 4-2

❑ Open drawing Practice-4-02.dwg and move to Layout1 in the **Drawing Mode**.
❑ Change the two dimensions shown in large font. Change 4.50 to 5.25 and the 2.25 to 3.00.
❑ Update both the part and the drawings.
❑ Change the names of Layout1 and Layout2 to Drawing One and Drawing Two.
❑ Create a new layout called Metric with an ISO A3 paper size.

CREATING BASE AND ORTHO VIEWS

Open drawing Example-4-02.dwg and switch to Layout1. You should have an ANSI-D size sheet with a title block. You now need to create a new view of the part. To create a new view, use the **AMDWGVIEW** command. To access this command, enter AMDWGVIEW or Q at the Command: prompt, pick **New View** from **Drawing** pull-down menu, or pick the **New View** button on the **Drawing Layout** toolbar.

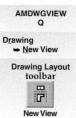

AMDWGVIEW
Q

Drawing
➥ New View

Drawing Layout
toolbar

New View

PROFESSIONAL TIP
You can also right-click anywhere in the drawing window or on Layout1 in **Browser** and select **New View** from the pop-up menu.

This accesses the **Create Drawing View** dialog box. See **Figure 4-6.** For the first view, you will need to select a base view of the active part. Pick **Base** in the **View Type:** drop-down list and **Active Part** from the **Data Set:** drop-down list. Other choices in this list (**Scene** and **Select**) are for assemblies of two or more parts and will be covered later in *Chapter 19.* The **Hidden Lines** tab controls whether hidden lines are calculated and displayed and we want both. Generally, tangencies are only displayed in isometric views and we always want to remove coincident (overlapping) edges. The **Section** tab controls cross hatching and symbols for a section view. This is covered later in this chapter.

The drawing Example-4-02.dwg is in metric and the outside diameter is 70mm. If you want to plot this on a D-size paper, you must adjust the *scale.* Enter a value of 0.1 in the **Scale** text box.

Figure 4-6.
The **Create Drawing View** dialog box is used to create a new view of a part.

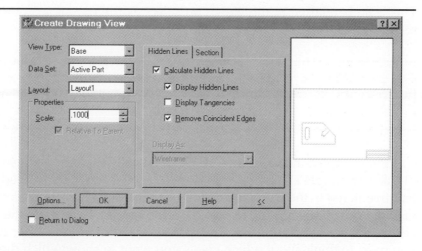

When you pick the **OK** button, the system switches back to the **Model Mode**. You are prompted to specify the direction you want to look at the part for the base view. The following prompt appears:

Select planar face, work plane or [Ucs/View/worldXy/worldYz/worldZx]:

At this prompt you can select a face on the part, a work plane, or one of the world coordinate system planes. If the base view is to be the top view showing the circles, then enter X for worldXy plane. Then, enter X again to specify the direction of the horizontal in the view. The system will reorient the UCS (user coordinate system) so that the view is looking down the Z-axis. If this is okay, press [Enter] to accept.

Now you are back in the **Drawing Mode** and you are prompted to pick a point for the center of the view. Pick points until the view is where you want it and press [Enter] to accept. The system will draw the view and put in the green parametric dimensions. The dimensions may not be where you want them or in the correct format. In *Chapter 11* you will be instructed on how to make the drawings look right. For now, please concentrate on creating the views.

Now create a front view. Access the **Create Drawing View** dialog box, pick **Ortho** in the **View Type:** drop-down list, and pick the **OK** button. *Ortho* is an abbreviation of *orthographic*, which in mechanical drawing means at right angles. There are only four choices for the location of the view—above or below and to the right or the left of the top view. Once you select **Ortho**, the scale value cannot be changed. You are prompted to pick the *parent view*. Make the top view the parent view by selecting anywhere on it. Locate the view below the top view and press [Enter]. The view will be drawn with some parametric dimensions. This is a *dependent view*, which means if the top view is moved the front will move with it.

PROFESSIONAL TIP The front view could also have been created by making it another base view. However, in mechanical drawing the orthographic views should generally be dependent.

Now create a right side view as an ortho view based on the front view. Then create an isometric view based on the front view. To do this, pick **Iso** in the **View Type:** drop-down list. The defaults for the isometric view are as follows:
- Hide all hidden lines.
- Display the tangencies of circular features.
- Remove coincident (duplicate) edges.

Generally, the view will look better if it is a little smaller than the ortho views. The scale is relative to the view you select as the parent so set this scale to 0.75. The result of creating these views is shown in **Figure 4-7.**

NOTE These same four views could have been created with one command by selecting **Multiple** in the **View Type** drop-down list.

Parametric dimensions will be automatically placed on at least the first two views. Think of a simple shaft with two dimensions—a diameter placed on the sketch and a length entered in the **Extrusion** dialog box. The diameter dimension will show in the top view and the length in the front view. See **Figure 4-8.** Because the part is fully defined, additional views will not show any parametric dimensions. Prior to creating a drawing, you need to think about which views you want to show which dimensions.

Figure 4-7.
Shown is the result
of generating two
Ortho and one Iso
views.

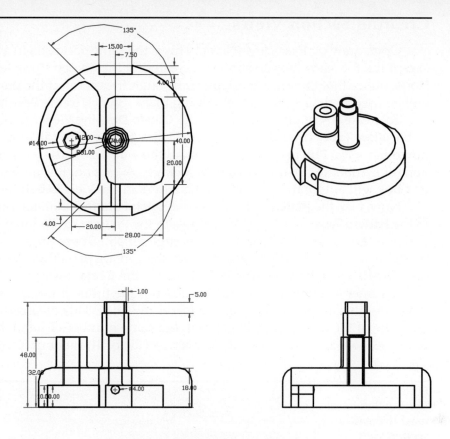

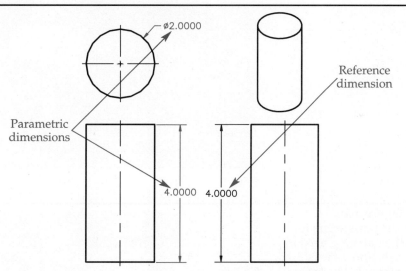

Figure 4-8.
The parametric
dimensions are
green by default.

Parametric
dimensions

Reference
dimension

Ø2.0000

4.0000   4.0000

Reference dimensions (the cyan ones) can be added to any view. The 4.00 dimension in the side view is a reference dimension. It is possible to create the cylinder without fully constraining the sketch. Thus, the parametric diameter would not appear in the drawing view. This would need to be added as a reference dimension in the MDT sense of the word.

CAUTION

The way MDT uses the term *reference dimension* should not be confused with how it is used in standard drafting practices. In standard drafting practices, a reference dimension is one that is already represented by one or more dimensions and the word "REFERENCE" follows the value.

Creating Section Views

Open drawing Example-4-03.dwg. This is the same part used in a previous example, except the top view has already been created in the Section Views layout. In the **Model Mode**, notice that there is a work plane through the center of the part. This work plane will be used to define the section-cutting plane. Switch to the **Drawing Mode**.

To create a section view, access the **Create Drawing View** dialog box. Select **Ortho** in the **View Type:** drop-down list. Next, pick the **Section** tab and in the **Type:** drop-down list select **Full** to create a section all the way across the part. See **Figure 4-9.** The default symbol for a section view is always "A." Changing the letter in the **Symbol:** text box will automatically change the letters in the **View Label:** text box.

Now, pick the **Pattern...** to access the **Hatch Pattern** subdialog box. See **Figure 4-10.** In the **Pattern Type** area, select Predefined. In the **Pattern:** drop-down list, select ANSI31, which is single continuous lines at an angle of 45°. There are more than 50 different predefined patterns. As you select them, they are displayed in the **Pattern Type** image tile. The distance between the lines is set in the **Scale:** text box. A scale value of 1.0 sets the lines 0.125 units apart. These are not the units of the Layout, but the units of the part—the larger the number the fewer the lines. This means that no matter what the scale of the drawing view you get, the same number of hatch lines are displayed in the section. Since this is a metric part, a scale of 10 is about right.

Figure 4-9.
The **Section** tab is used to create a section view.

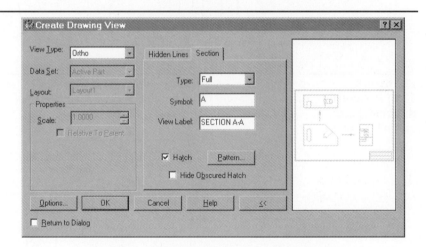

Figure 4-10.
The parameters for the hatching used in a section view is controlled by the **Hatch Pattern** dialog box.

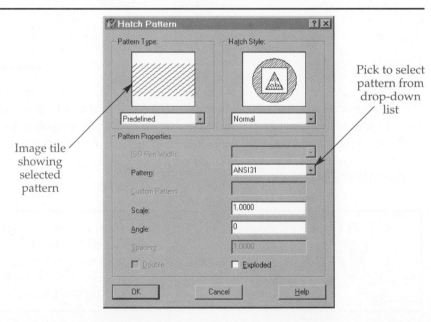

Image tile showing selected pattern

Pick to select pattern from drop-down list

Click **OK** to return to the **Create Drawing View** dialog box. Access the **Hidden Lines** tab and notice that the **Display Hidden Lines** check box is not checked. See **Figure 4-11.** This is the default for a section view. Touch **OK** and pick the top view as the parent view. Position the section view below the top view and press [Enter]. You are now prompted for location of the cutting plane:

> Enter section through type [Point/Work plane]<Work plane>: *(press [Enter] to accept the default)*
> Select work plane in parent view for the section: *(in the top view, pick the horizontal work plane shown in an edge view)*

If you wish to change the size of the text "SECTION A-A," double-click on it. This accesses the **Properties** dialog box. See **Figure 4-12.** Double-click on the value of 0.25 next to **Height** in the **Text** area and input the height you want. Try 0.5 for example.

Figure 4-11.
Make sure the **Display Hidden Lines** check box is not checked.

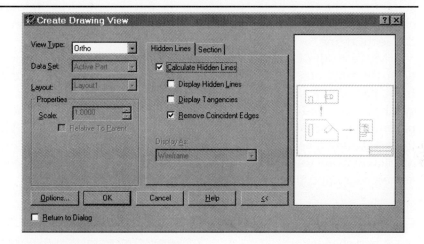

Figure 4-12.
A—The height of the text is changed in the **Properties** dialog box.
B—Shown is the result of the section view and the text change.

A

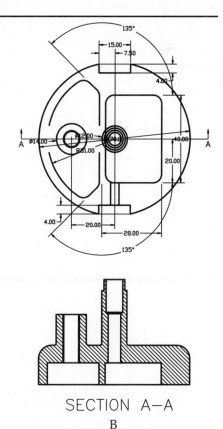

SECTION A—A

B

Creating an offset section

To create an offset section view, you need to construct a cutting line sketch in the **Model Mode**. Open drawing Example-4-04.dwg and go to the sketch view by entering 9. Draw a polyline as shown in **Figure 4-13A.** This polyline represents the cut line for the section view. Also remember, using a polyline requires only one pick when profiling. The line must be a continuous line type, so do not use a construction line. Since it will be dimensioned, you do not have to construct it accurately. Use the **AMCUTLINE** command to profile the line as a cut line, not as a sketch. To access this command, enter AMCUTLINE at the Command: prompt, select **Cut Line** from the **Sketch Solving** cascading menu in the **Part** pull-down menu, or pick the **Cut Line** button on the **Part Modeling** toolbar.

After profiling, dimension the cut line as shown in **Figure 4-13B.** Two dimensions with zero values locate the cut line through the centers of the circles. (This could also be accomplished with the **Y-Value** constraint described in **Chapter 10**.) The dimensions from the edge of the part keep the cut line outside the part if the part length is changed. This is not necessary for section view, but for the Parent (Top) view showing the section line.

Move to the Layout1, which has the top view already constructed. As you did with the previous example, access the **Create Drawing View** dialog box, pick **Ortho** from **View Type:**, and access the **Section** tab. Select **Offset** as the **Type** and enter a symbol, the default is A, in the **Symbol:** text box. Now, pick the **OK** button. Select the parent view and place the section view below it. The mode will automatically change to **Part.** You will be prompted to select the cut line, since there could be more than one. MDT will draw the section view. See **Figure 4-14.** The cut line can have as many steps as necessary and they don't have to be at right angles.

AMCUTLINE

Part
↳ **S**ketch Solving
↳ **C**ut Line

Part Modeling
toolbar

Cut Line

Figure 4-13.
A—A part showing
a polyline that will
be used as a cut line.
B—Shown are the
dimensions for the
cut line.

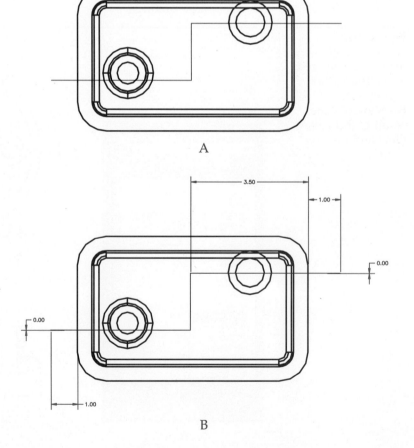

A

B

Figure 4-14.
Shown is the result
of an offset section
views.

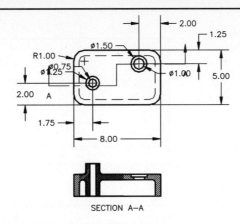

SECTION A—A

Making changes to the section view

For any view of the various properties, such as the linetype of hidden lines, the section symbol, the cross hatching, the relationship to the parent view, and the display of parametric dimensions, can be changed. Change this section view so that the parametric dimensions are displayed. In the **Browser**, right-click on the view name (Section A for this example). Then, select **Edit** in the pop-up menu. Access the **Display** tab, check the **Parametric Dimension** check box, and pick **OK**.

Creating Detail Views

Open the drawing Example-4-05.dwg as seen in **Figure 4-15A.** This simple block has a small hole in one corner. In the drawings, we want to construct a detail view of this hole shown at a larger scale than the main views. The height of the dimension font must remain the same size. MDT will do both of these in the same step. To do so, switch to the Layout1. This shows a front and top view at full scale. Access the **Create Drawing View** dialog box and input the information for a detail view at a scale of 3.0. See **Figure 4-15B.**

After picking the **OK** button, you will be prompted to pick a vertex in the parent view to attach the detail view. A *vertex* is an endpoint of an edge (line) or the center of an arc or circle. Do not use osnaps, because no matter what you select— midpoint, quadrant, etc.—MDT will default to the center or the endpoint. Select the small circle in the top view. Now select the small circle again for the center of the circular area defining the view; you can have other shaped areas. Then pick a screen position to set the size of the area you want in the detail view. Next, pick the location for the view and press [Enter]. See **Figure 4-16.** A reference dimension was added to the view to demonstrate the automatic sizing of the font.

Figure 4-15.
A—A block with a small hole used to create a detail view. B—Select Detail as the view type, set the scale, and change the detail symbol.

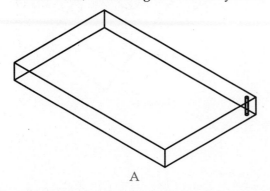

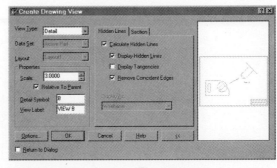

A   B

Figure 4-16.
Shown is the result
of creating a detail
view.

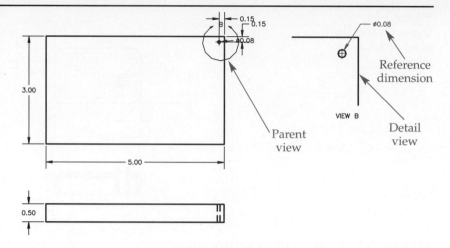

Creating Auxiliary Views

Open the drawing Example-4-06.dwg. This part has an angle cut on one corner and a hole pattern drilled into the face. In the drawing, a view showing the true size and shape of face and holes needs to be added. Switch to the Layout1, which shows a front and top view at full scale. Access the **Create Drawing View** dialog box. Select **Auxiliary** from the **View Type:** drop-down list. Also uncheck the **Display Hidden Lines** check box on the **Hidden Lines** tab. Click on **OK** and you are then prompted as follows:

Select first point for projection direction or [Workplane]:

This prompt is somewhat confusing because it does *not* say that you can pick the edge of face that you want to project. Carefully pick near the edge AB. Then, press [Enter] to use the selected edge and pick a point to locate the new view. See **Figure 4-17.**

In descriptive geometry, this view would be called the *first auxiliary view*. The procedure for creating second auxiliary views of planes that are at a compound angle to the view is in the discussion on sketch planes in *Chapter 7.*

Figure 4-17.
Shown is the result
of creating an
auxiliary view.

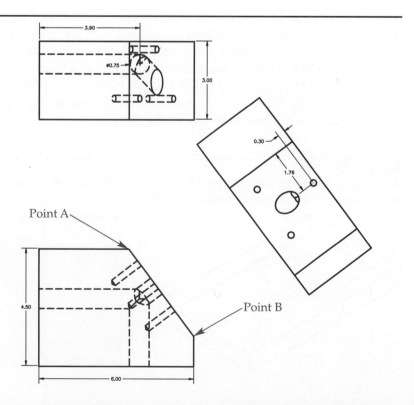

Chapter Test

Answer the following questions on a separate sheet of paper.

1. Describe the differences between **Model Mode** and **Drawing Mode**.
2. What command is used to switch between **Model Mode** and **Drawing Mode**?
3. What layer are parametric dimensions placed by default?
4. Why shouldn't parametric dimensions be deleted?
5. What command is used to edit parametric dimensions?
6. *True or False?* Once a parametric dimension is edited, the change is made on screen automatically.
7. What is the maximum number of layouts that can be created?
8. What command is used to create a new view of a part?
9. What is a *dependent view*?
10. Describe a situation where the **AMCUTLINE** command is used.

Chapter Exercises

Exercise 4-1. Open the drawing Exercise-4-01.dwg. Construct the four drawing views and the title block as shown.

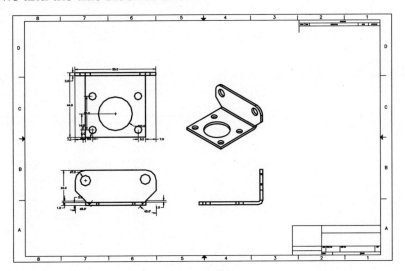

Exercise 4-2. Open the drawing Exercise-4-02.dwg. Construct the six drawing views as shown.

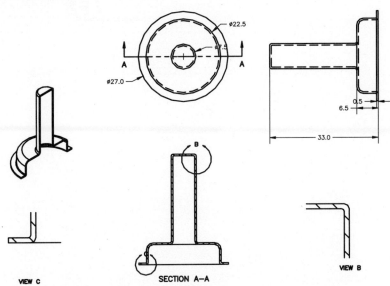

Exercise 4-3. Open the drawing of the injection molded valve body Exercise-4-03.dwg and construct the six views as shown. The Ampardim layer that shows the parametric dimensions has been frozen for clarity. The scale of the hatch pattern is set at 10 and the scale of the detail view is three.

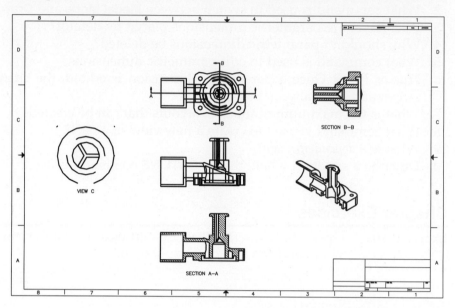

Exercise 4-4. Open the drawing of the weldment part Exercise-4-04.dwg and construct the front, top, side, isometric, and auxiliary views as shown.

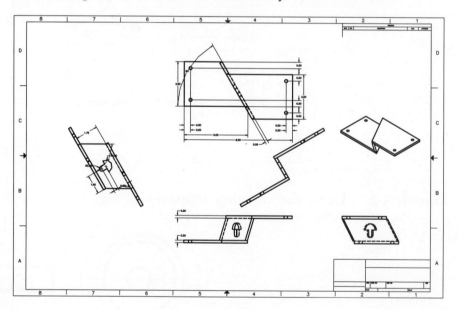

CHAPTER 5

Constructing Holes

Objectives

After completing this chapter, you will be able to:

* Construct holes in faces concentric to arcs and circles.
* Understand the term *planar face*.
* Place and edit hole notes in the drawing views.
* Construct threaded holes.
* Move features in the **Browser** to correct problems.
* Construct counter-bored and chamfered holes.
* Construct holes in planar faces from two edges.
* Construct holes in planar faces from an existing hole.
* Select multiple features in the **Browser**.
* Change the color of holes to highlight in the shaded view.

Drawings Needed

*The following is a list of files that you will need to work through this chapter. These files can be found on the **User's Drawings** CD included with this text.*

Examples	Practices	Exercises
Example-5-01.dwg	Practice-5-02.dwg	Exercise-5-01.dwg
Example-5-02.dwg	Practice-5-03.dwg	Exercise-5-02.dwg
Example-5-03.dwg		Exercise-5-03.dwg
Example-5-04.dwg		Exercise-5-04.dwg
Example-5-05.dwg		
Example-5-06.dwg		
Example-5-07.dwg		

CONSTRUCT HOLES IN THE PART

Holes can be *drilled* through any planar face on the part. A *planar face* is a flat plane. Open the part Example-5-01.dwg. This part has seven planar and five cylindrical faces as shown in **Figure 5-1.** (The technique for placing holes on cylindrical and conical faces using work points will be discussed in *Chapter 8*.) The specifications for

Figure 5-1.
This part has seven
planar and five
cylindrical faces.

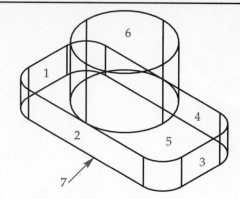

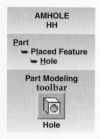

holes are input into a very useful dialog box. The first example is to use the **AMHOLE** command to place through holes concentric with arcs and circles. Then, you will put notes on these holes in the drawing views. To access the **AMHOLE** command, enter AMHOLE or HH at the Command: prompt, select **Hole** from the **Placed Feature** cascading menu of the **Part** pull-down menu, or pick the **Hole** button on the **Part Modeling** toolbar.

This opens the **Hole** dialog box that has six main areas—**Operation**, **Termination**, **Placement**, **Drill Size**, **Tapped**, and **C'Bore/Sunk** size. See **Figure 5-2**. Input the information for a 2.0 unit diameter hole drilled through and concentric with some feature in the part. Note that the image tile showing the hole changes as you change the **Operation:** and **Termination:** selections. Various options in other areas of the dialog box are grayed out based on the choices selected in the **Operation:**, **Termination:**, and **Placement:** drop-down lists.

Once you click the **OK** button, you are prompted to select a work plane or planar face. Select the top of the large cylinder—face 6 in **Figure 5-1**. Pick the face on the circumference of the circle or, since the surface is visible, pick within the face. The only indication that you have made a proper selection is that you are prompted to select a concentric edge. Select the edge of the large cylinder and the hole is *drilled*. This is a two-step process. First, select the face, and then the concentric edge. The result is shown in **Figure 5-3**.

Figure 5-2.
The **Hole** dialog box is used to input the information for constructing holes.

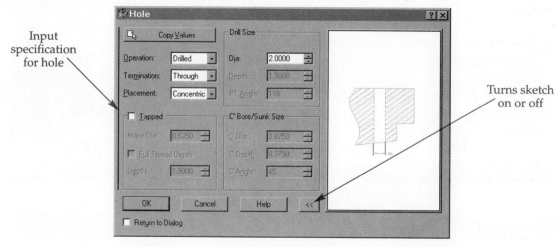

Figure 5-3.
Shown is the result of adding the concentric through hole.

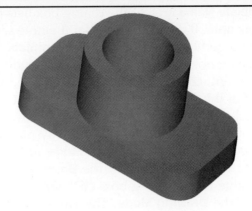

PROFESSIONAL TIP An easy procedure is to pick the circle twice; the first pick sets the face to drill into and the second the concentric edge.

Placing Hole Notes on the Drawing

Switch to the **Drawing Mode**. The views update showing the hole, but without dimensions. Holes require a hole note that specifies the size and the operations. To place a hole note, use the **AMHOLENOTE** command. To access this command, enter AMHOLENOTE at the Command: prompt, pick **Hole Note** from the **Annotation** cascading menu of the **Annotate** pull-down or right mouse menu, or pick the **Hole Note** button on the **Drawing Layout** toolbar. After entering the command you are prompted as follows:

Enter an option [Edit/New] <New>:

To create a new note, press [Enter] to select the default New. Entering E (for Edit), lets you edit an existing note. This may be difficult to remember since most editing in MDT is done by a unique editing command. Now, select the circle representing the hole in the top view. This accesses the **Create Holenote** dialog box. See **Figure 5-4.**

AMHOLENOTE
HH

Annotate
↪ A**n**notation
↪ **H**ole Note

Drawing Layout
toolbar

Hole Note

Figure 5-4.
The **Create Holenote** dialog box is used to place notes dealing with holes.

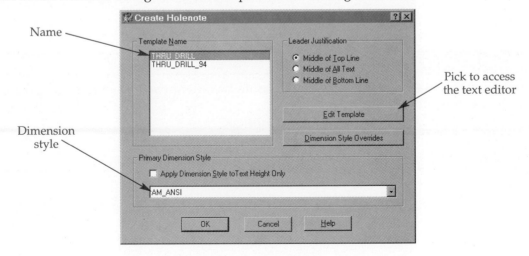

Name

Dimension style

Pick to access the text editor

The **Template <u>N</u>ame** area of the dialog box lists the name of the templates used to create the hole note. It does *not* list what the note says. The words in the template are shown and changed by selecting the **Edit Template** button. This accesses the **Multiline Text Editor** dialog box. See **Figure 5-5.** You can edit the text on a one-time basis here. To change the font, color, or text size of the note, first create a new dimension style. Then, instead of the default standard, select what you wish.

Then, click **OK** and place the note as shown in **Figure 5-6.** The hole note is attached to the circle and the size is one-way parametric. This means that to change the hole size you must go to the **Model Mode** and edit the hole. Remember, the easiest way is to right-click on the Hole in the **Browser** and select **Edit** in the pop-up menu. Change the value in the dialog box, click **OK**, and press [Enter]. Update the part and switch back to the **Drawing Mode.** The value and the location of the arrowhead will change. If you move the view, the note will move also.

Figure 5-5.
The **Multiline Text Editor** is used to make changes to hole notes.

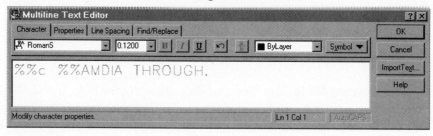

Figure 5-6.
The hole note is attached to the circle and the size is one-way parametric, which means changing the hole size must be done in **Model Mode** by editing the hole.

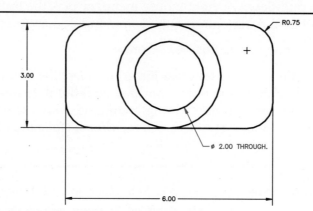

PRACTICE 5-1

❑ Use **Example-5-01.dwg.** This drawing should be on your screen. If it is not, please open it now.
❑ Change the diameter of the hole to 1.0 units.
❑ Update the part and check the drawing view.
❑ Change the diameter back to 2.0 units.

CREATING MULTIPLE HOLES

Still using Example-5-01.dwg, switch to the **Drawing Mode**. We are going to place a hole on each corner of the base. See **Figure 5-7**. The procedure used here will generate four holes that are independent of each other. You can change the size or characteristics of, or even delete, any one of the holes without affecting the others.

<table>
<tr>
<td>
NOTE</td>
<td>There are ways to create *patterns* (previously called arrays) of dependent holes. Creating patterns is covered in *Chapter 7*.</td>
</tr>
</table>

In the **Hole** dialog box, input the information for a 0.75 diameter, drilled through, and concentric hole. Click on each arc at each corner twice—once for the planar face and once for the concentric arc. The positions of these holes are related to the size of the base. This means that the holes will always remain concentric with the corner fillets no matter how the base changes.

Holes have a direction—they are *drilled* into the part from the selected planar face. When you select the *planar face* for a hole, the system sets the direction of the hole into the part. This can cause unexpected results, especially for through holes. A through hole drilled into the lower-corner of the top of the inclined plate in **Figure 5-8** goes through the lower plate as well. This problem can be eliminated by selecting the bottom planar face of the inclined plate and drilling *up*, or by selecting a blind hole of the correct depth, or by selecting the **To-Plane** termination.

Figure 5-7.
The holes, one at each corner of the base, are independent of each other.

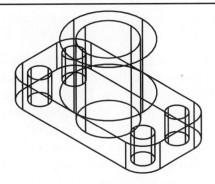

Figure 5-8.
When selecting a *planar face* for a hole, the system sets the direction of the hole into the part. For through holes, this can cause unexpected results.

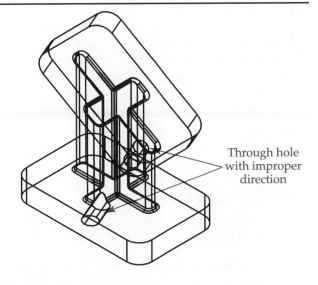

Through hole with improper direction

❏ Open part Practice-5-02.dwg, which is shown on the left here.
❏ Drill 4.0 diameter holes, four in the base and four in the inclined plate as shown on the right here.

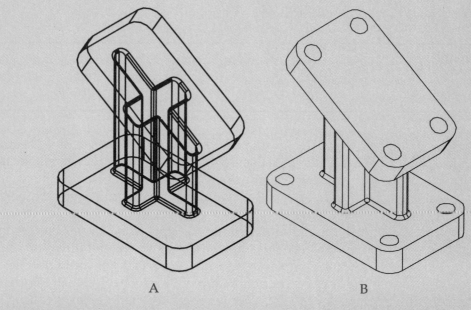

A   B

❏ Create a top drawing view.
❏ Put in a note that has the diameter and: DRILL THROUGH 8 PLACES

THREADED HOLES

To create a threaded hole, the **Tapped** check box in the **Hole** dialog box is checked. This feature will make no difference in the **Model Mode**, but it will put the symbolic representation for threads in the **Drawing Mode**. You must input the diameters you want for the major and minor (drill size) diameters. For example, a two-inch diameter 4.5 threads per inch National Course internal thread has minor diameter of 1.7113. This four-place accuracy is not required in symbolic thread representation.

Open the drawing Example-5-02.dwg as shown in **Figure 5-9A**. Issue the **AMHOLE** command and input the values as shown in **Figure 5-10**. These values are for a through hole, tapped to full depth with a 1.7 minor and a 2.0 major diameter. The results are shown in **Figure 5-9B**.

You can create your own standard hole notes by picking the **Hole Note Template** button on the **Drawing Layout** toolbar. It is right under the **Hole Note** button. This accesses the **Hole Template** dialog box. In the **Templates:** drop-down list, select **Thru Drill**. Type in the desired name in the **Name:** text box. Click the **Save** button and the new template is created. The **Edit Template** button is activated. This will let you type in whatever note you want. This template will be available the next time you input a hole note in this drawing. Release 5 also lets you put a reference dimension on the major diameter on the drawing.

PROFESSIONAL TIP If you want these notes available all the time, put them in your template drawings.

Figure 5-9.
A—A threaded hole
needs to be added
to this part.
B—The results of
adding the threaded
hole are shown with
a symbolic
representation and
as a thread note.

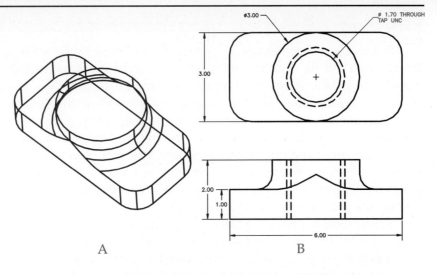

A   B

Figure 5-10.
The **Tapped** area of
the **Hole** dialog box
is used to add
threads to a hole.

Use this
area for adding
threads to hole

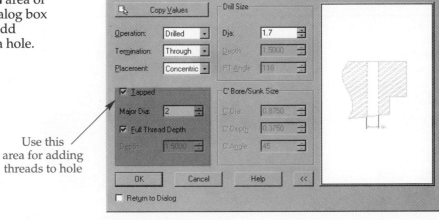

PRACTICE 5-3

❏ Open the part Practice-5-03.dwg as shown here.

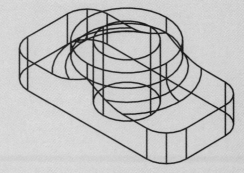

❏ Put four drilled through, 0.5 diameter 20 threads per inch UNF tapped holes concentric to the four corners radii. The minor thread diameter is 0.415.

❏ Create a top view.

❏ Put in a hole note calling out the threads and a reference dimension on the major diameter.

MOVING FEATURES IN THE BROWSER

Some times one feature is incorrectly affected by another feature that is added later. Open the file Example-5-04.dwg. The web that was added last in this example goes through the hole. See **Figure 5-11**. In most cases like this, you can reorder a feature in the **Browser**. That is, move Large Hole below the Web by dragging and dropping the Large Hole in the **Browser**. To do this, click and hold on Large Hole in the **Browser** and simply drag the feature below the Web. See **Figure 5-12A**. As you drag the feature, note the line showing where you can place the feature. Also note that a small circle with a line through it appears as you drag across places where you cannot place the feature. Once you have reached the location you wish to place the feature, release the mouse key. See **Figure 5-12B**.

In addition to changing the order in the **Browser,** you have also changed the part. The result of this drag-and-drop process is shown in **Figure 5-13**. In this case, you could also have dragged the Web above the Large Hole. Generally, any feature except the base can be moved. However, a feature that is constrained, dimensioned, or in any way related to a feature above it cannot be moved above that feature.

Figure 5-11.
The result of a feature that has been incorrectly affected by one that is added later.

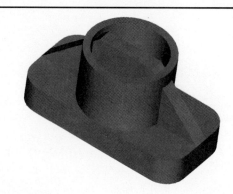

Figure 5-12.
A—In the **Browser**, left-click and hold on the Large Hole.
B—Drag-and-drop the Large Hole so it is located below the Web.

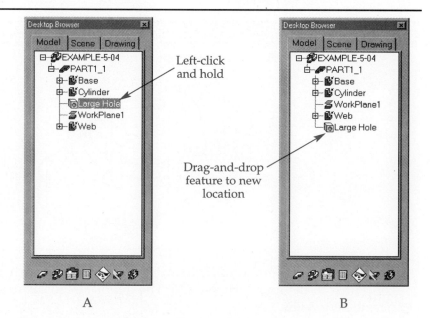

Figure 5-13.
The result of
moving the Large
Hole below the Web.

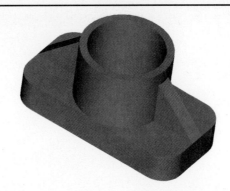

COUNTERBORED AND COUNTERSUNK HOLES

Open Example-5-05.dwg and switch to **Drawing Mode.** See **Figure 5-14.** As you can see the hole on the left, Hole-CB, is a counterbored hole. Now, switch to **Model Mode** and right-click on Hole-CB in the **Browser.** Select **Edit** in the pop-up menu to display the **Hole** dialog box.

In this dialog box, the options in the **Operation:** drop-down list are **Drilled, C'Bore,** and **C'Sink.** See **Figure 5-15.** If you select **C'Bore** and **C'Sink,** you must add diameters and depths or angles. These are added in the **C'Bore/Sunk** size area. The drill size diameter must be smaller than the value for **C'Dia.** The **C'Depth** value is the distance the larger hole is bored or sunk into the part. As an example, change the counterbore diameter to 20.0 and the counterbore depth to 5.0. Click on **OK,** press [Enter], and update the part.

The hole on the right (Hole-CS) is a countersunk hole with a drill diameter of 10.0 and a 60° angle. The circle at the top of the countersink, or the **C'Dia** value, is 15.0 diameter. This is also a blind hole. This means it is not drilled all the way through the part and the depth and angle of the drill point have to be specified. This information is shown in **Figure 5-16.** Change the countersunk diameter to 20 and the angle to 30°. Click on **OK,** press [Enter], and update the part.

Figure 5-14.
The part shown in
the **Drawing Mode.**

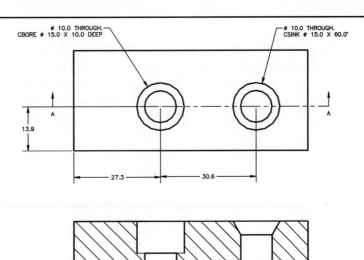

Figure 5-15.
Select **C'Bore** or **C'Sink** from the **Operation:** drop-down list to create counterbore and countersunk holes, respectively.

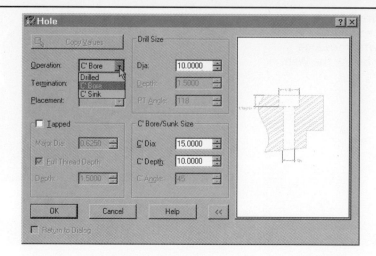

Figure 5-16.
This dialog box shows the values needed to produce the countersunk hole.

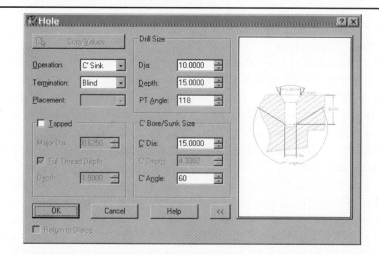

HOLES LOCATED FROM TWO STRAIGHT EDGES

A hole can also be located from any two nonparallel, straight edges of a planar face on the part. Open the part Example-5-06.dwg. A 0.5 diameter through hole needs to be drilled at point D and relative to two edges. See **Figure 5-17**.

To add this hole, issue the **AMHOLE** command. In the **Placement:** drop-down box of the **Hole** dialog box, one of the options is **2 Edges**. Select **Drilled**, **Through**, and **2 Edges**, input the correct diameter, and pick **OK**. See **Figure 5-18**.

Figure 5-17.
The point where a hole is to be located using the **2 Edges** option.

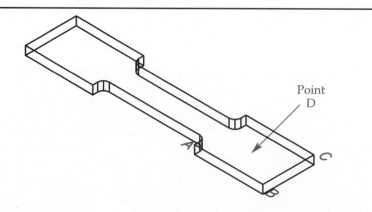

Point D

Figure 5-18.
This dialog box
shows the values
needed to produce
the hole at point D.

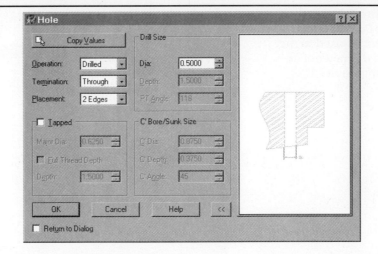

To locate the hole, pick any two nonparallel edges of any planar face—in this case edges AB and BC. After selecting the second edge, a right angle cursor will appear. Pick the approximate hole location with this cursor. Then, input the exact distances from each of the two edges—1.5 from edge AB and 1.65 from edge BC. The results are shown in **Figure 5-19.**

Figure 5-19.
The results of
placing a hole at
point D using the **2
Edges** option.

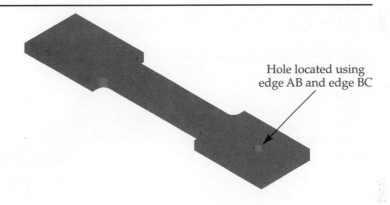

Hole located using
edge AB and edge BC

CAUTION

Note that the curved edge at the fillet in the long notch may *not* be selected for this type of hole feature.

HOLES AT A DISTANCE FROM EXISTING HOLES

The last option in the **Placement:** drop-down box is the **From Hole** option. With this option you can locate a hole at an X and Y distance from an existing hole. Open the part Example-5-07.dwg. This is the example you just finished in the previous section. The top of the part is the sketch plane. Note the orientation of the UCS—the X-axis is along the part.

Now, place a hole of the same diameter as the existing hole and with the same Y coordinate, but at 11.0 units in the negative X direction. To do this, issue the **AMHOLE** command and select **Drilled**, **Through**, and **From Hole**. See **Figure 5-20.** Click on the **Copy Values** button in the upper-left corner and then select the existing hole, Hole-D. This will copy the hole diameter into the dialog box. The diameters of two holes are not parametrically related, which means a change in the diameter of Hole-D will *not* change the new hole.

Figure 5-20.
The **Copy Values**
button is used to the
hole diameter of an
existing hole.

Click to get back to
drawing screen

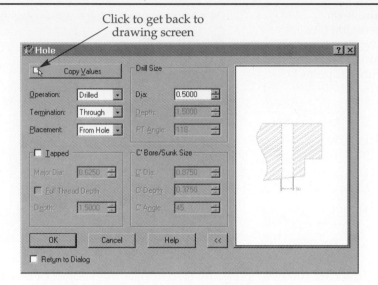

Click **OK** and select the top face as the face to drill the hole into. When prompted to rotate the X-axis, select edge AB to place it parallel to the edge. Now, select Hole-D for both the X and Y reference directions. Specify the approximate hole location with the cursor. Enter the 11.0 distance for the X direction. This is *not* a negative number, because you pointed the X direction with the approximate location. Enter zero for the Y distance. The results are shown in **Figure 5-21.**

Figure 5-21.
The result locating a
hole at an X and Y
distance from an
existing hole using
the **From Hole**
option.

New hole

Original Hole-D

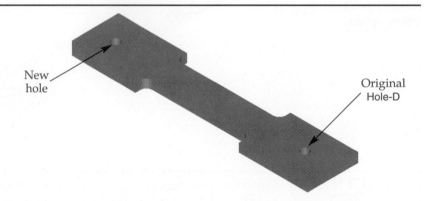

PROFESSIONAL TIP

As a test, edit the diameter of the existing hole to 0.75 units and see if the second hole changes after you update the part. Then change the distance of existing hole from edge AB to 2.0 units and update the part. The second hole should not be affected by these edits.

CHANGING THE COLOR OF HOLES

You can change the color of holes, and actually any features of the part, to give them emphasis in the shaded or rendered view. In the current drawing, Example-5-07.dwg, highlight both holes in the **Browser** by holding down the control [Ctrl] key as you select them. Then right-click to access the shortcut menu. Select **Color**, then **Apply...**, and then pick color number 40 from **Select Color** dialog box. Shade and then render the part to see the effect.

Chapter Test

Answer the following questions on a separate sheet of paper.

1. What is a *planar face*?
2. What command is used to add notes dealing with holes?
3. What does it mean when it is stated *a hole has direction*?
4. What differences are seen when viewing a threaded hole in **Model Mode** and in **Drawing Mode**?
5. In most cases, any feature except the base can be moved in the **Browser**. However, there are some limitations. Identify these limitations.
6. When adding a counterbore or countersink, the drill size diameter must be _____ than the value for **C'Dia**.
7. When adding a counterbore or countersink, what does **C'Depth** value mean?
8. Describe how the **2 Edges** option in the **Placement:** drop-down box of the **Hole** dialog is used.
9. *True or False?* When using the **2 Edges** option, it is okay to use a fillet as an edge reference.
10. When using the **From Hole** option in the **Placement:** drop-down box, the diameters of original and copied holes are not parametrically related. What does this mean?

Chapter Exercises

Exercise 5-1. Open the drawing Exercise-5-01.dwg. Drill a 0.5 diameter hole concentric with the rounded end and a 0.5 diameter hole at the center of the top. Use the dimensions from the drawing to locate the center of the hole.

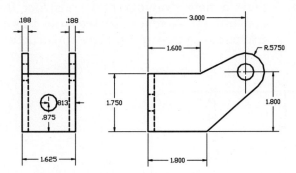

Exercise 5-2. Open the drawing Exercise-5-02.dwg. Drill an 8 diameter hole at the dimensions shown from the counterbored hole. Change the color of both holes to color 40 and shade the part.

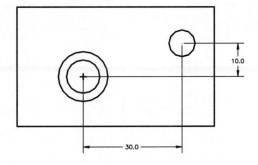

Exercise 5-3. Open the drawing Exercise-5-03.dwg. Put four 8mm through holes on the corners and a 30mm tapped hole concentric with the large radius as shown. Create two or more drawing views and put hole notes on one through hole and on the tapped hole.

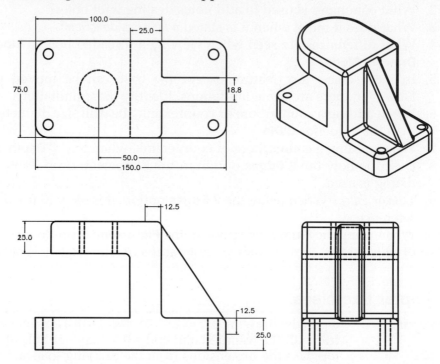

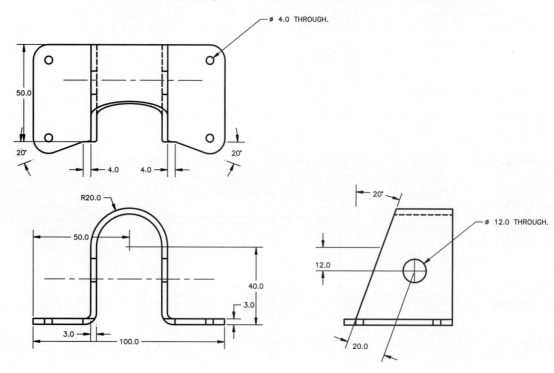

Exercise 5-4. Start a new drawing and construct the part shown using fully-constrained sketches. To see how this part was constructed, open Exercise-5-04.dwg and review construction with **AMREPLAY**.

CHAPTER 6

Dimensions, Equations, and Design Variables

Objectives

After completing this chapter, you will be able to:

* Understand he concept and format of the d0 variables.
* Display dimensions as equations.
* Use the d0 variables to logically constrain a sketch.
* Understand the concept and format of design variables.
* Understand the difference between local and global design variables.
* Use the global design variables to logically constrain a sketch.
* Export, import, and link a parameter file.
* Use the **Equation Assistant** to assign variables to dimensions.
* Change the values of design variables to change the part and drawing views.
* Create an **Excel** spreadsheet of the design variables.
* Understand how global variables control multiple parts in a file.

Drawings Needed

*The following is a list of files that you will need to work through this chapter. These files can be found on the **User's Drawings** CD included with this text.*

Examples	Practice	Exercises
Example-6-01.dwg	Practice-6-01.dwg	Exercise-6-01.dwg
Example-6-02.dwg		Exercise-6-02.dwg
Example-6-03.dwg		Exercise-6-03.dwg
Example-6-03.xls		
Example-6-04.dwg		
Example-6-05.dwg		
Example-6-06.dwg		

THE D0 VARIABLE NAMES

When you profile a sketch, each object (line, arc, or circle) is assigned a number starting with zero. These numbers are displayed when you display, edit, or add constraints. Constraints, such as tangent, also use these numbers to relate the objects. It can be seen in the sketch in **Figure 6-1** that there are six objects. The object

Figure 6-1.
This part consists of six objects that are assigned a number starting with zero. Constraints use these numbers to relate to the parts.

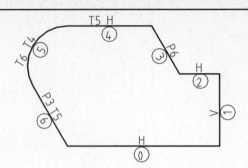

numbered zero is the horizontal line at the bottom. The arc number 5 is tangent to both line 4 and line 6 and line 3 is parallel to line 6. Every time you add and profile a sketch for a new feature, the objects in the new sketch are numbered again starting with zero.

When you add dimensions to a profiled sketch, each one is given a name consisting of the letter "d" and a number. The first dimension placed is named d0, the next is d1, the next d2, and so on. Unlike the numbers assigned to objects in sketches, the d0 numerical sequence continues for all features added to the part. This includes extrusions distances, revolutions angles, and taper angles.

Open the drawing Example-6-01.dwg. This is a fully constrained sketch with dimensions. The d0s can be displayed on the screen using the **AMDT_DIMASEQU** command. To access this command, enter AMDT_DIMASEQU at the Command: prompt, select **Dimensions As Equations** from the **Dimensioning** cascading menu in the **Part** pull-down menu, or pick the **Dimensions as Equations** button in the **Part Modeling** toolbar.

Display the constraints by selecting **Show Constraints** from the **2D Constraints** cascading menu in the **Part** pull-down menu. Note that the d0s are *not* related to the object numbers. They are sequenced based on the order the dimensions were applied. See **Figure 6-2**.

Go to the isometric view and extrude the sketch downward 1.0 units with a zero draft angle. Now right-click on ExtrusionBlind1 in the **Browser** and select **Edit**. Pick **OK** in the **Extrusion** dialog box. Two additional dimensions are shown on the screen—d7 the extrusion distance and d6 the draft angle. See **Figure 6-3**. Press [Enter] and update the part if necessary.

Now draw a 1.0 diameter circle on the sketch plane concentric to the radius in the upper-left corner. Profile the circle and apply the concentric constraint to it and the radius. The circle is numbered 0 and the radius is 1 and the N shows they are concentric. Add the diameter dimension to the circle and it is named d8. If you erase a dimension, the name is still retained and will not be used again. The dimension and constraints are shown in **Figure 6-4**. Extrude this circle **Cut**, **Through**, with a zero draft angle. There is no distance input for the extrusion and d9 becomes the draft angle. This is displayed when you edit the feature. Dimensions must be input to get a name. If you had not put the diameter dimension on the circle, the draft angle would have been d8.

Sidebar:

ADMT_DIMASEQU

Part
➥ Dimensioning
  ➥ Dimensions
     As Equations

Part Modeling
toolbar

Dimensions as
Equations

Figure 6-2.
D0s are sequences based on the order the dimensions were applied.

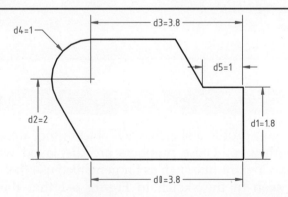

Figure 6-3.
Two additional
dimensions are
shown when the
extrusion is edited.

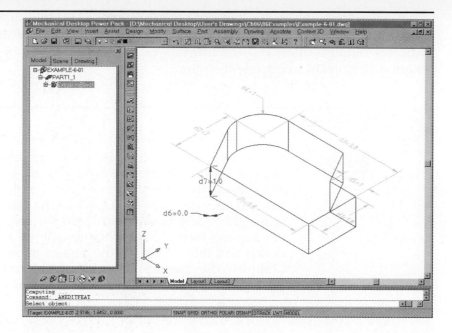

Figure 6-4.
The concentric
constraint and the
diameter dimension
for the circle.

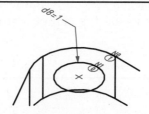

USING THE D0 NAMES TO CONTROL PART SIZE

The d0 names can be used to control the part size and the location and size of features by writing formulas for new dimensions. You can build arithmetical and trigonometric operations into these formulas. For example, design a rectangular block with a cylinder centered on the block. You want the width of the block to always be one-half the length, the thickness one-quarter the length, and the diameter of the cylinder 80% of the width.

Start a new drawing and construct a rectangle, profile it, and input the length dimension of 5.0. Place the width dimension, and for the value enter d0/2.0 at the command line or in the **Power Dimensioning** dialog box. This equation means d0 divided by 2.0 and will fix the width as one-half (1/2) of the length (d0). This is shown in **Figure 6-5** with dimensions shown as equations and the constraints displayed.

Figure 6-5.
The part with
dimensions as
equations.

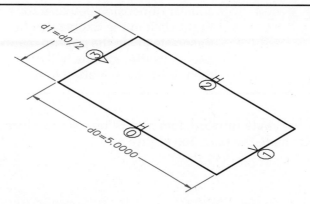

Now extrude the profile and flip the arrow until it is pointing downward. In the **Distance:** field of the **Extrusion** dialog box, enter the equation 0.25*d0. The asterisk indicates multiplication and the formula relates the extrusion distance to the part length. Click on **OK** to extrude the part.

PROFESSIONAL TIP

In simple parts, it is easy to remember what the d*x* name is for a dimension. In more complex parts with fifty or more dimensions, it is difficult. Here is how you can quickly associate a new dimension to an existing one. First use the **UNDO** command to undo the extrusion. Then extrude the profile again, but this time right-click in the **Distance:** field and select **Associate to<** in the pop-up menu. Now click on the d0 dimension on the sketch and this will place the correct name (d0) in the field. Complete the equation by entering *0.25 after the d0.

The top face of the part should be the current sketch plane. Construct a circle on the part in the top face and profile it. Input on a diameter dimension with a value of d1*0.75—that will always keep the diameter smaller than the part width. Input two linear dimensions to locate the center of the circle using equations d0/2 and d1/2 as shown in **Figure 6-6**.

Figure 6-6.
Shown is the part with the diameter dimension and two linear dimensions to locate the center of the circle.

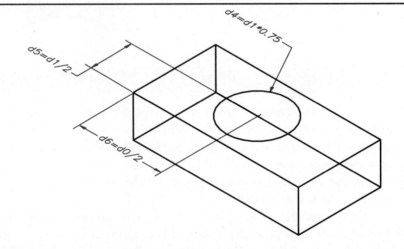

PROFESSIONAL TIP

Undo the linear dimensions. Use **Power Dimensioning** to place a linear dimension between the part edge and the circle center. Right-click in the **Expression** field of the dialog box and select **Associate to<** in the pop-up menu. Click on the edge of the box and all its dimensions will display. Pick the one you want and then finish the equation in the **Expression** field.

Extrude the circle upward **Join** and **Blind** using a distance of d3, the thickness of the part, and shade the part. See **Figure 6-7**.

Figure 6-7.
The result of the
extrusion and the
shading.

DESIGN VARIABLES

There are two problems with using the d0 variables. First, the values of dimensions can only be changed by editing the feature. Second, each part has its own set of names starting with d0 and you cannot relate dimensions of one part to another, such as a shaft diameter to a hole diameter. Using design variables can solve both of these problems. *Design variables* are special variables with characteristics as follows:

- Have individual names that you think up and input; i.e. Length_of_base.
- Have values based on numbers or equations.
- Have descriptions of what they are or represent.
- Can be used for any dimension or distance in sketches or operations.
- Can be used to set positions of work planes.
- Can be changed in a dialog box and the model and drawings will change automatically.
- Can be used on multiple parts.
- Can be exported to a file and imported from a file.
- Can be related in an Excel spreadsheet for families of parts.

So far we have only created one part in an MDT drawing but there can be more than one part and such drawings will be discussed in Chapter 14. *Global design variables* apply to all the parts in a drawing. *Local design variables* only apply to the active part.

Global Design Variables

AMVARS
WW

Part
↳ Design
Variables...

Part Modeling
toolbar

Design Variables

We will start out using the global variables. Open the drawing Example-6-02.dwg. This drawing does *not* have a part created in it, but some design variables have been input. To view these variables, enter AMVARS or WW at the Command: prompt, select **Design Variables...** from the **Part** pull-down menu, or pick the **Design Variables** button on the **Part Modeling** toolbar.

This accesses the **Design Variable** dialog box. See **Figure 6-8.** The design variables in this example are set up for a simple base block. In the **Design Variables** dialog box, the **Global** tab is displayed. After a part is created, there will be a **Local** tab in addition to the **Global** tab. The four variables shown in the dialog box have short names, good descriptions, and values for equations. The name of a variable can be up to 72 characters long, must start with a letter, may *not* contain spaces or any punctuation other than the underscore, and is not case sensitive.

CAUTION	It is good practice *not* to give variables names such as D3 or Line. D3 is a dimension name that is created automatically and Line is an AutoCAD command.

Figure 6-8.
The **Design Variable** dialog box is used to create, edit, and delete global design variables.

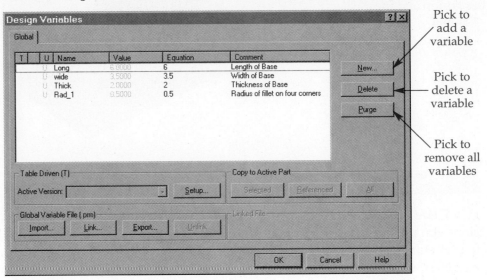

Pick to add a variable

Pick to delete a variable

Pick to remove all variables

Choosing a name is a compromise between long names that are descriptive and easy to understand, such as The_Length_of_the_Base, and short names that are easy to type but are more cryptic. The optional comment field can be used to describe variable names, which makes it easier to use shorter names.

PROFESSIONAL TIP

The **Equation Assistant** dialog box, which is covered later in this chapter, allows you to use the variable by picking the name from a list. This eliminates any chance of entering errors, and thus leads to the use of longer names. As you work with these features, you will develop your own method of using and naming design variables.

The letter U in front of the name indicates the variable is not in use. The value is the numeric result of the equation. Since each of the equations is a simple number, the value is the same as the equation. The information (not grayed out) can be changed by double-clicking and editing the entry as desired. For example, double-click on the name wide and edit it in place by changing it to Wide.

To remove a variable, select the variable you wish to remove and select the **Delete** button. To remove all of the variables, pick the **Purge** button. To add a variable, pick the **New** button. This accesses the **New Part Variable** dialog box. See **Figure 6-9.** Enter the desired information in the **Name:**, **Equation:**, and **Comment:** text boxes. Press the **OK** button and the **Design Variable** dialog box reappears with the new variable listed.

The equation can be a number, a calculated number, or a mathematical relationship between other variables. In this example, the equation for the width of the base is simply the number 3.5. However, it could be as follows:

- **0.65*Long.** This is a mathematical relationship with Long (the length of the base). The asterisk means multiply.
- **Long/2.** This is a mathematical relationship with Long (the length of the base). The slash means divide.
- **Long/wide.** This is a mathematical relationship with Long (the length of the base) divided by wide (the width of the base).

Figure 6-9.
The **New Part Variable** dialog box is used to add a variable.

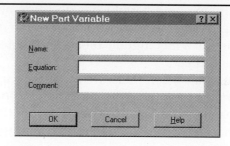

- **Long–1.25.** This is a mathematical relationship with Long (the length of the base). The hyphen (dash) means subtract.
- **Sqrt(7.125).** This is a calculated number meaning the square root of 7.125.

Excel spreadsheets are going to be used to create *table-driven parts*, so do *not* put an equal sign in front of the equation. The equal sign is not necessary and will confuse Excel if it is used. The *math operators* that can be used in the equations are shown in **Figure 6-10.**

Figure 6-10.
This table shows the math operators that can be used in equations.

Operator	Description	Example	Result
Arithmetic			
^	Exponent	3.0^3	27.0
+	Add	3.5+2.0	5.5
-	Subtract	3.5-2	1.5
*	Multiply	3.5*2	7.0
/	Divide	3.5/2	1.25
Mod	Modulus (remainder)	Mod(3.5/2)	?
sqrt	Square root	sqrt(3.5)	1.8708
log	Logarithm		
ln	Natural Logarithm		
exp(x)	exponent		
e	Base number for natural logarithm		
floor	Rounds down to nearest integer	floor(3.5)	3.0
ceil	Rounds up to nearest integer	ceil(3.5)	4.0
Trigonometry			
sin	Sine	sin(30)	0.500
cos	Cosine		
tan	Tangent		
asin	Arcsin ($\sin^{-1}$)		
acos	Arccos ($\cos^{-1}$)		
atan	Arctang ($\tan^{-1}$)		
sinh	Hyperbolic sine		
cosh	Hyperbolic cosine		
tanh	Hyperbolic tangent		
pi	pi		

The difficulty in using variables is deciding how you can use and relate them to make your design easier to create and change. For example, the four variables already created could define a rectangular block with holes on all four filleted corners and a large hole in the center.

You will find that when there are multiple parts in a file, it is a good idea to give each part a name other than the automatic Part_1, Part_2, etc. To do this in the drawing Example-6-02.dwg, select **New Part** from the **Part** cascading menu in the **Part** pull-down menu, or right-click in the **Browser** and select **New Part**. Name the part BASEBLOCK.

In the drawing, sketch an accurate rectangle 6.0 by 3.5 units with a 0.5 radius fillet and profile it. Use **AMPARDIM** to place a dimension across the bottom of the sketch. The command line shows the value. In this example, it is 6.000 since the sketch was drawn accurately. When prompted, you can enter in the variable Long or right-click in the *drawing area* and select **Equation Assistant...** from the pop-up menu. This accesses the **Equation Assistant** dialog box.

Using the Equation Assistant Dialog Box

Once the **Equation Assistant** dialog box is accessed, click on the **Global Variables** tab to show the names and equations. Double-click on the variable name and it will be shown in the calculation window. See **Figure 6-11**. Pick the **OK** button to assign the results to the dimension. Of course, the calculator can be used as well.

CAUTION

Note that you cannot edit the characteristics of design variables in this dialog box, but you can create new ones. Right-click in the **Global Variables** area and select **New.**

Repeat the process for the width and the radius. The sketch should look like **Figure 6-12**, where the dimensions are set to display as equations.

Figure 6-11.
The equations are added using the **Equation Assistant** dialog box.

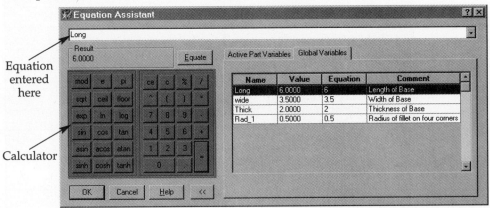

Figure 6-12.
The sketch with the dimensions as equations.

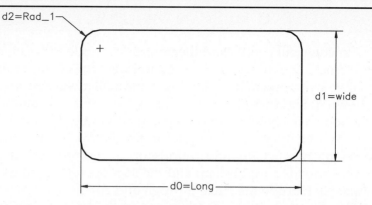

The **Equation Assistant** is also available in the **Extrusion** dialog box. Extrude the profile and flip the direction arrow so the part will be extruded downward. Set termination at **Blind**, the draft angle to zero, and right-click in the **Distance:** text box. From the pop-up menu select **Equation Assistant**, double-click on the Thick variable, and then pick the **OK** button. Of course, you could have entered Thick directly into the **Distance:** text box.

When you open the **Equation Assistant**, it defaults to the **Active Part Variables** tab. See **Figure 6-13.** In the example, this only shows the d0 automatic variables, because all the design variables were defined as global. The check box at the bottom controls the display (on or off) of the d0 automatic variables.

Design variables can be exported to a text file having the .prm extension. In early versions of MDT, the design variables were called parameters, which is the reason for the .prm extension. To export, issue the **AMVARS** command, select the **Global** tab, and click on the **Export** button. The file can then be imported into other drawings and linked. The variables will relate dimensions of parts in different drawings. Any changes in the file will affect all related drawings when the .prm file is changed and the drawing files reopened. The .prm files can also be edited using WordPad or some similar software.

Figure 6-13.
The d0 automatic variables, because all design variables were defined as global.

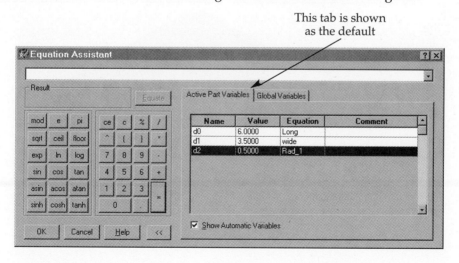

USING DESIGN VARIABLES IN EQUATIONS

Now we will profile and then extrude a circle to create a hole through the center of the block. Equations will be used to locate the center of the circle. The diameter of the circle will be related to the width and will always be an integer. This can be done as shown in **Figure 6-14** using the design variables in equations. The circle is centered in the top of the block with dimensions dividing the length by two (Long/2) and the width by two (Wide/2). The equation for the diameter takes the value of wide (3.5), divides by two (=1.75), and rounds up to the nearest integer (2). The ceil (ceiling) operator rounds a number up and the floor operator rounds it down. The equation is ceil(wide/2). Extrude the circle through and cut.

The design variables can be used in the **Hole** dialog box as well. The **Equation Assistant** can be recalled by right-clicking in the **Dia:** text box under **Drill Size**. Put a hole in each corner concentric with the fillet and with a diameter of Rad_1. Use the **Toggle Shading/Wireframe** button to view the results. See **Figure 6-15**.

Figure 6-14.
The ceil (ceiling) operator ensures that the circle's diameter will always be an integer.

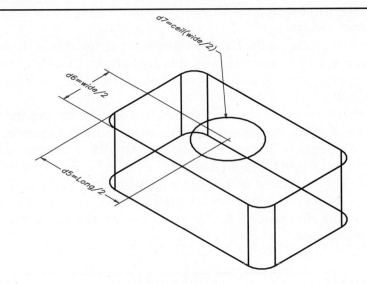

Figure 6-15.
Shown are the results of using the design variables in equations.

Changing the Equations for Design Variables

One of the great features of the design variables is that when you open the **Design Variables** dialog box and change any or all of the equations, as soon as **OK** is picked the model is automatically updated. When you switch to the **Drawing Mode**, the views are updated.

PRACTICE 6-1

❏ Open the drawing Practice-6-01.dwg. This drawing is set up for small metric parts using US standards.
❏ Input three global design variables as:
 Length: 134mm
 Dia_1: 75mm
 Dia_2: 0.50*Dia_1
❏ Draw a circle, profile it, and give it one dimension of Dia_1.
❏ Extrude it up a distance Length/2.
❏ Draw a second circle, profile it, constrain it concentric to the bottom of the part, and give it one dimension Dia_2.
❏ Extrude the circle in the up direction, **Join**, and **Blind** a distance of Length.
❏ Construct two drawing views of the shaft.
❏ Return to the part mode and change the formulas for the design variables to:
 Length: 100mm
 Dia_1: 60mm
❏ Return to the **Drawing Mode**.

CAUTION   You *must* have Microsoft Excel on your computer to complete the following section.

USING EXCEL TO CREATE TABLE-DRIVEN PARTS

In MDT, *table-driven parts* means a drawing file containing a part or parts that are dimensioned with design variables that have multiple values controlled by a linked Excel spreadsheet. An example spreadsheet is shown in **Figure 6-16**. The names of the *design variable* are the column heads (across the top) and the *version names* of the various sets of values (Generic, Small, Medium, and Large) are in the rows (down the left side). The values for the variables are in cells B3 through B5. The cells can contain numbers or equations. For example, cell C3 is the width of the small part and is calculated by the equation =0.75*B3. The Generic values were the values of the variables when the spreadsheet was created.

Open the drawing Example-6-03.dwg. For a part with a linked spreadsheet, the table and the version names are listed in the **Browser**. See **Figure 6-17A**. This part is linked to a spreadsheet called Example-6-03.xls, which should be in the same directory as the part file. If by error you get a highlight on **Global Table** in the **Browser**, select **Design Variables...** in the **Part** pull-down menu to access the **Design Variables** dialog box. In the **Global** tab, pick the **Setup...** button, pick the **Update Link** button, and then **OK** twice.

Currently the values for the table name Generic are the active ones. Double-click on Large in the **Browser** and it will become active and the part and the drawings will change. See **Figure 6-17B.** You may also right-click on one of the names and select Activate to make the change.

Figure 6-16.
This Excel spreadsheet shows design variables, version names, and variable values.

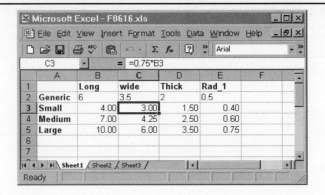

Figure 6-17.
A—Shown is the Generic version of the part. B—This is the Large version of the part. Notice the size difference.

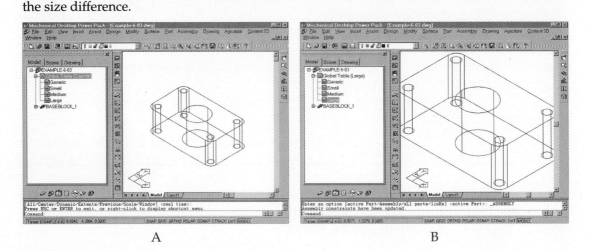

A   B

Setting Up a Linked Spreadsheet

A linked spreadsheet is very easy to set up and is created directly from the **Design Variables** dialog box. Open the drawing Example-6-04.dwg. Open the **Design Variables** dialog box and show the global variables. Click the **Setup...** button. The default setup will have *variable names* across the top and the *version names* down the left side. See **Figure 6-18.**

Now, select the **Create...** button. Change the file name to Example-6-04.xls. This creates and opens an Excel spreadsheet with the current values for the variables in the row labeled Generic. Enter version names in column A and numbers in the cells for the three variations. Refer to **Figure 6-16.** Then click **Update Link** button in the **Table Driven Setup** dialog box. The version names should now display in the **Browser**.

You can change the numbers and names in the spreadsheet at any time by selecting **Edit...** in the **Table Driven Setup** dialog box. When you are finished with the changes, press the **Update Link** button. Press **OK** twice to return to the drawing.

Figure 6-18.
A linked spreadsheet is set up using the **Table Driven Setup** dialog box.

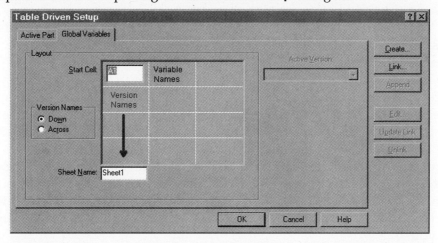

WORKING WITH ADDITIONAL PARTS IN A FILE

A large number of parts can be created in one drawing. You will use this feature in later chapters to create assemblies. In this example, observe how the global design variables can be used to relate and control the size of several parts at once. The example is to create a stepped pin that fits in the large hole, regardless of the set of variables selected (Large, Medium, or Small). Open the file Example-6-05.dwg and review the design variables. Make one of the other version names active and note the change in size of both parts. See **Figure 6-19.**

Open the file Example-6-06.dwg. This is the block from the previous example. To create the shaft, a second (new) part, either enter AMNEW or N at the Command: prompt and enter P for part when prompted to enter an option. You can also pick **New Part** from the **Part** cascading menu in the **Part** pull-down menu, select the **New Part** button on the **Part Modeling** toolbar, or right-click in a blank area of the **Browser** and select **New Part** from the pop-up menu.

It is highly recommended that you take the time to type in the part name, because this will put the same name in the **Browser** and the **Catalog** (used in assembly modeling). Call this part SHAFT. This will put the name SHAFT_1 in the **Browser** and make it the active part. It is also recommended that the part be placed on its own

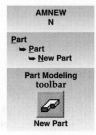

Figure 6-19.
The global design variables are used to relate and control the size of several parts at once.
A—Two related parts shown as generic versions. B—The change of versions changes both parts.

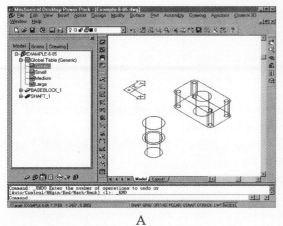

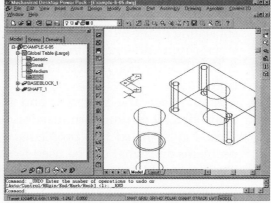

A   B

layer. The layer called Shaft exists. Make it current. The XY plane of the UCS when the part was created becomes the sketch plane. So, it is good practice to put the UCS back to the World (UCS W) before creating the part.

It does not matter in what location or orientation the new part is created. When the parts are positioned into assemblies, **3D Assembly Constraints** to move and position them. This is discussed in *Chapter 15*.

The larger diameter of the shaft is drawn first. Draw a circle in a blank area of the screen, profile it, and dimension with a value of 0.50 + Ceil(Wide/2). Remember how you dimensioned the hole in the block. This will give a step of 0.5 units larger than the hole in the center of the block. Extrude the circle downward, blind, and a distance of Thick. Notice that pressing [Enter] flips the arrow.

Now, draw a second smaller circle inside the first circle. Profile it, constrain it concentric, and dimension it with the equation of Ceil (Wide/2) . Extrude the circle downward, **Join**, and a distance of 2*Wide. See **Figure 6-20.** There are no tolerances for type of fit between the shaft and the hole. These can be edited into the dimensions of the layouts.

The next step is to create a second layout and input the views of the shaft. Select **New Layout** in the **Drawing** pull-down menu. Now select **New View...** in the **Drawing** pull-down menu. Make sure you have a Base View of the Active Part. Now selecting the various tabular parts (Large, Medium, etc.) in the **Browser** will change both the baseblock and the shaft and their respective drawings.

Figure 6-20.
Shown are the results of creating and relating two parts in a file.

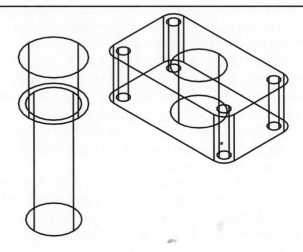

Chapter Test

Answer the following questions on a separate sheet of paper.

1. When dimensions are added to a profiled sketch, each dimension is given a name consisting of the letter _____ and a(n) _____.
2. What command is used to see the dimensions displayed as equations?
3. If the d0s are not related to the object numbers, how are they labeled on a drawing?
4. *True or False?* The part size and the location and size of features cannot be controlled by writing formulas for new dimensions.
5. This chapter listed two problems with using the d0 variables. What are these problems and what can be used to solve them?
6. Design variables are special variables with nine characteristics. List any five of the nine.
7. Describe the difference between *global design variables* and *local design variables*.

8. The equation for a design variable can be a number, a calculated number, or a mathematical relationship between other variables. Explain the following equations:
   A. 0.65*Long
   B. Long/2
   C. Long/wide
   D. Long–1.25
   E. Sin(thick)
   F. Sqrt(7.125)
9. *True or False?* The design variables cannot be changed in the **Equation Assistant** dialog box.
10. Describe how and why table-driven parts are used.

Chapter Exercises

Exercise 6-1. Open the drawing Exercise-6-01.dwg and construct the wheel using fully-constrained sketches, the revolve sketched feature, and the design variables shown. One way of constraining the sketch is also shown. Save the file for use with *Exercise 6-2*.

Design Variables		
Name	**Equation**	**Comment**
OD_Wheel	18	Outside Diameter of Wheel
ID_Wheel	4	Inside Diameter of Wheel
Thick_Rim	1.5	Thickness of Rim
Thick_Hub	1.5	Thickness of Hub
Thick_Web	1	Thickness of Web
Wide_Rim	5	Width of Rim

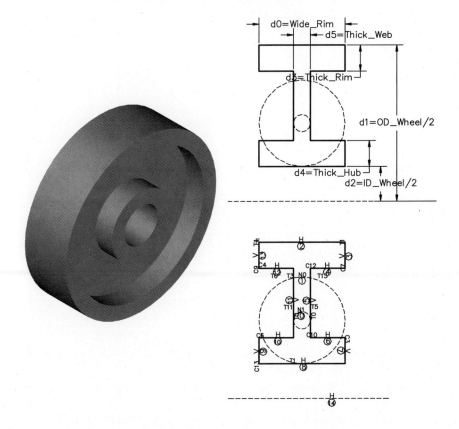

Exercise 6-2. Open the saved drawing created in *Exercise 6-1*. If you have not done *Exercise 1*, please do it now. Open the **Design Variable** dialog box and select the **Global** tab. Export the variables to a global variable file and name it Exercise-6-01-2.prm. Link the file to this drawing and keep the drawing file open.

Open the empty file Exercise-6-02.dwg and open the **Design Variable** dialog box. Import the .prm file and link it. Construct the shaft as shown, where the dimensions are shown with their design variables.

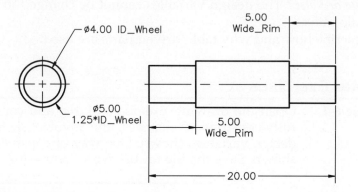

Open the .prm file in NotePad or some similar text file editor. Change the value of Wide_Rim to 3.0. Import the file using the **Design Variable** dialog box and update the part. Do this for both the Wheel and the Shaft. The same changes would take place by closing the files and then reopening them.

Exercise 6-3. Start a new drawing in metric units and construct the two parts as shown. Use the global design variables here to relate all the dimensions of the parts. Save the drawing as Exercise-6-03.dwg. Create a separate 2D layout for each part.

Global Design Variables		
Name	**Equation**	**Comment**
Wide	260	Width of base block
High	150	Height of base block
Long	300	Length of base block

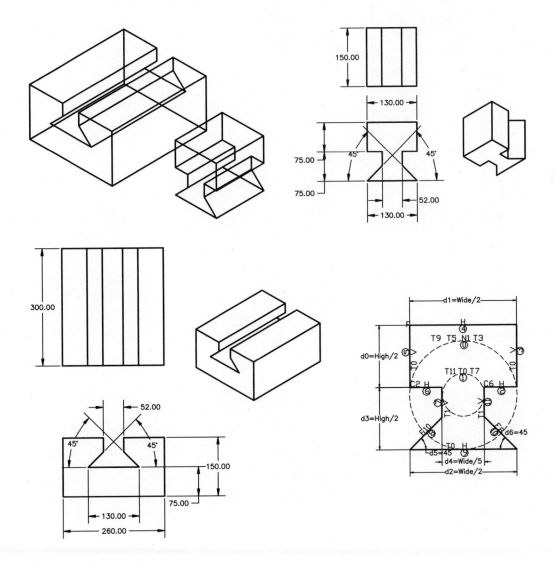

Exercise 6-4. Start a new drawing and construct the part as shown. Use seven design variables to control the part: Long, Wide, Thick, Large_Radius, Small _Radius, Large_Hole_Dia, Small_Hole_Dia.

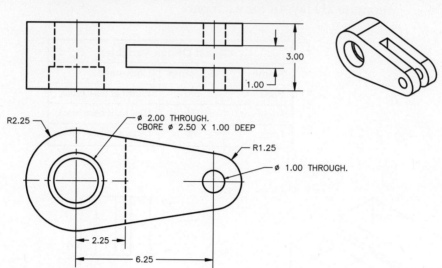

R2.25

Ø 2.00 THROUGH.
CBORE Ø 2.50 X 1.00 DEEP

R1.25

Ø 1.00 THROUGH.

3.00

1.00

2.25

6.25

CHAPTER 7

The Sketch Plane and More Features

Objectives

After completing this chapter, you will be able to:

* Understand the concept of a sketch plane.
* Change the sketch plane on a model.
* Change the viewpoint to the sketch plane.
* Use the **Shell** feature.
* Build rectangular patterns of features.
* Create a work axis on circular features.
* Create work points on the sketch plane.
* Build polar patterns of features.
* Copy sketches to the current sketch plane.

Drawings Needed

*The following is a list of files that you will need to work through this chapter. These files can be found on the **User's Drawings** CD included with this text.*

Examples	Practices	Exercises
Example-7-01.dwg	Practice-7-01.dwg	Exercise-7-01.dwg
Example-7-02.dwg	Practice-7-02.dwg	Exercise-7-02.dwg
Example-7-03.dwg	Practice-7-03.dwg	Exercise-7-03.dwg
Example-7-04.dwg		Exercise-7-04.dwg
Example-7-05.dwg		Exercise-7-05.dwg
Example-7-06.dwg		Exercise-7-06.dwg
Example-7-07.dwg		Exercise-7-07.dwg
Example-7-08.dwg		Exercise-7-08.dwg
Example-7-09.dwg		

THE SKETCH PLANE

When you create a new part, the current User Coordinate System (UCS) is the default sketch plane. In a new drawing file where the two coordinate systems, the UCS and the World Coordinate System (WCS) are coincident, this would also be the WCS. That is, all the objects in the sketches have a zero Z coordinate. This is how you built models in the first few chapters, from the ground up, extruding everything up or down from the XY plane of the UCS or what is called in MDT the *sketch plane*.

When you inserted holes, the first selection you made on the part was to select a planar face to drill the hole into. This face was temporarily changed to the sketch plane. What we want to do now is create features other than holes on different faces of the part. To do this we must first change the sketch plane permanently.

Moving the Sketch Plane

Open part Example-7-01.dwg, a block with two fillets, all extruded up from the WCS from the same sketch. Design variables are used to control the size of the part and fillets. The base sketch, with the dimensions as equations is shown in **Figure 7-1A**. The extruded part is shown in **Figure 7-1B**. The distance of the extrusion is set to the variable Thick.

Use the **AMVARS** command to access the **Design Variables** dialog box. Select the **Global** tab and note that there are the four variables of Long, Wide, Thick, and High. See **Figure 7-2**. The variable High has a U in front of it—that means it has not been used. Pick **OK** to return to the drawing.

Figure 7-1
A—A base sketch with dimensions shown as equations.
B—The extruded part.

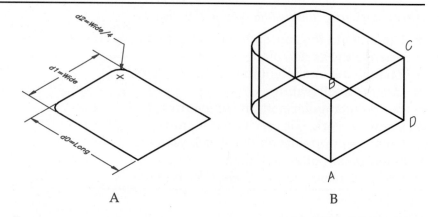

Figure 7-2.
The **Design Variables** dialog box showing the **Global** tab and the four variables.

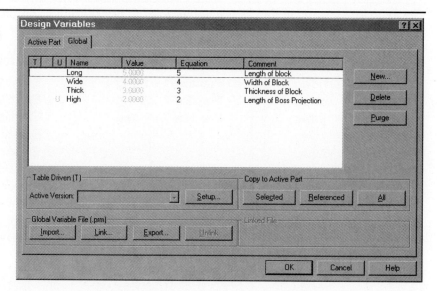

The letters ABCD show the corners of a plane, or *face* on the part. The part has six planar (flat) faces and two curved faces that are the fillets. The sketch plane is on the bottom of the part since the sketch was extruded upwards (+Z direction). We can move the sketch plane to any of the other five planar faces and thereby create features on them. We cannot move the sketch plane to a curved face because there is no clear definition of the X-and Y-axis.

CAUTION If you did 3D work in AutoCAD R14 or 2000, you may be tempted to move the UCS to the face using three points or by rotation. Do *not* do this in MDT. The sketch plane will not be parametrically related to the part! Use the following procedure and MDT will move the UCS.

Adding a Boss

AMSKPLN
SS

Part
→ New Sketch
Plane

Part Modeling
toolbar

New Sketch Plane

Create a boss at the center of face ABCD with a diameter of Thick/2 and a length of High. The **AMSKPLN** command is used to select a face for the new sketch plane. Enter AMSKPLN or SS at the Command: prompt, select **New Sketch Plane** from the **Part** pull-down menu, or click the **New Sketch Plane** button in the **Part Modeling** toolbar.

To change the sketch plane, put the cursor on the visible face that is inside the edges ABCD. The edges of the face will be highlighted. See **Figure 7-3.** Move the mouse around the part and you will see that you can select any of the planar faces. To select a hidden face such as the bottom of a part, pick on edge AD. The small mouse icon is displayed and you can left-click to choose the next face.

Highlight and click on face ABCD and accept this face by right-clicking or pressing [Enter]. The screen will display a plane, an XYZ coordinate system icon, and the mouse icon. See **Figure 7-4.** A left-click will rotate the XY axis around the Z. This has four different positions. For viewing reasons, we want the X-axis horizontal along AD and the Y-axis vertical. Right-click or press [Enter] to accept. Note that the UCS in the lower-left corner of the screen changed orientation.

Construct a circle on face ABCD, profile it, and dimension as shown in **Figure 7-5A.** Extrude the profile out from the part **Join** and a distance of High. Pick **OK** and the result is shown in **Figure 7-5B.**

Figure 7-3.
Face ABCD is
shown highlighted.

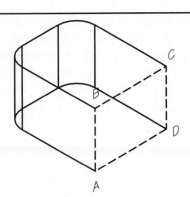

Figure 7-4.
Once the face has
been selected, the
screen will display a
plane and an XYZ
coordinate system
icon.

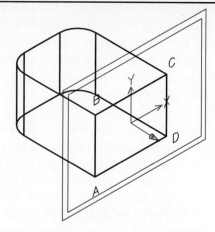

Figure 7-5.
A—A circle on face
ABCD that has been
profiled and
dimensioned.
B—The circle has
been extruded.

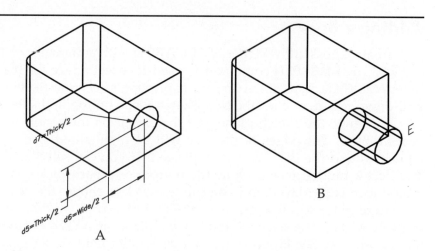

Adding a Taper to the Boss

We want to extrude the boss further, but with a taper on it. Move the sketch plane
to the face at the end of the boss, labeled E in **Figure 7-5B.** First use the **AMSKPLN**
command and move the cursor in the circular face on the end of the boss—this is a
planar face. Once this face is highlighted, left-click once and right-click twice to
accept.

You can use the edges of the part in the sketch plane to define a fully constrained
sketch. In this case, it is the circle that makes up the boundary of face E. Profile the
sketch and when prompted to select objects for the sketch, press [Enter]. Then when
you are prompted to select a part edge to close the profile, select the circle. Press
[Enter] and you have a fully constrained sketch. Extrude the profile **Join, Blind,** a
distance of **High,** and with a –7.0° draft angle. The results are shown in **Figure 7-6.**

Figure 7-6.
The extrusion has
been extended and
tapered.

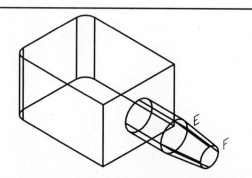

GETTING A PLAN VIEW OF THE SKETCH PLANE

Sometimes it is helpful in creating sketches to view the part with the sketch plane parallel to the screen. This is called the *plan view* of the current UCS in AutoCAD. It is especially helpful when you want to use the edge of a cylinder, called the *silhouette curve*, to close or align a profile. Remember, to use the silhouette curve to close a profile, the system variable DISPSILH must be set to 1.

Open the drawing Example-7-02.dwg. This block has six planar faces and one of them is labeled ABCD. See **Figure 7-7.** Make this face the sketch plane and if the X-axis is not parallel with edge AB, rotate it until it is and then accept it. The direction of the X-axis is important, because it will determine the orientation of the part in the view of the sketch plane.

To change the viewpoint to the plan view of the sketch plane, enter 9 at the Command: prompt and press [Enter]. You can also select **Current UCS** from the **Plan View** in the **3D Views** cascading submenu of the **View** pull-down menu, or click the **Sketch View** button in the **Part Modeling** toolbar.

Note that edge AB is horizontal and parallel to the X-axis. See **Figure 7-8.** Sketches drawn in this view will appear in their true size and shape.

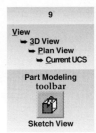

Figure 7-7.
Shown is a block
with six planar faces
and one of them
labeled ABCD.

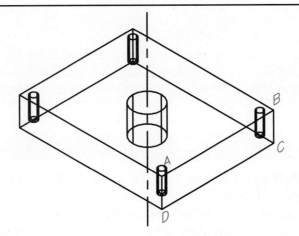

Figure 7-8.
Shown is the plan
view of the sketch
plane.

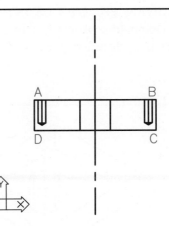

❑ Open the drawing Practice-7-01.dwg.

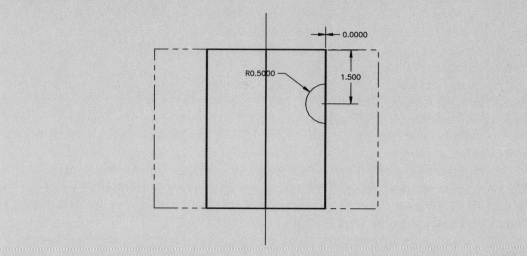

❑ Use the **AMSKPLN** command to make the red work plane the new sketch plane.
❑ Set the system variable **DISPSILH** to 1.
❑ Get the plan view of the sketch plane.
❑ Construct the arc shown here.
❑ Profile the arc and use the edge of the cylinder to close the profile.
❑ To fully constrain the sketch, apply the three dimensions as shown.
❑ Revolve the profile, **Cut**, and 360° about the vertical red work axis.

USE THE SHELL FEATURE

To *shell* a part means to make a part hollow with a specified wall thickness and to indicate which faces of the part are to be left open. The part in **Figure 7-9A** shows a solid part and the shelled part in **Figure 7-9B**. The top of the part was *excluded*, or left open, and the face ABCD has a different wall thickness. Note the hole in the part gets a wall around it as well.

Open the drawing Example-7-03.dwg. To access the **AMSHELL** command, enter AMSHELL at the Command: prompt, select **Shell** from the **Placed Features** cascading menu in the **Part** pull-down menu, or pick the **Shell** button on the **Part Modeling** toolbar.

AMSHELL

Part
↳ Placed Features
  ↳ Shell

Part Modeling
toolbar

Shell

Figure 7-9.
A—Shown is a solid part. B—The part with the top shelled and one end of a different thickness.

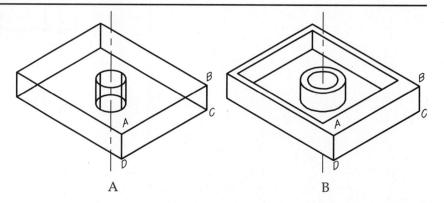

Figure 7-10.
The **Shell Feature**
dialog box has three
choices for the
direction in which
the walls will be
formed.

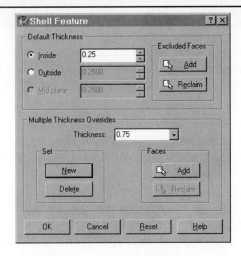

This accesses the **Shell Feature** dialog box. See **Figure 7-10.** There are three choices for the direction in which the walls will be formed. These choices are **Inside**, **Outside**, and **Mid-plane** and are found in the **Default Thickness** area. Select the **Inside** radio button and set the wall thickness at 0.25. To exclude a face, click the **Add** button in the **Excluded Faces** area, pick the top face of the box, and press [Enter] twice. Now, click on the **New** button in the **Multiple Thickness Overrides** area and enter 0.75 for the thickness. Click on the **Add** button in the **Multiple Thickness Overrides** area, pick face ABCD, and press [Enter] twice. Click **OK** and the part will be shelled.

CAUTION	A part can only have one shell feature! There is a work-around. Create two parts, shell each one, and then combine them as described in *Chapter 14*.

PRACTICE 7-2

❏ Open the drawing Practice-7-02.dwg. This is a 3D pipe. The construction of this type of part is described in *Chapter 13*.
❏ We want to shell the pipe outward so that the existing face becomes the inside of the pipe as shown here.

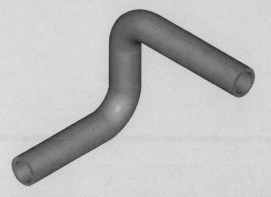

❏ Issue the **AMSHELL** command. Set the default thickness to **Outside** and 0.20.
❏ Click on the **Add** button in the **Exclude Faces** area and pick the ends of the pipe at A and B. Make sure that when you pick the face at A, only the circle is highlighted. Do the same at B.
❏ Press **OK** in the dialog box to shell the pipe.

CREATING RECTANGULAR PATTERNS OF FEATURES

Almost any feature or set of features on the part can be used to create a pattern of features. In previous versions of MDT and in AutoCAD 2000i, patterns are called *arrays*. A pattern is created in the current sketch plane, which can be changed before using the **AMPATTERN** command. **Figure 7-11** shows a pattern of four rows that go along the Y-axis of the UCS and are labeled 1 through 4. The five columns, labeled A through E go along the X-axis. This may also be a linear edge of the part or at a specified angle, in this case 60°. The example also shows in the upper right that a pattern can "fall off the edge" of a part.

Open the drawing Example-7-04.dwg. This plate has a rod welded to it at an angle. There is a drilled hole in the rod. Set the sketch plane to the top of the plate. To create a pattern of these rods, use the **AMPATTERN** command. To access this command, enter AMPATTERN at the Command: prompt, select **Rectangular Pattern** from the **Placed Features** cascading menu in the **Part** pull-down menu, or pick the **Rectangular Pattern** button on the **Part Modeling** toolbar. You are now prompted to select the features to be patterned. Select the rod and the drilled hole and press [Enter].

AMPATTERN

Part
→ Placed Features
→ Rectangular Pattern

Part Modeling toolbar

Rectangular Pattern

Figure 7-11.
A rectangular pattern of a circle feature set in four rows, five columns, and at a specified angle of 60°.

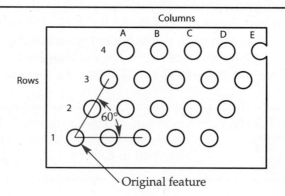

PROFESSIONAL TIP

Multiple features such as the rod and the drilled hole can be selected by holding down the [Ctrl] key, picking each feature in the **Browser**, and then right-clicking to access a shortcut menu. Then select **Pattern** and **Rectangular...** from the shortcut menu.

This accesses the **Pattern** dialog box. On the part display there are arrows showing the direction of the columns and rows. Each element in the pattern is called an *instance*. Any edits to the original feature changes every instance in the entire pattern. A specific instance can be suppressed. Input 5 for the number of column instances and 1.75 for the incremental column spacing. Input 3 for the number of row instances and 4.00 for the included row spacing. See **Figure 7-12**. Note the difference between the incremental and the differential spacing.

The **True Preview** choice is unselected in **Preview** pop-up menu. See **Figure 7-13**. This type preview can be time-consuming due to the generation of the actual pattern. The **Dynamic Preview** shows simple boxes for all instances. Click on **OK** and the pattern will be created. See **Figure 7-14**.

Figure 7-12.
The **Pattern** dialog box is used to input the number of columns and rows used to create rectangular arrays.

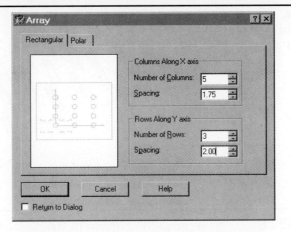

Figure 7-13.
A—The **Preview** options.
B—The dynamic preview of the rectangular pattern.

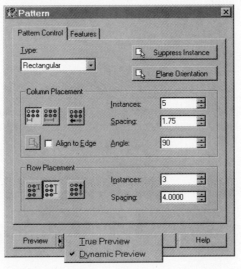

A

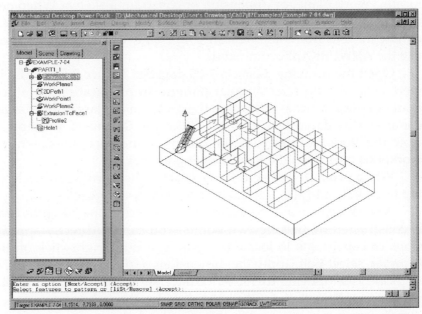

B

Figure 7-14.
The result of the
rectangular array.

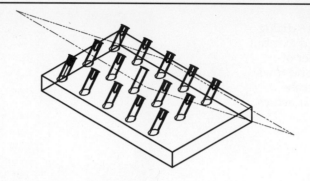

NOTE	To send the pattern in the negative direction along an axis, use the **Flip Row** or **Flip Pattern Direction** button in the dialog box.

PROFESSIONAL TIP	Features can be selected on the part after issuing the command. Features can also be added and deleted in the features tab of the **Pattern** dialog box. Features can include existing patterns resulting in "Patterns of Patterns" and features based on silhouette curves such as the groove created in *Practice 7-1*. Several types of features, most importantly fillets and chamfers, cannot be patterned individually.

CREATING A WORK AXIS AND A WORK POINT

A work axis can be used to revolve a sketch, create a polar (circular) pattern, locate a work plane, or dimension or constrain a sketch. When created, the axis goes on the AM_WORK layer and you can set the color and linetype of this layer to differentiate the axis from other features. To create an axis on a cylindrical feature, we will use the **AMWORKAXIS** command.

Open the drawing Example-7-05.dwg. See **Figure 7-15A.** The holes in the corners were created with a rectangular pattern. In preparation for a polar pattern, we are going to put a work axis through the center of the large hole. Issue the **AMWORKAXIS** command by entering AMWORKAXIS at the Command: prompt, selecting **Work Axis** from the **Work Features** cascading menu in the **Part** pull-down menu, or picking the **Work Axis** button on the **Part Modeling** toolbar.

When prompted to select a cylinder, cone, torus, or sketch, click on the edge of the large hole and the work axis is created. See **Figure 7-15B.**

A work axis can also be created on the sketch plane using the **Sketch** option. The axis is constructed by picking two points. You cannot directly apply parametric dimensions or constraints to locate the axis. You must right-click on the work axis in the **Browser**, select **Edit**, apply the dimensions and constraints, and then update the part.

Open the drawing Example-7-06.dwg. See **Figure 7-16.** We want to create a bolthole pattern (a polar pattern) of the hole labeled A about the center of the top face of the part. A polar array requires a work axis, circular feature, or work point as the center of the array. The top of the plate is the sketch plane. Now locate a work point at its center.

To access the work point command, enter AMWORKPT at the Command: prompt, select **Work Point** from the **Work Features** cascading menu in the **Part** pull-down menu, or pick the **Work Point** button on the **Part Modeling** toolbar.

Figure 7-15.
A—A plate with four holes created by a rectangular array, a large center hole, and a small hole that will be used for a polar array.
B—A work axis is added to the part.

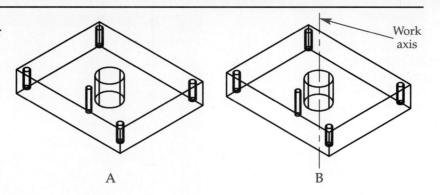

A

B

Work axis

Figure 7-16.
A plate with one hole that will be used to create a bolthole pattern.

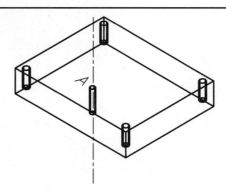

When prompted, pick at about the center of the top of the plate. Just as if it were a sketch, dimension the point using the **AMPARDIM** command as shown in **Figure 7-17**. This will fully constrain the work point. Note that the dimensions will remain on the drawing until the point is used for another feature.

Figure 7-17.
A work point is added and dimensioned.

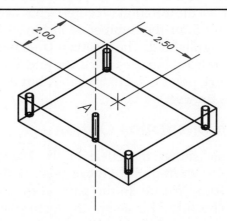

2.00

2.50

CREATING POLAR PATTERNS

Open Example-7-07.dwg, which is the same as the completed *Example 7-6*. Make sure the top of the plate is the sketch plane. The axis of rotation must be perpendicular to the sketch plane. Now we want to create a polar pattern of the hole A using the **AMPATTERN** command. To access this command, enter AMPATTERN at the Command: prompt, select **Polar Pattern** from the **Placed Features** cascading menu in the **Part** pull-down menu, or pick the **Polar Pattern** button on the **Part Modeling** toolbar.

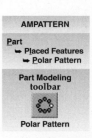

AMPATTERN

Part
→ Placed Features
→ Polar Pattern

Part Modeling
toolbar

Polar Pattern

Figure 7-18.
The **Pattern** dialog
box is used to input
the number of
instances and the
type of polar
pattern.

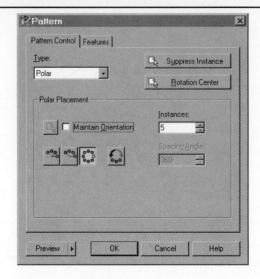

Figure 7-19.
The result of the
polar pattern.

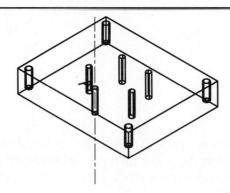

When prompted to select the feature to be patterned, select hole A and then select
the work point as the rotational center. This accesses the **Pattern** dialog box. See
Figure 7-18. Input 5 for the number of instances and make sure that the **Full Circle**
radio button is highlighted. The **Maintain Orientation** check box only makes a difference for noncylindrical features. Pick **OK** and the results are shown in **Figure 7-19.**
Note that the work point is set to invisible, but it is not deleted. You can see that it is
still listed in the **Browser**.

The **Maintain Orientation** Option

This example shows the effect of the **Maintain Orientation** option. Open the
drawing Example-7-08.dwg. See **Figure 7-20.** Both parts have polar patterns of the
same noncylindrical feature. In the part on the right the **Maintain Orientation** check
box was left unchecked. This means the feature orientation is *not* maintained so that
it is rotated as it is copied—much like the spokes on a bicycle wheel. In the part on
the left, the **Maintain Orientation** check box was checked. The orientation is maintained using the work point as a reference—much like the cars in a Ferris wheel.

**PROFESSIONAL
TIP**

The **AMPATTERN** command can be used to quickly mirror
symmetrical features across a part as shown in the *Practice 7-3.*

Learning Mechanical Desktop R5

Figure 7-20.
The left pattern was done with the **Maintain Orientation** check box unchecked. The right pattern was done with the **Maintain Orientation** check box checked.

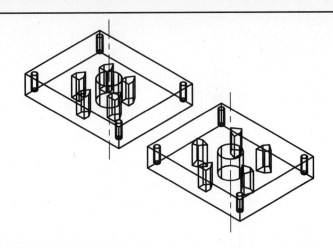

PRACTICE 7-3

❑ Open the drawing Practice-7-03.dwg. This is an injection molded plastic cap with one snap labeled FEATURE A. We want a mirror image of the snap on the other side of the slot to get the finished cap shown here.
❑ Set the sketch plane to the top face of the cylinder.

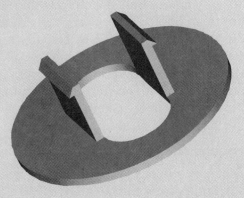

❑ Create a work axis using the outer edge of the cylinder.
❑ Issue the **POLAR PATTERN** command and select FEATURE A and then select the work axis for the center of the array.
❑ In the dialog box, set the number of instances at 2, pick the **Full Circle** radio button. Do not check the **Maintain Orientation** check box, and then pick **OK**.

COPY SKETCHES TO THE SKETCH PLANE

A sketch of a feature can be copied to a new sketch plane to create a feature that would be difficult to pattern or one that is not symmetrical. This process also copies most of the dimensions from the original sketch, which can save time and increase accuracy.

Open the drawing Example-7-09.dwg. We want to put a feature identical to Feature A on Face B. See **Figure 7-21.** To understand how Feature A was originally created, right-click on it in the **Browser** and pick **Edit Sketch**. Note that there are six dimensions, five *internal* that define the shape of the part, and one *external* (the 0.5 horizontal dimension) that locates the sketch to the left end of the Baseplate. If you display the constraints, you will see that the bottom of the sketch is collinear to the top of the Baseplate. The external dimension and the collinear constraint will *not* be copied. To prepare the part to copy the sketch, make Face B the sketch plane and align the X-axis with the top

Figure 7-21.
A—A plate showing Feature A and Face B. B—The feature has been copied to the new sketch plane, profiled, constrained, and extruded.

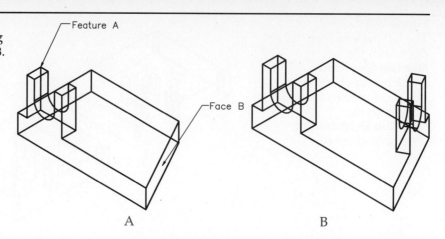

A   B

AMCOPYSKETCH

Part
  ↳ **Sketch Solving**
    ↳ Copy Sketch

Part Modeling toolbar

Copy Sketch

edge of Face B with the Y-axis pointing up. Also freeze the layer Text. This will hide the two leaders making the view less confusing.

To copy a sketch, use the **AMCOPYSKETCH** command. To access this command, enter AMCOPYSKETCH at the Command: prompt, select **Copy Sketch** from the **Sketch Solving** cascading menu in the **Part** pull-down menu, or pick the **Copy Sketch** button on the **Part Modeling** toolbar.

You are prompted to copy an existing sketch (one that has not been extruded) or the sketch of a feature. Input an F for feature and press [Enter]. Pick Feature A and the sketch and its internal dimensions are displayed. You are then prompted for a sketch center. Pick a point on Face B, the current sketch plane, to locate the sketch. You can keep picking points until you are satisfied and then press [Enter]. Profile the sketch with the **AMPROFILE** command and make sure you include the dimensions. Resolve the sketch by selecting **Re-Solve** from the **Sketch Solving** cascading menu in the **Part** pull-down menu and the command line should read:

> Solved under constrained sketch requiring 2 dimensions or constraints.

Make the bottom of the sketch collinear with the top of Face B and input a dimension to locate the sketch 0.5 units from the back edge. Extrude the sketch **Join**, **Blind**, 0.5 units into the part.

Chapter Test

Answer the following questions on a separate sheet of paper.

1. In MDT, what is the *sketch plane*?
2. What command is used to move the sketch plane?
3. What is a *plan view*?
4. What is a *silhouette curve*?
5. *True or False?* To shell a part means to make a part hollow with a specified wall thickness.
6. Each element in the array is called a(n) _____.
7. When creating a work axis, it goes on the _____ layer.
8. What command is used to insert a work point?
9. Describe the differences between a rectangular and polar array.
10. *True or False?* To create a polar array, the sketch must be perpendicular to the work axis.

Chapter Exercises

Exercise 7-1. Open drawing Exercise-7-01.dwg. As shown here, construct cylindrical bosses 7.0 units in diameter and 12 units long on the centers of the three faces.

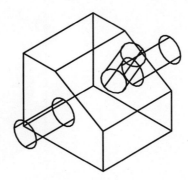

Exercise 7-2. Open drawing Exercise-7-02.dwg. As shown here, shell the part with a 3mm wall thickness.

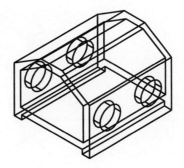

Exercise 7-3. Open drawing Exercise-7-03.dwg. As shown here, shell the part with a 0.75 wall thickness.

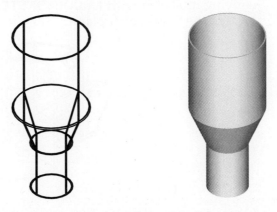

Exercise 7-4. Open drawing Exercise-7-04.dwg. As shown here, create a rectangular pattern of 0.5 diameter holes equally spaced across the part.

Exercise 7-5. Open drawing Exercise-7-05.dwg and review the part construction with **AMREPLAY** to see how a work point was used. As shown here, construct a polar pattern of 0.50 diameter holes on a 3.5 diameter circle in the inclined face.

Exercise 7-6. Open drawing Exercise-7-06.dwg. As shown here, mirror the feature using a work point and a polar pattern.

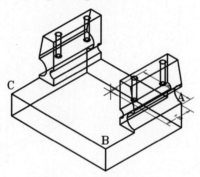

Exercise 7-7. Open drawing Exercise-7-07.dwg and review the part construction with **AMREPLAY**. Construct a mirror image of the BIG_EAR feature by copying its sketch with **AMCOPYSKETCH**. Note the orientation of the UCS in the new sketch plane.

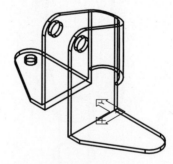

Exercise 7-8. Open drawing Exercise-7-08.dwg and review the part construction with **AMREPLAY**. The groove was created using a silhouette curve. As shown here, create 3 additional grooves spaced 30mm apart using a rectangular pattern.

CHAPTER 8

Work Planes

Objectives

After completing this chapter, you will be able to:

* Understand the concept of a work plane.
* Create work planes on the model.
* Set the sketch plane to a work plane.
* Use the **AMRIB** command to create a rib from a single line.
* Use work planes to terminate extruded and revolved features.
* Create a work axis on the sketch plane.
* Create a hole on a work point.
* Use the work features for parametric dimensions and constraints.
* Use the **AMPARTSPLIT** command with work planes to split a part.

Drawings Needed

*The following is a list of files that you will need to work through this chapter. These files can be found on the **User's Drawings** CD included with this text.*

Examples	Practice	
Example-8-01.dwg	Practice-8-01.dwg	Exercise-8-06.dwg
Example-8-02.dwg		Exercise-8-07.dwg
Example-8-03.dwg	**Exercises**	Exercise-8-08.dwg
Example-8-04.dwg	Exercise-8-01.dwg	Exercise-8-09.dwg
Example-8-05.dwg	Exercise-8-02.dwg	Exercise-8-10.dwg
Example-8-06.dwg	Exercise-8-03.dwg	Exercise-8-11.dwg
	Exercise-8-04.dwg	Exercise-8-12.dwg
	Exercise-8-05.dwg	

THE WORK PLANE

In *Chapter 7* you saw that you could put a sketch plane on any planar face of the part. However, what if you want to create a feature on a plane that is not represented by a face? This is the job for work planes.

A *work plane* is a permanent feature related to the part and is therefore listed in the **Browser.** There can be as many work planes as needed on the part. The sketch plane can be set to any planar face on the part or to any of the work planes. Of course you can only have one sketch plane at any one time. If you delete a work plane, all the features that used the work plane in their construction will be deleted as well.

Work planes are used to create sketches that are not on a planar face of the part, to terminate extrusions or revolutions, and to locate dimensions and constraints. Right-click on a work plane in the **Browser** to set the plane visible or invisible, rename it, delete it, suppress it, change its properties (color and layer), and to edit any numeric values that were associated with its creation. Since work planes are automatically created on the layer AM_WORK, it is probably not a good idea to change the layer of an individual plane. It is a good idea to set the linetype and color of AM_WORK in your template drawings. In a new drawing, AM_WORK does not exist until a work feature is created, so when you set up a template you will have to create a layer AM_WORK.

When you first start a drawing and before a part is created, you can make work planes that are related to the World Coordinate System. Work planes parallel to the world XY, YZ, and ZX axis can be created in one step with the **AMBASICPLANES** command. This is very useful for round parts such as shafts and pins that need features perpendicular to the axis. The details of this will be explained in *Example 8-3*. After you create a part, do not build work planes based on the UCS or the WCS. The features based on these planes will not be fully related to the part. An example of this is demonstrated in *Example 8-5*.

CREATING A WORK PLANE

Open part Example-8-01.dwg. See **Figure 8-1A.** This is a filleted plate with a vertical cylinder. The plate is extruded down from the starting sketch plane (the WCS) and the cylinder extruded up. A through hole is placed concentric to the top of the cylinder. Design variables are used to control the size of the part and features. Remember, **AMREPLAY** can be used to review the construction. We are going to add the three webs supporting the cylinder as shown in the finished part in **Figure 8-1B.** The first web will be constructed on a vertical work plane through the center of the cylinder. Then the web will be patterned about the center of the cylinder.

AMWORKPLN

Part
→ **Work Features**
→ **Work Plane...**

Part Modeling toolbar

Work Plane

The **AMWORKPLN** command is used for creating the plane. To issue this command, enter AMWORKPLN at the Command: prompt, select **Work Plane** from the **Work Features** cascading menu in the **Part** pull-down menu, or pick the **Work Plane** button on the **Part Modeling** toolbar.

Figure 8-1.
A—A filleted plate with cylinder.
B—The filleted plate with three webs between plate and cylinder.

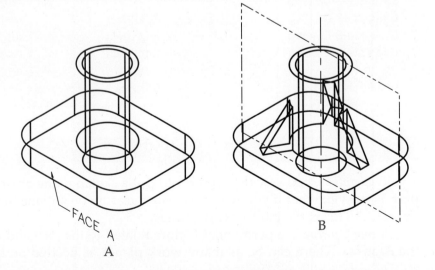

Figure 8-2.
The **Work Plane** dialog box showing settings for inserting a work plane.

Check to make work plan the new sketch plane

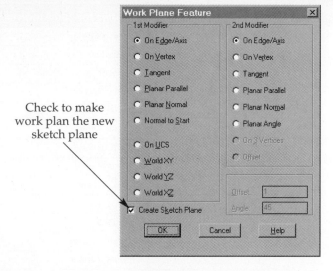

This accesses the **Work Plane Feature** dialog box. See **Figure 8-2.** Near the bottom of this dialog box there is the **Create Sketch Plane** check box. If this is checked, the new work plane will become the sketch plane. You will be required to set the orientation of the axis with the same procedure as putting a sketch plane on a planar face. If it is not checked, the sketch plane does not move and there will be no axis to orient. For this example, the **Create Sketch Plane** check box should be checked.

The **1st Modifier** and **2nd Modifier** areas have a number of choices. The last four choices in the **1st Modifier** area are for the UCS and the World, and do not require a second modifier. Click **On Ucs** and notice the second column is totally grayed out. Remember, do not use these four choices after you create the base sketch. The **Normal to Start** choice creates a sketch plane normal to the start point of a path. This is used to create a sketch to sweep along a 2D or 3D path. This will be discussed further in *Chapter 13.*

If you select **On Edge/Axis**, notice that you have six choices for the second modifier. An edge is an edge of a face on the part and an axis is a work axis. Think about the geometry required to define a plane. The geometry required is one of the following:
- Three noncollinear points.
- One line and one noncollinear point.
- One line and a tangent surface.
- Two noncollinear lines in the same plane.
- One line and a plane that is parallel, normal, or at an angle.

A map of the available second modifiers for each choice of first modifier is shown in **Figure 8-3.**

A *vertex* is a corner of any face on the part. Only two choices require numerical values: selecting **Offset** requires inputting an offset distance and selecting **Planar Angle** requires inputting an angle. Some of the choices in **Figure 8-3** are the same and these are shown in color. It does not matter whether you pick **Tangent** and then **Planar Parallel**, or **Planar Parallel** and then **Tangent**. So, there are really only twelve choices. Pick the **Cancel** button to close the dialog box.

Figure 8-3.
This table shows the 1st and 2nd modifier combinations.

Pick the 1st Modifier	Choice for 2nd Modifier	Value Required
On Edge/Axis	On Edge/Axis	None
	On Vertex	None
	Tangent	None
	Planar Parallel	None
	Planar Normal	None
	Planar Angle	Angle
On Vertex	On Edge/Axis	None
	Planar Parallel	None
	On 3 Vertices	None
Tangent	On Edge/Axis	None
	Tangent	None
	Planar Parallel	None
	Planar Normal	None
Planar Parallel	On Edge/Axis	None
	On Vertex	None
	Tangent	None
	Offset	Offset Distance
Planar Normal	On Edge/Axis	None
	Tangent	None

Creating a Work Axis Through the Cylinder

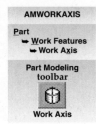

AMWORKAXIS

Part
→ Work Features
→ Work Axis

Part Modeling toolbar

Work Axis

For our example, we first need to create a work axis through the cylinder. Use the **AMWORKAXIS** command. To access this command, enter AMWORKAXIS at the Command: prompt, select **Work Axis** from the **Work Features** cascading menu in the **Part** pull-down menu, or pick the **Work Axis** button on the **Part Modeling** toolbar.

Now pick the outside circle of the top of the cylinder for the axis. Do not pick the circle for the hole because the web we create will interfere with the hole. We may want to be able to move the Hole1 feature downward in the **Browser**. If the new work plane was based on the edge of the hole, then the geometry would be related and that means we could not move the hole beneath the new work plane in the **Browser**.

The layer AM_WORK is created as soon as a work feature is created, in this case the work axis. Enter the **LAYER** command and change the color of AM_WORK to Red. To do this, click on the color square in the row and select Red. Then, change the line-type to PHANTOM2. The dialog box with changes is shown in **Figure 8-4.** These are arbitrary choices, but they seem to work well for good visibility and recognition.

Creating a Rib Feature

To create the rib, we need a work plane parallel to the front face of the base labeled FACE A and through the work axis. Use **AMWORKPLN** command and choose **Planar Parallel** and **On Edge/Axis** and check the **Create Sketch Plane** check box. See **Figure 8-5.** Click the **OK** button and select the front face of the base and press [Enter] to accept. Now select the work axis and rotate the displayed coordinate system with left-clicks until the Y is vertical and the X horizontal to the right. Then right-click or press [Enter] to accept. The part with the work plane is shown in **Figure 8-6**. Enter 9 to get the sketch view and then set the system variable **DISPSILH** to 1 so the silhouette curve can be used.

Construct a single line representing the top edge of the rib. Issue the **AMPROFILE** command, select the line and press [Enter] twice. This will create an "open profile." Use the project constraint to attach the lower-right end to the top of the base. Apply

Figure 8-4.
Layer Properties Manager dialog box showing the changes to the AM_WORK layer.

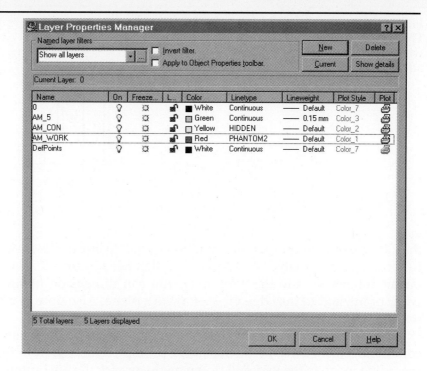

Figure 8-5.
The **Work Plane** dialog box showing settings for inserting a work plane.

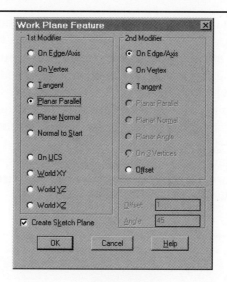

Figure 8-6.
The part with work axis and work plane, which is also the sketch plane. Notice the change in the UCS icon location.

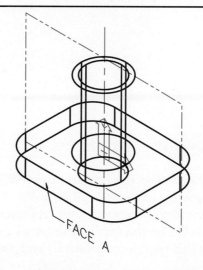

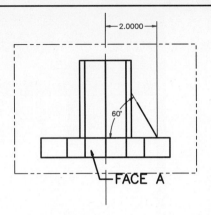

Figure 8-7.
The sketch of the
web with dimension
added.

the 2.0 (use the design variable equation Long/2.5) linear dimension and the 60-degree (use the design variable Web_Angle) angular. See **Figure 8-7.** Note that the horizontal linear dimension is to the work axis. You can dimension features to a work axis, a work point, and to the edge view of a work plane. Also note that the top of the line is not constrained to the silhouette curve.

The **AMRIB** command is used for creating the rib from the single line. To issue this command, enter AMRIB at the Command: prompt, select **Rib** from the **Sketched Features** cascading menu in the **Part** pull-down menu, or pick the **Rib** button on the **Part Modeling** toolbar.

This command displays the **Rib** dialog box shown in **Figure 8-8A.** Select **Midplane** in the **Type:** drop-down list and enter the design variable Web_Thick in the **Thickness** text box. Make sure the **Fill Direction** arrow is pointing toward the cylinder and click on **OK**. The rib automatically fills in to the curved cylinder and the flat base. Enter 8 to get a better view of the rib.

The next step is to pattern the web about the work axis. Issue the **AMPATTERN** command, select the web, and press [Enter]. In the **Pattern** dialog box, select Polar in the **Type:** drop-down list. Select the work axis as the rotational center. Input 3 for the number of instances and click **OK**. Finally, change the color of the Rib1 and the Polar Pattern1 for emphasis. Use the **Toggle Shading/Wireframe** button and view the results. See **Figure 8-8B.**

Figure 8-8.
A—The **Rib** dialog box.
B—The finished,
shaded part.

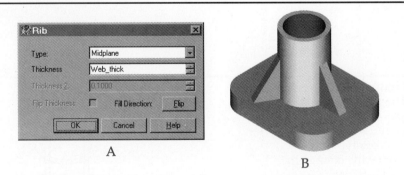

OFFSET WORK PLANES AND EDITING WORK PLANES

With **Planar Parallel** as the first modifier, **Offset** can be used as the second modifier. Open the drawing Example-8-02.dwg. We want to create a groove on the inside diameter as shown in the section views in **Figure 8-9**. One way to do this is to create a work plane parallel to the top of the small cylinder and offset downward. Then, sketch a circle on this plane, constrain it, and extrude it **Cut** and downward.

To create this groove on the inside diameter, enter **AMWORKPLN** and select **Planar Parallel** and **Offset**, enter 1.25 for the offset, and check **Create Sketch Plane**. See **Figure 8-10.** Select the top face of the small cylinder. A blue rectangle will appear with arrows pointing up. Change the direction by entering F for Flip, or by clicking anywhere on screen. Then press [Enter] or right-click to accept. With the arrows pointing down, the work plane will be below the top face. Press [Enter] again to accept the orientation of the coordinate system. Change the color and linetype of layer AM_WORK to Red and PHANTOM2. See **Figure 8-11.**

Entering 9 displays the plan view of the current sketch plane. In this case, this view is also the top view. Draw a 1.0 unit diameter circle, profile it, constrain it concentric to outside circle of the part, and dimension it. In **Figure 8-12,** the circle is highlighted. Extrude the circle **Cut**, **Blind**, a distance of 0.25 units in the downward direction.

Figure 8-9.
The finished part in Example-8-02.dwg shown in section view.

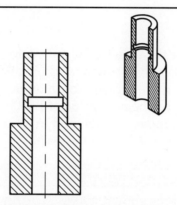

Figure 8-10.
The **Work Plane** dialog box showing settings for inserting a work plane.

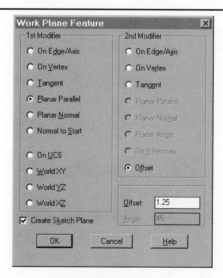

Figure 8-11.
The work plane that is offset from the top face of the small cylinder.

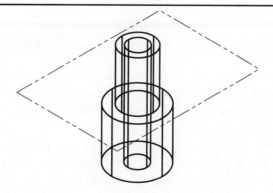

Figure 8-12.
The top view with
the profiled and
dimensioned circle
shown in color.

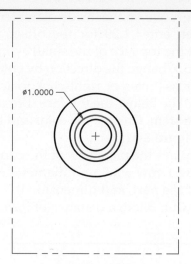

The amount of the offset used to create the work plane can be changed. Right-click on WorkPlane I in the **Browser** and pick **Edit** in the pop-up menu. The offset value of 1.2500 appears on the screen. You may need to zoom out to see the entire dimension. Select it and enter the new value of 1.5000 units. Press [Enter] twice and then update the part. Since the groove was created in a sketch plane based on the work plane, the groove will also move down on the part.

PRACTICE 8-1

❑ Construct a part as shown. Open the drawing Practice-8-01.dwg.
❑ Create a work plane using the **Planar Parallel** and **Offset** options with a distance of 1.25 from Face A.
❑ Set the color to Red for layer AM_WORK and the linetype to PHANTOM2.
❑ Create the sketch, profile, constrain, and dimension as shown. Note the use of the construction circle to center the sketch.
❑ Extrude the sketch 2.0 units, **Join**, and **Blind**.

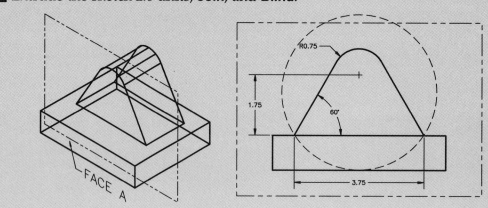

USING WORK PLANES TO TERMINATE EXTRUSIONS

Open the drawing Example-8-03.dwg. See **Figure 8-13A**. This is a plate with the sketch for a vertical cylinder profiled, constrained with construction circles, and dimensioned. In this example, the top of the cylinder should be at a 45° angle to the top of the base plate. It should have a height at the center of 3.5 units. See **Figure 8-13B**. This will be accomplished by extruding the circle to a work plane. The plan is to create a work plane at an angle to the base and then another work plane parallel and offset using a little trigonometry to calculate the offset.

Figure 8-13.
A—The plate with a sketch for a vertical cylinder profiled, constrained with construction circles, and dimensioned.
B—The part with a cylinder extruded to a work plane that is at a 45° angle to the top of the base plate.

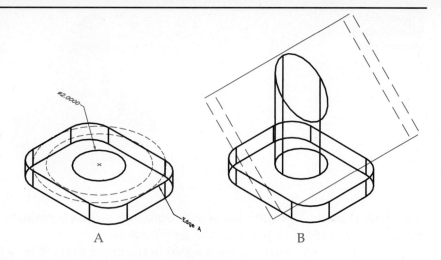

A

B

The offset calculation is done in the design variables shown in **Figure 8-14A.** The variable is named Offs. The geometry for the calculation is shown in **Figure 8-14B.** The details for the construction are to first create a work plane **On Edge/Axis** and **Planar Angle**. Do not create a sketch plane. Use the design variable Top_Angle for the angle. Select Edge A for the straight edge and the top face of the base for the planar face. Then, flip the plane if necessary.

Now create a second work plane **Planar Parallel** and **Offset** with Offs as the value for the offset. Do not create a sketch plane. Select the first work plane and flip until

Figure 8-14.
A—The offset calculation is done in the design variable Off.
B—The geometry for the calculation.

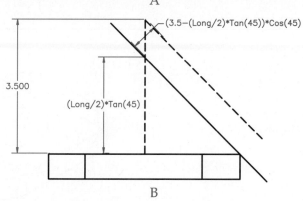

A

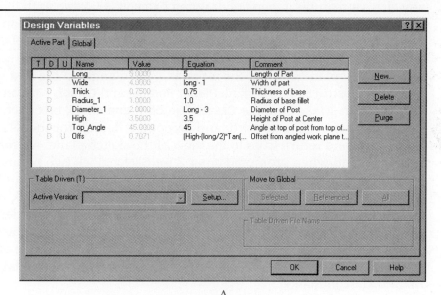

B

Figure 8-15.
The finished,
shaded part.

the offset is above. Extrude the sketch **Join** and with a termination of **To-Face/Plane**. Select the second work plane as the termination.

The work plane could be used again in putting a vertical hole though the cylinder. Sketch a circle on the top of the base that is still the sketch plane. Profile and constrain the hole. Extrude **Cut** and **From-To** using the bottom of the base as the first termination and the work plane as the second termination. Use the **Toggle Shading/Wireframe** button to view the results. See **Figure 8-15.**

CREATING WORK AXIS ON THE SKETCH PLANE

Another way to construct the feature just described, but with no trigonometric calculations, is to construct a work axis on the sketch plane. This can be done without any circular features. Open the drawing Example-8-04.dwg with the base and the circle to be extruded, sketched, and profiled. See **Figure 8-16.** The plan is to offset a work plane from the top face of the base, which is also the sketch plane, and construct a work axis on that plane that is parametrically related to the part. Then, create a second work plane through the work axis at an angle to the first work plane.

First construct a work plane **Planar Parallel** and **Offset**, with an offset of the design variable High. Make sure the **Create Sketch Plane** check box is checked. Pick the top face of the base, offset upward, and accept the default for the X- and Y-axis.

Enter 9 and return to go to the view of the sketch plane. Access the **AMWORKAXIS** command and when prompted enter S for Sketch. Pick two points (with **Ortho** on) on the screen to construct a vertical work axis. The position does not have to be accurate because you will next add a parametric dimension. Right-click on WorkAxis1 in the **Browser** and select **Edit** in the pop-up menu. This is required in order to place the dimension. With **AMPARDIM**, input a horizontal parametric dimension between the work axis and the right edge of the part. The dimension should have a value of Long/2 as shown in **Figure 8-17**. Press [Enter] to get out of the edit mode and update the part with the **AMUPDATE** command.

Figure 8-16.
The plate with a
sketch for a vertical
cylinder profiled,
constrained with
construction circles,
and dimensioned.

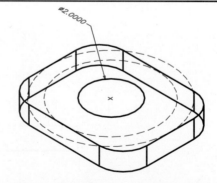

Figure 8-17.
Shown is the part with a horizontal parametric dimension between the work axis and the right edge of the part.

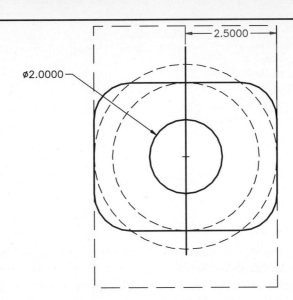

Go to the Iso view and construct a work plane **On Edge/Axis** and **Planar Angle** with an angle of Top_Angle. Do not create a sketch plane. Pick the work axis and then the work plane. Flip it if necessary. Extrude the circle **Join** and **To-Face/Plane** to the new work plane.

CREATING WORK PLANES FOR CYLINDRICAL PARTS

In cylindrical parts where there are no planar faces with straight edges, it is difficult to create work planes tangent to the cylinders without resorting to the world coordinate system. For example, constructing a drilled hole perpendicular to the cylinder axis would be difficult. See **Figure 8-18.**

Remember, we want to avoid resorting to the world coordinate system after we start a part. A technique used to avoid this situation is to create three work planes in a new drawing before creating a part. This can be done automatically with the **AMBASICPLANES** command. To access this command, enter AMBASICPLANES at the Command: prompt, select **Basic 3D Work Planes** from the **Work Features** cascading menu in the **Part** pull-down menu, or pick the **Create Basic 3D Work Planes** button on the **Part Modeling** toolbar.

This command creates three orthogonal work planes and a work point at a designated origin. On a new part it is good practice to put the origin at 0,0,0 of the world coordinate system.

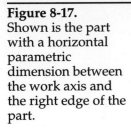

AMBASICPLANES

Part
➥ Work Features
➥ Basic 3D Work
Planes...

Part Modeling
toolbar

Create Basic
3D Work Planes

Figure 8-18.
A part with a drilled hole perpendicular to the cylinder axis. Without the basic work planes, this hole is difficult to construct.

Figure 8-19.
Three orthogonal
work planes and a
work point at the
designated origin
of 0,0,0.

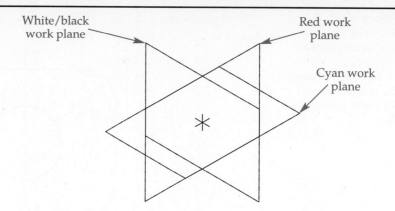

White/black
work plane

Red work
plane

Cyan work
plane

As an example, open a new drawing and go to the isometric view by entering 8. Issue the **AMBASICPLANES** command and when prompted to pick an origin enter 0,0,0. Use the **LAYER** command to change the color of the AM_WORK layer to Red and the linetype to PHANTOM2. Change the color of WorkPlane2 to White by right-clicking on it in the **Browser**, selecting **Properties** in the pop-up menu, and then selecting **Color...** to access the **Select Color** dialog box. Now, change the color of WorkPlane3 to Cyan. Note that the cyan work plane is by default the sketch plane. See **Figure 8-19.**

PROFESSIONAL TIP

Once you have reached this point, it is a good idea to save this file as a template for future use.

Now use this setup to construct a cylinder with a blind hole drilled perpendicular to the axis of the cylinder. To do this, you will first construct a work plane parallel to the red work plane, a work axis for the cylinder, and another work point. Then you will drill the hole on the work point.

To get started, construct a 2.0 diameter circle at 0,0. Profile and extrude it 3.0 units up. Now construct a work plane **Planar Parallel** to the red work plane and **Tangent** to the cylinder and make it the sketch plane.

Put a work axis through the cylinder. Go to the sketch plane view by entering 9. Put a work point on screen position and input a parametric dimension locating the point 0.90 units from the top of the cylinder. To locate the point at the center of the cylinder, use the AMADDCON command and the **Project** constraint option. Using **Endpoint** osnap, pick the work point and then pick the work axis. The command line should read: Solved fully constrained sketch.

Issue the **AMHOLE** command and select a drilled, blind hole on a point with a diameter of 0.50, a depth of 1.0, and an angle of 118. See **Figure 8-20.** Select **OK** and pick the work point, make sure the arrow is in the correct direction (into the part), and press [Enter]. The results are shown in **Figure 8-21.**

NOTE

Notice that the work point is made invisible when the hole is created. To use the point for some other feature, you must right-click on WorkPoint1 in the **Browser** and make it **Visible.**

Figure 8-20.
The **Hole** dialog box with the setting for creating the drilled hole from the work point.

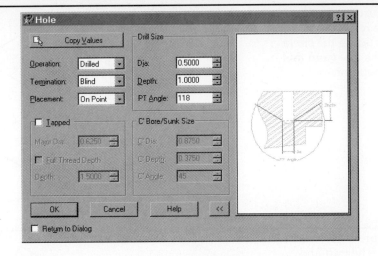

Figure 8-21.
The result of the drilled hole in the cylinder.

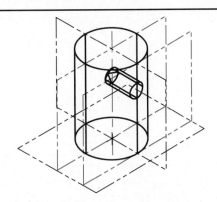

PROBLEMS CREATED BY USING THE UCS TO CREATE WORK PLANES

Open the drawing Example-8-05.dwg. See **Figure 8-22.** We want to create a circular feature into the inclined face of the part. However, we are going to do it the *wrong way*. That is, we will do it by moving the UCS instead of changing the sketch plane. We will then change the angle of the face and this will demonstrate that the feature is *not* parametrically related to the face.

The sketch plane is on the right front face of the part. Use the **UCS** command to move it to plane ABC. First, make sure **Osnap** is on and set to **Endpoint**. Then, enter UCS and then 3 for the **3point** option—this option is not listed, but it is still recognized. Select the endpoint at A for the origin, the endpoint at B for the X-axis, and the endpoint at C for the Y-axis.

Figure 8-22.
The base part. Notice the location and orientation of the UCS icon.

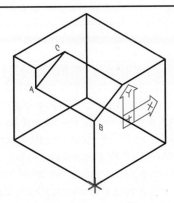

Figure 8-23.
The result of
extruding the circle
looks fine from this
viewpoint.

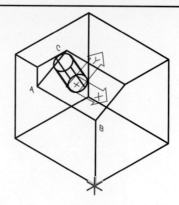

Figure 8-24.
A—The part and
extruded cylinder
also looks fine from
this viewpoint.
B—Changing the
angle of the face
shows that the
construction of the
cylinder is flawed.

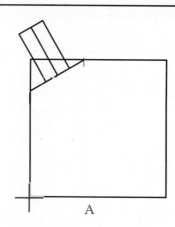 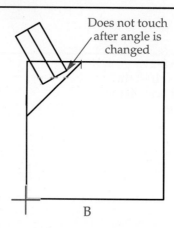

Does not touch
after angle is
changed

A   B

Now, draw a 1.0 diameter circle at the approximate center of face ABC. Profile the circle and a warning message is displayed that offers to create a nonparametric sketch plane based on the sketch object (the circle). Select **Yes** to continue. Dimension the circle with a diameter dimension on 1.0 units. Then extrude **Join** and **Blind** a distance of 2.0 units away from the part. The finished part is shown in **Figure 8-23.**

The problem will be shown most clearly if you go to the right side view. Do this by entering 7. See **Figure 8-24A.** Now change the global design variable Face_Angle. To do this, enter **AMVARS** and click on the **Global** tab in the **Design Variables** dialog box. Next, click on the 30 under equation, and enter 45. Press [Enter] and note that now the cylinder is *not* perpendicular to the face and does not even touch it. See **Figure 8-24B.**

PROFESSIONAL TIP  We have learned from this example to always use the **AMSKPLN** command to place the sketch plane on a face or work plane and keep the parts parametric.

USING AMPARTSPLIT WITH WORK PLANES

AMPARTSPLIT

Part
→ Placed Features
  → Part Split

Part Modeling
toolbar

Part Split

Open the drawing Example-8-06.dwg. See **Figure 8-25.** This part, called HOUSING, is a shelled part with a 0.20 wall. We want to cut the part in half creating a new part called TOP. To do this, we will first create a work plane and split the part along that plane with the **AMPARTSPLIT** command.

First create a work plane **Planar Parallel** and **Offset** a distance of 1.0 units. Pick the top face of the part and offset the work plane downward. Then issue the AMPARTSPLIT

Figure 8-25.
The original part that will be split.

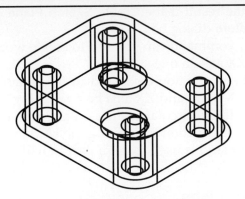

Figure 8-26.
The finished, shaded, and split part.

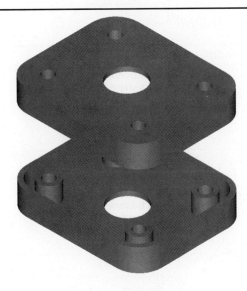

command by entering AMPARTSPLIT at the Command: prompt, selecting **Part Split** from the **Placed Features** cascading menu in the **Part** pull-down menu, or picking the **Part Split** button on the **Part Modeling** toolbar.

Now, pick the work plane. You are prompted to define the side for the new part. Flip the arrow so that it is pointing upward and press [Enter] to accept. Enter the name for the new part as TOP and press [Enter]. Use the **MOVE** command to move TOP upward 3.0 units. See **Figure 8-26.**

NOTE	As you will see in *Chapter 12* and *Chapter 14,* the **AMPARTSPLIT** command can also be used to turn features into parts that can then be copied and recombined.

Chapter Test

Answer the following questions on a separate sheet of paper.

1. How many work planes can be on a part?
2. *True or False?* The sketch plane can be set to any planar face on the part, but cannot be set to work planes.
3. What are the last four choices in the **1st Modifier** area used for?
4. A(n) _____ is a corner of any face on the part.
5. What does the **Normal to Start** choice in the **1st Modifier** area in the **Work Plane Feature** dialog box mean?

6. What is the result if the **Create Sketch Plane** check box is checked in the **Work Plane Feature** dialog box?
7. The _____ command can be used to create work planes parallel to the world XY, YZ, and ZX axis in one step.
8. For what construction discussed in this chapter would work planes parallel to the world XY, YZ, and ZX axis be a necessity?
9. When creating a circular feature on an inclined face of a part, why is it incorrect to do the construction by moving the UCS instead of changing the sketch plane?
10. What command is used to cut a part in half and create a new part?

Chapter Exercises

These 12 exercises use all 12 combinations of the 1st and 2nd modifiers of the **AMWORKPLN** *command, which are listed below.*

Pick the 1st Modifier	Choice for 2nd Modifier	Value Required	Exercise
On Edge/Axis	On Edge/Axis	None	Exercise-8-01
	On Vertex	None	Exercise-8-02
	Tangent	None	Exercise-8-03
	Planar Parallel	None	Exercise-8-04
	Planar Normal	None	Exercise-8-05
	Planar Angle	Angle	Exercise-8-06
On Vertex	On Edge/Axis	None	
	Planar Parallel	None	Exercise-8-07
	On 3 Vertices	None	Exercise-8-08
Tangent	On Edge/Axis	None	
	Tangent	None	Exercise-8-09
	Planar Parallel	None	Exercise-8-10
	Planar Normal	None	Exercise-8-11
Planar Parallel	On Edge/Axis	None	
	On Vertex	None	
	Tangent	None	
	Offset	Offset Distance	Exercise-8-12
Planar Normal	On Edge/Axis	None	
	Tangent	None	

Exercise 8-1. Open the drawing Exercise-8-01.dwg. Use the work plane modifiers **Edge/Axis** and **Edge/Axis**. Put a work plane through edges AB and CD and then construct a web as shown.

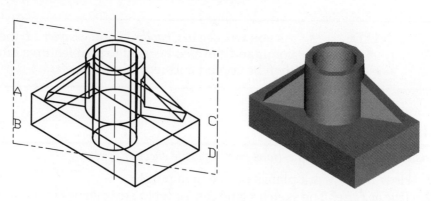

Exercise 8-2.   Open the drawing Exercise-8-02.dwg. Use the work plane modifiers **Edge/Axis** and **On Vertex**. Put a work plane through the work axis and vertex A and then construct features as shown.

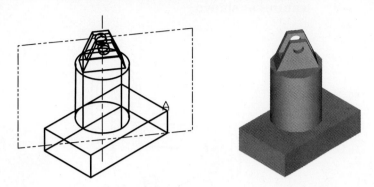

Exercise 8-3.   Open the drawing Exercise-8-03.dwg. Use the work plane modifiers **Edge/Axis** and **Tangent**. Put a work plane through the edge A and tangent to radius B. Then construct features as shown.

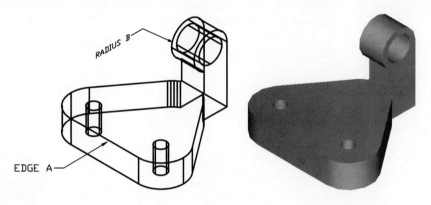

Exercise 8-4.   Open the drawing Exercise-8-04.dwg. Use the work plane modifiers **Edge/Axis** and **Planar Parallel**. Put a work plane through the work axis and parallel to face A and then construct features as shown.

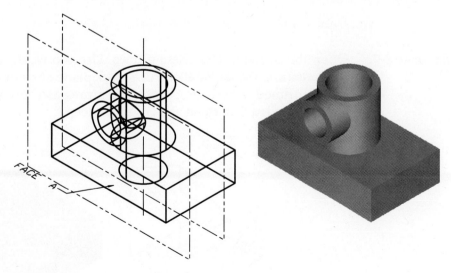

Exercise 8-5. Open the drawing Exercise-8-05.dwg. Use the work plane modifiers **Edge/Axis** and **Planar Normal**. Put a work plane through the work axis and normal (normal means perpendicular) to face A. Then construct features as shown.

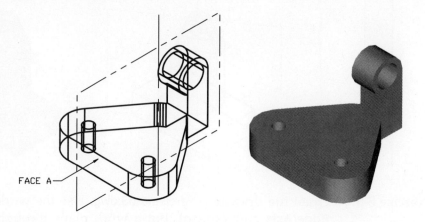

FACE A

Exercise 8-6. Open the drawing Exercise-8-06.dwg. Use the work plane modifiers **Edge/Axis** and **Planar Angle**. Put a work plane through the work axis and at a 65° angle to the horizontal (Cyan) work plane. Then construct features as shown.

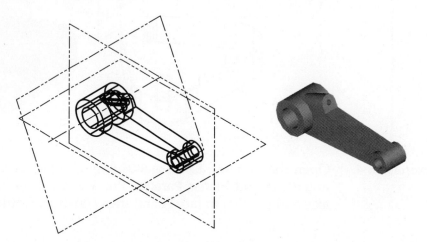

Exercise 8-7. Open the drawing Exercise-8-07.dwg. Use the work plane modifiers **On Vertex** and **Planar Parallel**. Put a work plane on vertex A and parallel to the inclined face. Then construct features as shown; 0.85 diameter circle, extruded **Cut, Blind** and 3.0 units deep.

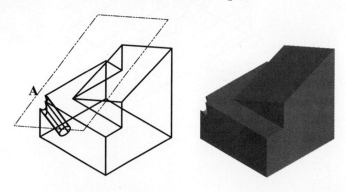

A

Exercise 8-8. Open the drawing Exercise-8-08.dwg. Use the work plane modifiers **On Vertex** and **On 3 Vertices**. Put a work plane through the vertices A, B, and C. Then construct features as shown.

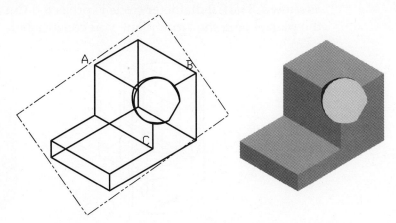

Exercise 8-9. Open the drawing Exercise-8-09.dwg. Use the work plane modifiers **Tangent** and **Tangent**. Put a work plane tangent to the visible side of the two cylinders. This part is shown in two views so you can see in the top view which of the four possible planes that you are getting. Then construct features as shown.

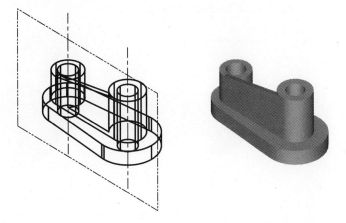

Exercise 8-10. Open the drawing Exercise-8-10.dwg. Use the work plane modifiers **Tangent** and **Planar Parallel**. Make sure **Create Sketch Plane** is checked. Put a work plane tangent to the cylinder and parallel to face A. Put a work plane that is *not* the sketch plane parallel to face A and on the work axis. Then construct features as shown.

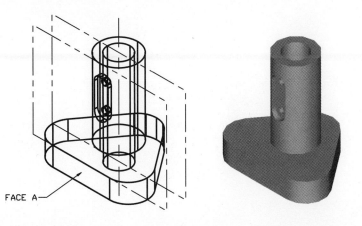

FACE A

Exercise 8-11. Open the drawing Exercise-8-11.dwg. Use the work plane modifiers **Tangent** and **Planar Normal**. Put a work plane tangent to the bottom of the slot and normal to the GreenWorkPlane1. Then construct features as shown. The 2.0 diameter circle is projected **Cut** and **To-Face/Plane**– the plane being the work plane you constructed.

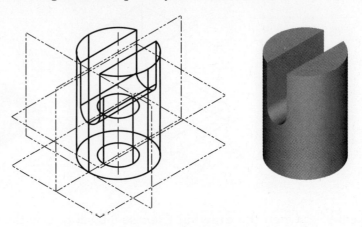

Exercise 8-12. Open the drawing Exercise-8-12.dwg. Use the work plane modifiers **Planar Parallel** and **Offset** parallel to WorkPlane6 (green) and offset 1.5 units. Then construct the 1.25 diameter boss. The **Boss** is projected **Join** and **To-Face/Plane**.

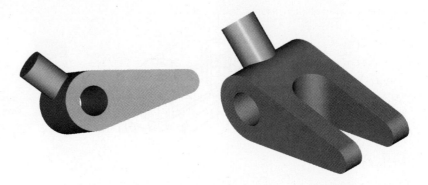

CHAPTER 9

Fillets and Chamfers

Objectives

After completing this chapter, you will be able to:

* Create constant radius fillets on parts.
* Edit the radii of fillets.
* Create several fillets with different radii with one command.
* Understand the automatic chaining feature of fillets.
* Know the common causes of fillet failure.
* Create fixed width fillets.
* Create variable radius fillets both linear and cubic.
* Create and edit chamfers on a part.
* Copy a feature to a different location on the part.
* Use the copy feature command to mirror a part.

Drawings Needed

*The following is a list of files that you will need to work through this chapter. These files can be found on the **User's Drawings** CD included with this text.*

Examples

Example-9-01.dwg
Example-9-02.dwg
Example-9-03.dwg
Example-9-04.dwg
Example-9-05.dwg
Example-9-06.dwg
Example-9-07.dwg
Example-9-08.dwg
Example-9-09.dwg

Example-9-10.dwg
Example-9-11.dwg

Practices

Practice-9-01.dwg
Practice-9-02.dwg
Practice-9-03.dwg
Practice-9-04.dwg

Exercises

Exercise-9-01.dwg
Exercise-9-02.dwg
Exercise-9-03.dwg
Exercise-9-04.dwg
Exercise-9-05.dwg
Exercise-9-06.dwg
Exercise-9-07.dwg
Exercise-9-08.dwg

THE TYPES OF FILLETS

In 2D AutoCAD, the **FILLET** command requires you to select two objects. These objects generally are lines or arcs. You have some control over the resulting trim of the selected objects. In MDT, the **AMFILLET** command requires you to select one or more edges on the part and there is no control on the trim. The fact that you can select more than one edge lets you create multiple fillets with one command and also edit the fillet radius for multiple fillets in one operation.

The most common type of 3D fillet is one where the fillet radius is the same along the entire length of the selected edge or edges. This is the *constant radius fillet* and it is the default. There are also three types of variable radius fillets. These types are *fixed width*, the *variable linear*, and the *variable cubic*. All of these will be discussed in this chapter.

THE CONSTANT RADIUS FILLET

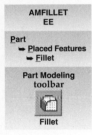

AMFILLET
EE

Part
➥ Placed Features
➥ Fillet

Part Modeling
toolbar

Fillet

Open drawing Example-9-01.dwg as shown in **Figure 9-1.** This is a rectangle extruded 2.5 units with two edges labeled A and B. To create a fillet with a constant radius, use the **AMFILLET** command. To access this command, enter AMFILLET or EE at the Command: prompt, select **Fillet** from the **Placed Features** cascading menu in the **Part** pull-down menu, or pick the **Fillet** button on the **Part Modeling** toolbar.

This accesses the **Fillet** dialog box. See **Figure 9-2.** A constant radius fillet with a value of 0.500 is the default. If you click in the **Radius** field, you can enter a different value or click on the spinner to change the value in increments of 1.000. In the lower-left corner is the **Return to Dialog** check box. If it is checked, you will return to the dialog box after creating one set of fillets. If you select the button with the double brackets in the lower right of the dialog box, then the dialog box expands to include an image tile of the proposed fillet. Accept the defaults, click **OK**, pick edges A and B, and press [Enter].

Note that the fillets are formed tangent to the edges in the top and bottom faces. See **Figure 9-3.** Also, even though there are two fillets, Fillet1 is only one operation listed in the **Browser**. Right-click on Fillet1 and select **Edit**. You do *not* get the **Fillet** dialog box and you cannot change the type of fillet. However, the radius appears on the part. Changing the radius will change both fillets at once. Change the radius to 0.75 units and press [Enter] twice. Update the part using the **AMUPDATE** command. Right-click on Fillet1 again, select **Delete**, and press [Enter] to remove both fillets.

Figure 9-1.
The base part used to create fillets.

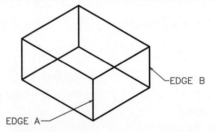

EDGE B

EDGE A

Figure 9-2.
The **Fillet** dialog box is used to create fillets.

Figure 9-3.
Two fillets created but only one fillet listing in the **Browser**.

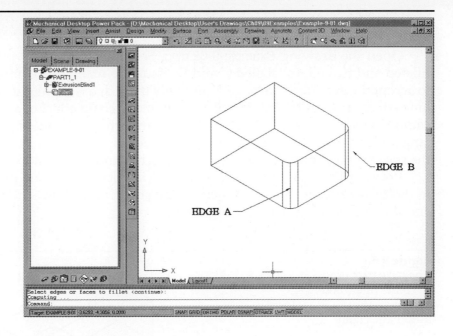

Filleting All the Edges of Face at Once

In drafting and machining terminology, a radius on an inside corner (concave) is called a fillet and on an outside corner (convex) a round. Rounds removed material and fillets added material. MDT calls both of these a fillet. If a face on the part has the same convexity on all its edges, all fillets or all rounds, then all the edges can be filleted with one pick. Issue the **AMFILLET** command, set the radius at 0.25, and click on **OK**. When prompted to *select edges or faces to fillet*, put the cursor anywhere on the top face of the part. Note that all the edges are highlighted. Select the face and press [Enter] twice and all edges of the face will be filleted. If a face has mixed convexity, this fillet will fail.

Adding Fillets with Different Radii

You may want one radius on edge A, a different radius on edge B, and only one fillet entry in the **Browser**. To do this, check the **Individual Radii Override** check box in the **Fillet** dialog box. See **Figure 9-4.** When you select the edges and press [Enter], the values of the two radii appear on the screen. You can select these radii on screen individually and change its value at the command line. Change the value of edge B to 1.00. Now if you right-click on Fillet1 in the **Browser** and select **Edit**, both values will appear on the screen.

Figure 9-4.
Using the **Individual Radii Override** option allows two fillets to be created with different radii, but still have only one fillet listing in the **Browser**.

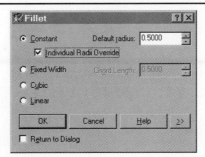

UNDERSTANDING THE USE OF THE AUTOMATIC CHAIN

Open the drawing Example-9-02.dwg. See **Figure 9-5.** This part has four edges filleted and two edges labeled C and D. Issue the **AMFILLET** command and click **OK** to accept the defaults. Pick edge C and only that edge is highlighted. Pick edge D and nine edges are highlighted. This is the automatic chain effect in action. An *automatic chain* is when an edge is selected and it is in a chain of edges where all corners are tangent, and the entire chain is selected. The chain continues in both directions from the selected edge until it reaches a sharp corner and it then stops.

Pressing [Enter] to accept the command will fillet the part and puts one entry, Fillet2, in the **Browser**. Note the differences in the shape of the fillets at the ends of edge D and the rest of the corners. See **Figure 9-6.**

Figure 9-5.
The base part used
to create fillets and
demonstrate an
automatic chain.

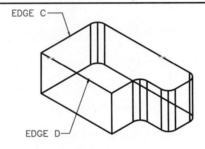

Figure 9-6.
The finished filleted
part.

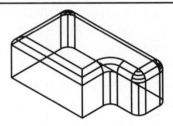

PRACTICE 9-1

❑ Open the drawing Practice-9-01.dwg.
❑ Fillet the vertical edges of the base with a 1.0 radius.
❑ Fillet the short edges of the inclined plate with a 1.25 radius.
❑ Fillet all the vertical edges of the ribs with a 0.125 radius.
❑ Fillet the top of the base and the bottom of the plate with a 0.125 radius.
❑ Fillet the intersection of the ribs with the base and the top with a 0.125 radius.
❑ The completed part is as shown.

Figure 9-7.
Shown is the base plate used to demonstrate how fillets fail.

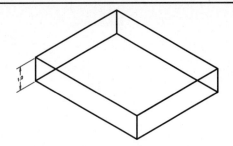

WHEN FILLETS FAIL

The filleting operation can fail for a variety of reasons. The most common is that the fillet is too large and interferes with an adjacent face. Open the part Example-9-03.dwg. See **Figure 9-7.** This plate is 1.0 unit thick. Attempting to place a fillet with a radius larger than 1.0 on all the edges of any of the six faces will fail. If you pick one of the four faces on the sides of the part, you will get no fillets and the command line will display the message:

    Failed to compute Fillet.
    Error: Unable to complete sheet – blend may be too big

If you pick:
- One or more of the short edges, the 1.5 fillet will work.
- One of the longer edges, it will work.
- The top or bottom faces, only one edge will be filleted.
- Two opposite long edges, it will work.
- Two adjacent long edges, only one edge will be filleted.

When you get errors like this, try a smaller fillet radius. In this example where the plate is 1.0 thick, it is easy to figure out that 1.0 is the largest fillet radius that will work.

NOTE	Linear, variable radius fillets that are larger than the plate thickness can be applied to individual edges. This is discussed in the section on variable radius fillets later in this chapter.

Delete all fillets and unsuppress ExtrusionBlind2 and ExtrusionBlind3 by right-clicking on them and selecting **Unsuppress** in the shortcut menu. Place a 1.0 fillet at the base of one cylinder. Attempting to place a 1.0 fillet at the base of the second cylinder will fail. Delete all fillets. Place 1.0 fillets on the bases of both cylinders in the same command and this will work. The largest fillet that can be placed on both cylinders is approximately 1.65 units where the two fillets meet on the long edge of the part.

PRACTICE 9-2

❏ Open the drawing Practice-9-02.dwg.
❏ Find the largest fillet that will work on all edges of the part.
❏ The completed part is as shown.

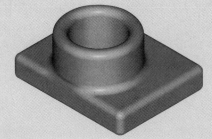

Figure 9-8.
The base plate used
to demonstrate how
edge selection
affects fillet results.

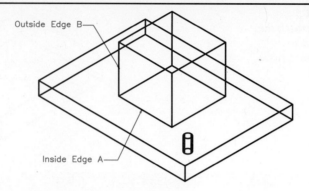

Fillets and a Convexity Problem

Fillets can also fail because of wrong choices of edges, which will cause a *convexity problem* as mentioned earlier in the chapter. Selecting an inside edge and an outside edge in the same **AMFILLET** command will result in an error message. For example, open the drawing Example-9-04.dwg. See **Figure 9-8.** Issue the **AMFILLET** command and set the radius to 1.0. Pick both inside edge A and outside edge B. The system will respond with the following message:

Error: Can only mitre edges of the same convexity.

Problems with Edge Order Selection

The order that you pick edges is also important. A different selection order will result in slightly different fillets. Use the **AMFILLET** command and select the four inside edges (starting with edge A) on the square boss for a 0.75 unit radius fillet. Again issue the **AMFILLET** command and select the four vertical edges using the same radius. The result is shown in **Figure 9-9A.** Now, delete the two fillets.

Redo the fillets in the opposite order. First select the four vertical edges to fillet and then select the horizontal edges with one pick. The result of this is shown in **Figure 9-9B.** Compare the differences between the two orders of fillet selection.

PRACTICE 9-3
❏ Open the drawing Practice-9-03.dwg.
❏ Put a 1.0 unit fillet on the boss by first selecting the three inside corners and then the four outside corners.
❏ Delete the fillets.
❏ Put a new 1.0 unit fillet on the boss. This time reverse the selection order.

Figure 9-9.
Note the different results of selecting edges to fillet.

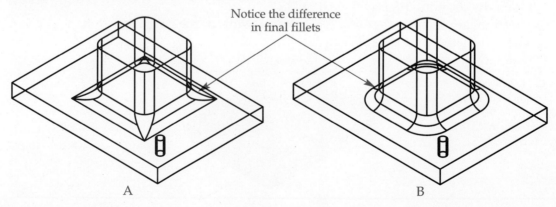

A   B

Figure 9-10.
The base part with fixed width fillets.

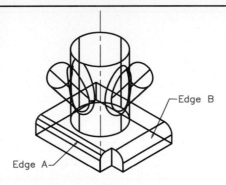

Figure 9-11.
The **Fillet** dialog box with settings for creating fixed width fillets.

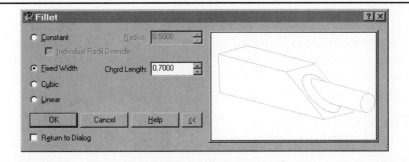

THE FIXED WIDTH FILLETS

Open the drawing Example-9-05.dwg. See **Figure 9-10.** This part has four fillets. The Fillet1 on edge A is a 0.35 unit constant radius fillet. The Fillet2 on edge B is a fixed width fillet with a chord length of 0.70 units. The **Fillet** dialog box for creating Fillet1 is shown in **Figure 9-11.** On a straight edge the fixed width is identical to a constant radius of half its value. This means Fillet1 with a radius 0.35 is identical to Fillet2 with a width 0.70.

On complex edges, such as where the inclined cylinders meet the vertical cylinder, the results are quite different. Fillet3 is a constant 0.30 radius fillet and Fillet4 is 0.30 fixed width. One of the uses of the fixed width fillet is to represent welds.

THE VARIABLE RADIUS FILLETS

There are two types of variable radius fillets. These fillets are the linear and the cubic type. They differ by the type of edge they may be applied to.

The Linear Fillet

The linear fillet can only be applied to an open, single segment edge. It requires two radii—one at each end of the edge. It also can only be applied to one edge per command.

Open the drawing Example-9-06.dwg. See **Figure 9-12.** The linear fillet *cannot* be applied to edge A, because it will chain around the left corner and contain multiple

Figure 9-12.
The base part used to demonstrate linear, variable radius fillets.

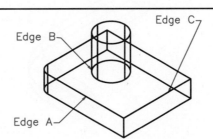

Figure 9-13.
The finished part
with a linear,
variable radius
fillet.

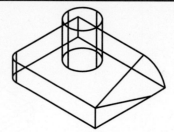

segments. It *cannot* be applied to edge B because it is closed. Issue the **AMFILLET** command, select the **Linear** radio button in the **Fillet** dialog box, pick **OK**, and select edge C.

Two R=0 equations appear at each end of the line. Click on the equation closest to edge A. At the command line, enter a value of zero. This value can be any number greater or equal to zero. Click on the second equation and set it to 1.25. This value may seem to be too large, but the value can be greater than the thickness. Press [Enter] and the fillet is formed. See **Figure 9-13.**

The Cubic Fillet

The cubic fillet can be applied to multiple segments—closed or open edges. All of the radii must be greater than zero. Open the drawing Example-9-07.dwg. See **Figure 9-14.** This part was created by sweeping a shape along a path. This process is discussed in detail in *Chapter 13.* We will use the top edge of this part to demonstrate the cubic variable radius fillet. Issue the **AMFILLET** command, select the **Cubic** radio button in the **Fillet** dialog box, pick **OK**, and select edge B.

Since this edge has three segments, four equations will appear, one at the end of each segment. The command line allows addition of more equations or vertices. Enter an A to add a vertex. Pick near the arrow on leader edge B. The command line shows the percentage of distance to the nearest vertex—set this to 50%. Starting at the front vertex set the values at 0.05, 0.20, 0.40, 0.20, and 0.05 as shown in **Figure 9-15A.** Once all the values have been input, press [Enter] to form the fillet as shown in **Figure 9-15B.**

Figure 9-14.
The base part used
to demonstrate
cubic, variable
radius fillets.

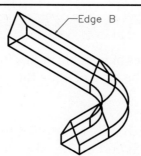

Figure 9-15.
A—The part with
the different radii
applied.
B—Shown is the
finished, shaded
part.

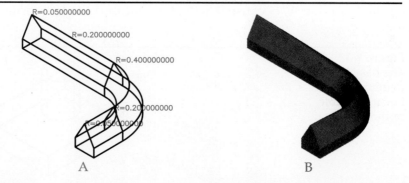

THE CHAMFER

Open the drawing Example-9-08.dwg. See **Figure 9-16.** This part has fillets on two corners. To create a chamfer use the **AMCHAMFER** command. To access this command, enter AMCHAMFER at the Command: prompt, select **Chamfer** from the **Placed Features** cascading menu in the **Part** pull-down menu, or pick the **Chamfer** button on the **Part Modeling** toolbar.

This accesses the **Chamfer Feature** dialog box. See **Figure 9-17.** To control the size of the chamfer, select the **Equal Distance** radio button and set **Distance1:** to 0.20. Pick edge A and note that, just like in the fillet command, the whole chain is selected. Now select the edge C, select edge B, and then press [Enter]. You could have selected the entire top face of the base as chamfers do not have the convexity problems that fillets do.

Put a 0.80 equal distance chamfer on the top of the 1.60 diameter cylinder. With a 1.60 diameter cylinder, a chamfer value greater than 0.80 would fail to make the chamfer. The results are shown in **Figure 9-18.**

AMCHAMFER

Part
↳ Placed Features
↳ Chamfer

Part Modeling
toolbar

Chamfer

NOTE: In the **Chamfer Feature** dialog box, there is no individual dimension override, as there is in the **Fillet** dialog box.

Figure 9-16.
The base part used to demonstrate chamfering.

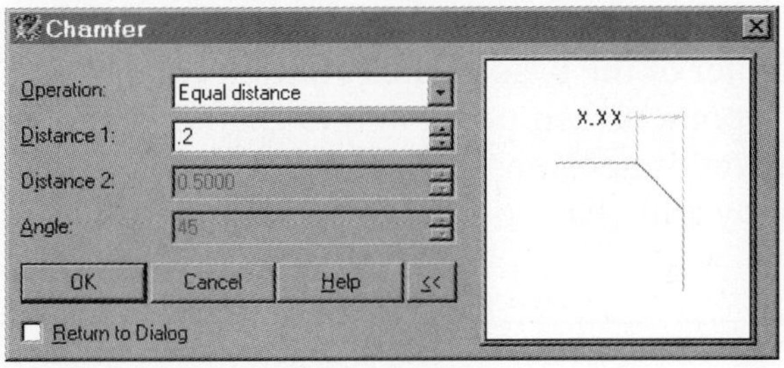

Edge C
Edge B
Edge A

Figure 9-17.
The **Chamfer** dialog box is used to create chamfers.

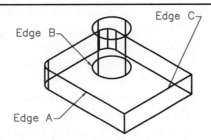

Figure 9-18.
The finished, shaded part.

You can also edit an existing chamfer. To edit Chamfer1, right-click on Chamfer1 in the **Browser** and then click on **Edit** in the pop-up menu. The **Chamfer Feature** dialog box appears on the screen. This is different than editing fillets where only the radius appears on the screen. The dialog box shows the **Equal Distance** chamfer with 0.5 in the **Distance1:** text box. This distance is the only information you can change. Change it to 0.75, click **OK**, and press [Enter]. Now use **AMUPDATE** to update the part.

PRACTICE 9-4

❑ Open the drawing Practice-9-04.dwg.
❑ Put a 0.80 radius fillet at the top and bottom edge of cylinder B.
❑ Put a 0.80 distance chamfer at the top and bottom of cylinder C.
❑ Go to the front view and note the differences.

COPYING A FEATURE

Open the drawing Example-9-09.dwg as shown in **Figure 9-19**. The **AMCOPYFEAT** command lets you copy a feature, like cylinder A and its counterbored hole, to another location on any face or sketch plane of the part. The copied feature can be dependent or independent. *Dependent* in this command means if you edit and change the original then the copy also changes. For example, the height of cylinder A is 2.0 units and has the variable d6. The height of a dependent copy would be assigned the value d6. *Independent* means that changes in the original would *not* affect the copy. The height of the independent copy would be assigned a value of 2.0.

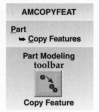

AMCOPYFEAT

Part
➥ Copy Features

Part Modeling
toolbar

Copy Feature

To access the **AMCOPYFEAT** command, enter AMCOPYFEAT at the Command: prompt, select **Copy Feature** in the **Part** pull-down menu, or pick the **Copy Feature** button on the **Part Modeling** toolbar.

You are prompted to select one feature to copy. This is a little confusing because the feature and all its dependent features, like the counterbored hole, will be copied. Select cylinder A (ExtrusionBlind2) and it includes Hole1. When prompted for a location, enter P for parameters. Now you are prompted whether the copy is dependent or independent. Enter D for dependent. The current sketch plane is on the top of the part and the screen shows the X-axis and Y-axis. Use the **Reference From** osnap and pick the center of the base of the cylinder, enter @3.0,0.5, and press [Enter]. You could also pick a point screen position and then edit the sketch for the copied feature and input location dimensions or constraints. Edit cylinder A and change its diameter to 2.0. The copy will change as well. See **Figure 9-20**.

Figure 9-19.
The base part used
to demonstrate how
to copy a feature.

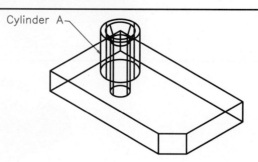

Cylinder A

Figure 9-20.
Result of copying
the feature and then
editing the original.

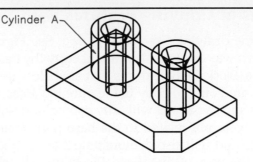

Cylinder A

Copying a Feature to a Different Sketch Plane

This next example shows how to copy a feature to a different sketch plane and some of the issues in copying. Open the drawing Example-9-10.dwg. See **Figure 9-21A.** Use the **AMREPLAY** command to review how this part was created.

Make face A the sketch plane. See **Figure 9-21B.** Construct a diagonal line on face A from endpoint to endpoint. We will use the line to locate the center of the face. Issue the **AMCOPYFEAT** command, set the parameters to independent, and select the feature called Top_Cylinder. Pick the midpoint of the line to locate the independent copied feature.

Note that the cylinder and fillet have been copied, but the extruded hole was not. See **Figure 9-22.** In general, placed features such as chamfers and fillets will copy. However, extruded features and features based on cylinder edges or work planes will not copy.

Figure 9-21.
A—The base part
used to demonstrate
how to copy a
feature to a different
sketch plane.
B—A line has been
constructed that
will be used to
locate the copied
feature.

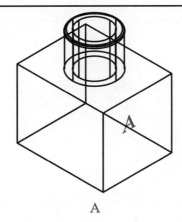

A

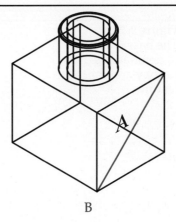

B

Figure 9-22.
Result of copying
the feature. Notice
that the hole is not
copied.

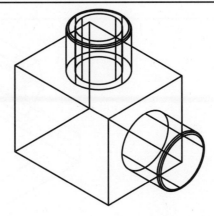

Problems of Copying Features with Subfeatures

Open the drawing Example-9-11.dwg. See **Figure 9-23**. Use **AMREPLAY** to review how this part was created. You cannot copy the base feature of a part—in this example named Base_Block. Only features that are below the base feature in the **Browser** can be copied. We are going to copy the feature Big_Boss, which has several subfeatures. The subfeatures such as Slot_B will only copy along with Big_Boss if their dimensions and constraints were not related to the base part Base_Block. This example demonstrates this problem and will also demonstrate a technique for mirroring features.

Make the top of the Base_Block the sketch plane. Issue the **AMCOPYFEAT** command and click on the Big_Boss. Note that Big_Boss and all its subfeatures are highlighted. Pick a point screen position on the top face of the Base_Block. Enter R for rotate and the copy will rotate 90°. Enter R again to rotate the copy an additional 90°. Thus, the copy is the mirror of the original. Press [Enter] to place the copied feature. See **Figure 9-24.**

The hole called Slot_A did not copy, whereas Slot_B did. Edit Slot_A and you will see that this feature was located from two edges, one of which was the edge between the top edge of the Base Block. This relationship prevented the hole from copying. The location of Slot_B was based solely on the top and side edges of the Big_Boss. Therefore, Slot_B was copied.

The copy is listed as ExtrusionBlind1 in the **Browser**. To locate the copy exactly on the base part, edit its sketch. Use constraints to make the front edge collinear to edge AB of the Base_Block and the short edge collinear to edge BC. Use the **Toggle Shading/Wireframe** button to view the completed part. See **Figure 9-25.**

Figure 9-23.
The base part used to demonstrate how to copy and mirror a feature.

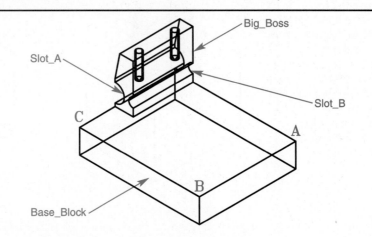

Figure 9-24.
The Big_Boss has been copied and mirrored to a random point on top of base. Notice that Slot_A was not copied.

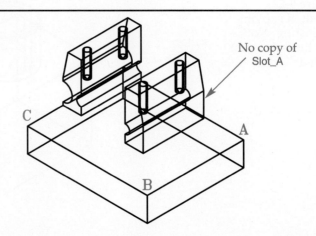

Figure 9-25.
The Big_Boss has been relocated by constraining it to the edges of the base. Shown is the finished, shaded part.

Chapter Test

Answer the following questions on a separate sheet of paper.

1. Describe the difference between the **FILLET** command in 2D AutoCAD and the **AMFILLET** in MDT.
2. The _____ _____ fillet is one where the fillet radius is the same along the entire length of the selected edge or edges.
3. Describe the procedure for adding fillets with different radii, but only one fillet entry in the **Browser.**
4. A(n) _____ _____ is when an edge is selected and it is in a string of edges where all corners are tangent, and the entire string is selected.
5. *True or False?* When adding fillets, the order in which edges are selected has no effect on the final result.
6. *True or False?* On a straight edge, the fixed width fillet is identical to a constant radius fillet of half its value.
7. What are the two types of variable radius fillets and how do they differ?
8. What command is used in MDT to construct a chamfer?
9. When copying a feature, what do the terms *dependent* and *independent* mean?
10. How could Slot_B in *Example 9-11* have been constructed so that it would copy along with Big_Boss?

Chapter Exercises

Exercise 9-1.   Open the drawing Exercise-9-01.dwg. Put a 0.5 unit fillet on the four vertical edges, one of them is labeled "A". Put a 0.25 unit fillet on all other edges on the top of the part. Put a 0.25 by 2.0 chamfer on the edge of the hole labeled "B". The completed part is as shown.

Exercise 9-2. Open the drawing Exercise-9-02.dwg. Fillet both ends of the pin with the maximum size fillet that will fit. The completed part is as shown.

Exercise 9-3. Open the drawing Exercise-9-03.dwg. Put 0.25 unit fixed width fillets at the intersection of both of the small tubes with the large tube. The completed part is as shown.

Exercise 9-4. Open the drawing Exercise-9-04.dwg. The bottom face, the top of the large cylinder, and the slot are machined in this casting. The large hole is bored and smaller holes drilled. Put a 0.02 fillet on all other edges. Chamfer the slot at 0.25 and the top of the bored hole at 0.125. The completed part is as shown.

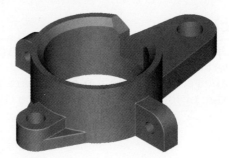

Exercise 9-5. Open the drawing Exercise-9-05.dwg. The 5mm thick aluminum plate needs a bend radius on the inside corners of 5mm and on the outside of 10mm. Fillet each corner and relief as shown.

Exercise 9-6. Open the drawing Exercise-9-06.dwg. Put 1.0 radius fillets on the four short edges around the holes. Put a 0.50 radius along the edge where the inclined face meets the vertical face and 0.25 radius on the other front edges as shown.

Exercise 9-7. Open the drawing Exercise-9-07.dwg. This part is the start of a windshield wiper arm bracket. Put a cubic fillet on both curved top edges. Add a vertex at the center. The radius at the ends is 9.0mm and at the center 3.0mm. The completed part is as shown.

Exercise 9-8.   Open the drawing Exercise-9-08.dwg. Review how the part was created with **AMREPLAY**. Copy Wing_1 as shown. Add fillets 0.125 on inside and 0.25 on outside. Use the bottom face labeled face A as the sketch plane for copying the feature. Rotate and flip the copy to get it into the correct position. The completed part is as shown.

Exercise 9-9.   Open the drawing Exercise-9-09.dwg. Review how the part was created with **AMREPLAY**. Place a 0.20 fillet on all the edges of the cylinder at once by selecting the face at point Face A.

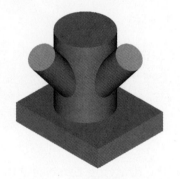

CHAPTER 10

More 2D Constraints

Objectives

After completing this chapter, you will be able to:

* Display the existing constraints on a sketch.
* Change the size of constraint display.
* Identify the names and symbols for the 15 2D constraints.
* Understand which constraints can be applied to work objects.
* Remove constraints on a sketch.
* Apply **Vertical** and **Horizontal** constraints to lines.
* Apply and understand the use of **Fix** constraints.
* Apply **Perpendicular**, **Parallel**, **Collinear**, and **Length** constraints to lines.
* Apply **Radius** and **Concentric** constraints to arcs and circles.
* Apply **Tangent**, **Xvalue**, and **Yvalue** constraints to lines, arcs, and circles.
* Apply and understand the use of the **Mirror** constraint and mirror axis.
* Use the **Project** constraint to position lines, arcs, circles, and work points on the sketch.
* Practice the use of construction lines and circles to constrain geometry in the sketch.

Drawings Needed

*The following is a list of files that you will need to work through this chapter. These files can be found on the **User's Drawings** CD included with this text.*

Examples

Example-10-01.dwg	Example-10-10.dwg	
Example-10-02.dwg	Example-10-11.dwg	
Example-10-03.dwg	Example-10-12.dwg	
Example-10-04.dwg		
Example-10-05.dwg	**Practices**	
Example-10-06.dwg	Practice-10-01.dwg	
Example-10-07.dwg	Practice-10-02.dwg	
Example-10-08.dwg	Practice-10-03.dwg	
Example-10-09.dwg	Practice-10-04.dwg	

Exercises

Exercise-10-05.dwg
Exercise-10-06.dwg
Exercise-10-07.dwg

THE 2D CONSTRAINTS

There are 15 geometric constraints that you can apply to profiled sketches and also use to locate work axis and work points. Refer to the *Appendix*. We have used some of them already on a variety of sketches. In this chapter, we will review those already presented and study all of the rest. They are called 2D constraints because they are applied to 2D sketches. They are different than the 3D assembly constraints that will be discussed in *Chapter 14* and *Chapter 15*. In this chapter, we will also study the use of construction geometry, both lines and circles, in more detail.

THE CONSTRAINT DISPLAY—THE SIZE AND LETTERS

Open part Example-10-01.dwg. See **Figure 10-1.** This is a rough sketch of a brake shield, which is comprised of five lines and two arcs. Using the **AMPROFILE** command, profile the sketch by individually selecting the objects starting with line A and working your way counterclockwise around the part. Be careful not to select the leaders. Even though they are on a locked layer, they are selectable.

The command line will display the message that this sketch requires 11 dimensions, or constraints, to fully define it. Remember that this means that some combination of dimensions and 2D constraints is required. In general, the more constraints used, the fewer dimensions that show on the drawings, and the more predictable the sketch is when changed.

There are five commands that will temporarily display the constraints, which were automatically applied to a sketch. These are as follows:

- **AMPARDIM.** Used to add dimensions to the sketch.
- **AMADDCON.** Adds any of the 15 constraint commands.
- **AMDELCON.** Used to delete constraints.
- **AMSHOWCON.** Used to show constraints on one or more objects in the sketch.
- **AMDT_SHOW_CONS.** Displays all constraints on the sketch.

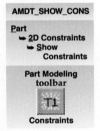

AMDT_SHOW_CONS

Part
↳ 2D Constraints
↳ Show
Constraints

Part Modeling
toolbar

Constraints

To display all the constraints, use the **AMDT_SHOW_CONS** command. To access this command, enter AMDT_SHOW_CONS at the Command: prompt, select **Show Constraints** from the **2D Constraints** cascading menu in the **Part** pull-down menu, or pick the **Show Constraints** button on the **Constraints** flyout on the **Part Modeling** toolbar.

When the constraints of the object are displayed, the object numbers are also displayed. They are numbered starting with zero and continue numerically in the order they were selected. See **Figure 10-2.** The letters representing the constraints are shown on

Figure 10-1.
A rough sketch of a brake shield, which is comprised of five lines and two arcs.

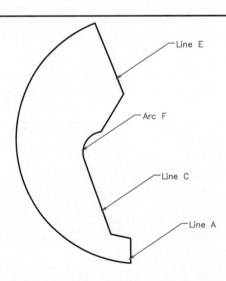

Line E

Arc F

Line C

Line A

Figure 10-2.
The constraints of the object are displayed, as well as the object numbers.

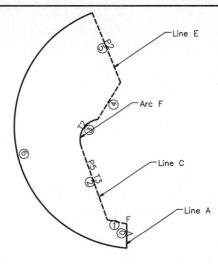

Line E
Arc F
Line C
Line A

the object either as a single letter or with a number. A letter **V** on line zero means it's **Vertical**, and a **T3** on line 2 means it's **Tangent** to arc 3. The four constraints automatically applied to this sketch are one **Vertical** (**V** on line 0), one **Fixed** (**F** at the upper end of line 0), one **Parallel** (line 2 is parallel to 5), and one **Tangent** (arc 3 is tangent to line 2).

Letters on the objects will also represent any additional constraints that you apply. Except for **Vertical**, **Horizontal**, and **Fixed**, the letters come in pairs showing the constraint between two objects.

The size of the letters and numbers is controlled with the **AMOPTIONS** command. To access this command, enter AMOPTIONS at the Command: prompt, select the **Part Options** in the **Part** pull-down menu, or pick the **Part Options** button on the **Part Modeling** toolbar.

This accesses the **Desktop Options** dialog box. See **Figure 10-3A.** Pick the **Part** tab and use the **Constraint Size...** to access the **Constraint Display Size** dialog box. See **Figure 10-3B.** Use the scroll bar to change the size, click the **Apply & Close** button, and click **OK**.

AMOPTIONS

Part
↳ Part Options

Part Modeling
toolbar

Part Options

Figure 10-3.
A—To change the constraint size, pick the **Constraint Size...** button. B—Use the scroll bar to make the constraint size larger or smaller.

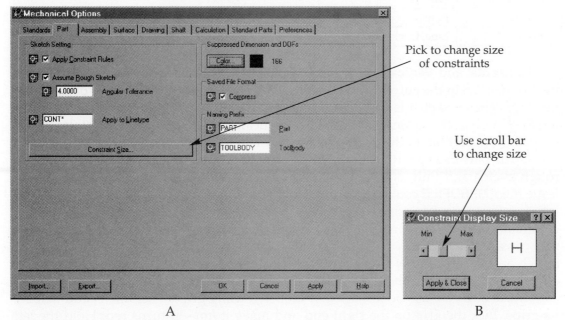

Pick to change size of constraints

Use scroll bar to change size

A

B

Figure 10-4.
While the constraints are displayed, a transparent zoom (**'Z**) will display the text larger or smaller. Redisplaying the constraints will set them back to the viewport related size.

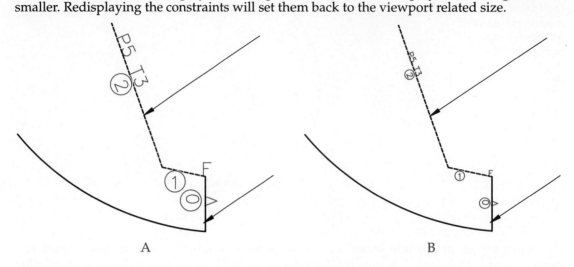

A   B

The constraints will appear in all viewports. The size is based on the current viewport size and is somewhat independent of the zoom factor. While the constraints are displayed, a transparent zoom (**'Z**) will display the text larger or smaller. Redisplaying the constraints will set them back to the viewport related size. See **Figure 10-4.**

The constraint names, their shortcut, and the types of applicable objects are summarized under each of the following headings. The shortcut for the constraint is capitalized and boldface. It is not always the first letter. Most of these constraints can also be applied to ellipses and spline segments. Because these objects are not widely used, they will not be covered.

USING **HORIZONTAL**, **VERTICAL**, AND **FIXED** CONSTRAINTS

Name	1st Choice—Object to be Reoriented	2nd Choice—Reference Object
Horizontal	Line	(None)
Vertical	Line	(None)
Fix	Line, Circle, Arc	(None)

Horizontal and **Vertical** constraints can only be applied to lines. They orient the line to the UCS in the current sketch plane. Therefore, a second reference object is *not* needed. When a sketch is profiled, if lines are within the angular tolerance angle of the X- or Y- axis, these constraints are automatically applied. The angular tolerance is set using the **Part** tab in the **Desktop Options** dialog box. By default, it is 4°.

Open part Example-10-02.dwg. See **Figure 10-5.** This four-segment polyline is a rough sketch and only one segment, side A, is within the angular tolerance of 4°. Issue the **AMPROFILE** command and pick the polyline on line A. The sketch will turn into four lines requiring five dimensions, or constraints. Use the **AMDT_SHOW_CONS** command to show the constraints and note that side A is vertical. When the polyline was constructed, side D was drawn first. Therefore, side D is the zero segment of the sketch and the left end of side D is fixed. Any dimensions placed on the sketch will adjust the sketch about the fixed point.

The sketch can also be reshaped or adjusted using grips. Pick line D and display its grips. Pick the grip on the right end and make it *hot*—it turns red. Hold the left

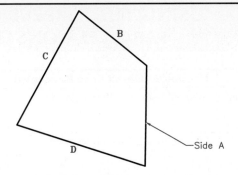

Figure 10-5.
A four-segment polyline that is a rough sketch with only one segment, side A, within the angular tolerance of 4°.

mouse key down, move the mouse, and the sketch dynamically changes. If you pick the midpoint grip of D and move it, you will see the angle between D and C changes. If you pick the grip at the left end, the fixed point, it will *not* move. You can move the entire profiled sketch in the coordinate system with the **MOVE** command and occasionally this is useful.

Use the **AMDELCON** command to remove the fixed constraint (the letter F) and the number of required constraints goes from five to seven. Apply a **Fix** constraint to line A. If you pick line A at an endpoint, both the X and Y coordinates of the point are fixed and only the length of the line can be changed. If you pick the line near the midpoint, then only one coordinate is fixed—in this case the X because line A is vertical. Pick near the lower end of line A.

You can apply as many fixed constraints as you wish until the sketch is fully constrained. Generally this is *not* recommended because it prevents you from modifying the sketch in any way. **Fix** constraints may also be applied to sketches for features, but this is *not* recommended either, as it limits the relationship of the feature to the base. In some sketches, adding a dimension or constraint grossly distorts the sketch. For example, place a vertical dimension of 9.0 units on line B as shown in **Figure 10-6.** Adding a fixed point to the upper end of line A before applying the dimension holds the basic shape of the sketch. This is probably the major use of the fixed point constraint.

Figure 10-6.
A—Fixed point only at lower end of line A.
B—Fixed points at both ends of line A.

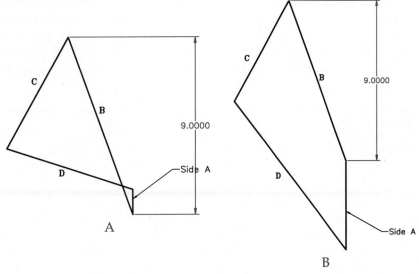

USING THE **PERPENDICULAR AND PARALLEL** CONSTRAINTS

Name	1st Choice—Object to be Reoriented	2nd Choice—Reference Object
Perpendicu**L**ar	Line, ellipse, or spline segment	Line, ellipse, spline segment, linear edges, and objects that appear as lines in the sketch view
Parallel	Line, ellipse, or spline segment	Line, ellipse, spline segment, linear edges, and objects that appear as lines in the sketch view

All the rest of the constraints are applied to pairs of objects. For **Perpendicular** and **Parallel** constraints, the first object has to be a line, ellipse, or spline segment. Once one of these constraints has been initiated, you will be prompted as follows:

> Valid selections: line, ellipse or spline segment
> Select object to be reoriented: *(Pick an object)*
> Valid selections: line, ellipse or spline segment
> Select object to be made parallel to:

Although this states that the second object can only be a line, ellipse, or spline segment, this is not so. Actually the second object can be any linear edge of the part, work planes that are perpendicular to the sketch plane, work axes that are not perpendicular to the sketch plane, circular edges of the part that are perpendicular to the sketch plane, and silhouette edges of cylinders. It can be almost anything that looks like a line in the sketch view. To pick some of these you must be in the sketch view. Examples and practices to demonstrate these features are presented later in this chapter.

To see these constraints applied to segments of an ellipse, open part Example-10-03.dwg. Objects A and C are trimmed ellipses and B and D are lines. Display the constraints as shown in **Figure 10-7**. Object C is parallel to A, B is parallel to D, and B is perpendicular to both C and A.

Open part Example-10-04.dwg. This six-sided polyline is a rough sketch. Profile it and it turns into six lines requiring seven constraints. Show the constraints. MDT has applied many of them as shown in **Figure 10-8**. Object 4 is vertical, object 3 is horizontal, objects 1 and 0 are perpendicular, all the endpoints join, and the intersection of lines B and C is fixed in relation to the UCS. Make line F (object 2) vertical, and make line B (object 0) perpendicular to line C (object 5). This also makes lines A and C parallel even though there is no **Parallel** constraint displayed or applied.

Figure 10-7.
A part with constraints displayed.

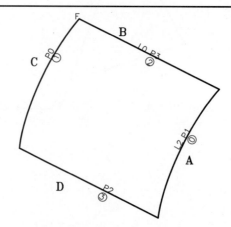

Figure 10-8.
A six-sided polyline with constraints displayed.

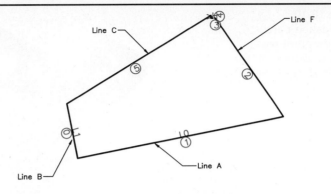

Figure 10-9.
The six-sided polyline after addition of and changes to the constraints.

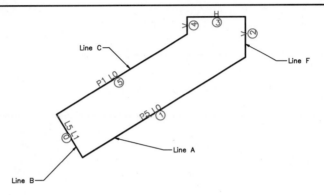

Apply a **Parallel** constraint between lines A and C. Note that this does *not* change the required number of dimensions, or constraints. It still requires five, since the **Parallel** constraint is implied by the perpendicular. What does this mean? It means that some redundant constraints that do not help constrain the sketch can be applied. The results of all the constraints are shown in **Figure 10-9.**

USING THE **COLLINEAR** AND EQUAL **LENGTH** CONSTRAINTS ON LINES

Name	1st Choice—Object to be Reoriented	2nd Choice—Reference Object
Collinear	Line or spline segment	Line, spline segment, linear edges, and objects that appear as lines in the sketch view
F Equal Length	Line or spline segment	Line or spline segment

The **Length** constraint can only be applied to two lines. For the **Collinear** constraint, the first choice must be a line or spline segment, but the second choice can be almost anything that looks like a line in the sketch view. The objective of the next example is to put two rectangular ears on the base that do not interfere with the slot. The final part is shown in **Figure 10-10.** There are several ways to create this feature. We are going to create it with two profiles to emphasize the use of constraints. Open part Example-10-05.dwg. The sketch for the ears is made up of two polylines that are shown in color.

Issue the **AMPROFILE** command and pick the two red polylines. Press [Enter] and pick edges B and C to close the profile. The two polylines are turned into six lines when the sketch is profiled. Increase the size of the constraint symbols with **AMOPTIONS** and display the constraints with **AMSHOWCON** as in **Figure 10-11.** MDT applied 10 constraints; lines 0, 2, 3, and 5 are **Horizontal** and **Projected** to the part edges, and lines 1 and 4 are **Vertical.**

Figure 10-10.
The finished part.
The viewpoint was
changed to present
a better view of the
part.

Figure 10-11.
Part and profiled
sketch with original
constraints
displayed.

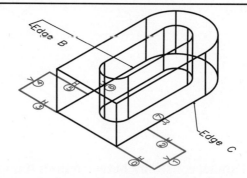

Following the procedures we learned in previous chapters, apply the major dimensions to hold the part shape. Dimension line 0 as 2.5 units and line 1 as 2.0 units. To line the sketch up with the front of the part, use the **Collinear** constraint. The command line prompts you to select the object to be reoriented. Pick line 0. The command line then prompts you to select the object to be made collinear to. Pick edge A. This is the common sequence of steps in the application of constraints applied to pairs of objects. Now make line 3 collinear to line 0.

Make line 5 collinear to line 2. Because line 2 is constrained by a dimension, you must select line 5 first. Use the equal length constraint and make line 3 the same length as line 0. This results in a fully-constrained sketch. All the constraints and dimensions are shown in **Figure 10-12.** Extrude the sketch **Join**, **Blind**, and 1.25 units up.

Figure 10-12.
Part with additional
and changed
constraints
displayed.

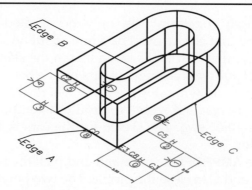

❏ Open the drawing Practice-10-01.dwg.
   ❏ Profile the two colored polylines.
   ❏ Use edges A, B, and C to close the profile.
   ❏ Put a horizontal dimension on the profile at edge A on 1.50 units and a vertical dimension of 2.00 units.
   ❏ Make the two vertical colored lines collinear and equal length.
   ❏ Extrude the profile **Cut**, **Blind**, and 1.00 units down.
❏ The results are shown here.

USING **COLLINEAR** ON ROUND PARTS

The edges of cylinders and arcs can be used to close sketch profiles, but the **DISPSILH** (DISPlay SILHouette curves) system variable must be set to 1. Silhouette curves are the nonquadrant edges of the cylinder that make it appear round. The default lines on a cylinder are *isolines*. The default for **DISPSILH** in a new drawing is zero, which means *off*. **Figure 10-13** shows a cylinder (A) with **DISPSILH** = 0 and **ISOLINES** = 6. The same cylinder (B) is shown with **DISPSILH** = 1 and **ISOLINES** = 0.

Open part Example-10-06.dwg as shown in **Figure 10-14.** The shaft has an annular (round) groove around the right end and a work axis. The objective is to put a 1.0 wide slot across the shaft with a depth equal to the radius of the shaft. The sketch for the slot is shown in color.

First enter 9 to go to the sketch view and then set **DISPSILH** to 1. Then, profile the sketch using edge A to close the profile. The edge will temporarily turn red. Put a 0.75 dimension on the width of the slot and locate the left edge 1.00 from the end of the cylinder.

Figure 10-13.
A—A cylinder with
DISPSILH = 0 and
ISOLINES = 6. B—A
cylinder with
DISPSILH = 1 and
ISOLINES = 0.

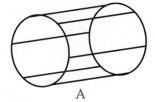
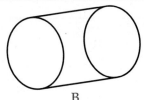

A   B

Figure 10-14.
The original part
with polyline that
will be extruded as
a slot through the
part.

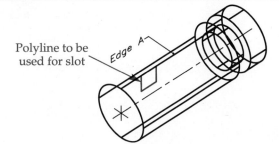

Polyline to be used for slot
Edge A

Figure 10-15.
A—Finished part
with slot. B—
Finished part after
shading.

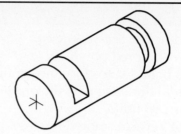

To make the bottom of the slot collinear with the work axis, use the **Collinear** constraint and pick the bottom of the slot. The command line states that the valid selections are a line of spline segment, but you can also use a work axis. Pick the work axis and extrude the sketch **Cut** and **Mid-Through**. Enter 8 to get a better view of the finished part. See **Figure 10-15**.

Creating a Pattern of Slots

Issue the **AMPATTERN** command and select the slot, make sure you only get the slot and not the entire shaft. Input five columns, 1.25 units spacing, and one row as shown in the dialog box in **Figure 10-16**. The results are shown in **Figure 10-17**.

Figure 10-16.
A part that will be
used to create an
array of slots
through the part.

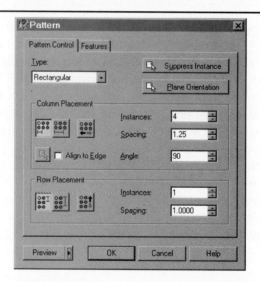

Figure 10-17.
Finished part with
the array of slots.

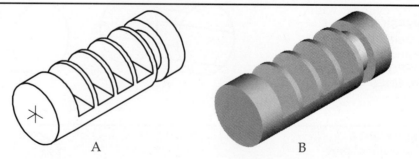

A

B

USING THE RADIUS, CONCENTRIC, AND TANGENT CONSTRAINTS

Name	1st Choice—Object to be Reoriented	2nd Choice—Reference Object
Radius	Circle or arc	Circle, arc, or circular edge
CoNcentric	Circle or arc	Circle, arc, or circular edge
Tangent	Line, circle, or arc	Line, circle, arc, edge, perpendicular work plane, non-perpendicular work axis

The **Radius** constraint makes either a circle or an arc have the same radius as another circle, arc, or circular edge. The **Concentric** constraint makes the same kinds of objects concentric (have the same center). Open part Example-10-08.dwg. See **Figure 10-18**. This casting is to have five 1.0 unit high bosses on the top surface. Profile all of the circles at once. This will result in 15 required constraints. Put a 0.75 diameter dimension on circle A. Use the **Radius** constraint to make the three circles in the other corners the same radius as circle A.

Now, use the **Concentric** constraint to make circle A concentric to corner radius B. Make each of the other circles concentric to their corresponding corner radii. The large circle should be in the center of the part and that will be done in the next example. For now, extrude the profile 1.0 units up. The results are shown in **Figure 10-19**.

The first object selected in the **Tangent** constraint can be either a line, arc, or circle. The second object can be anything that looks like a line, arc, or circle in the sketch view—this includes part edges. Open part Example-10-09.dwg. See **Figure 10-20**. There are three circles in the sketch plane. The two outside circles (circles A and B) are construction circles. Construction geometry has any linetype other than Continuous. If construction geometry is drawn with the **CONSTRUCTION LINE** or **CIRCLE** command it will be on layer AM_CON, which is yellow by default. In this example the color of that layer has been changed to red. Remember, the **Construction Circle** icon is on the **2D Sketch** flyout.

Figure 10-18.
The original part with circles that will be profiled, constrained, and extruded.

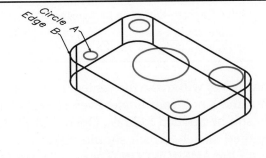

Figure 10-19.
The finished part.

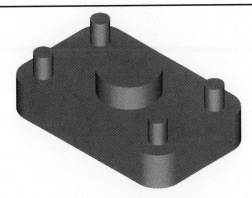

Figure 10-20.
The original part with circles and construction circles.

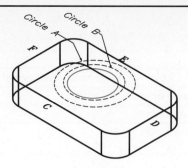

Figure 10-21.
The finished part.

This technique will position the smallest circle in the exact center of the rectangle without dimensions, equations, or design variables. This will help reduce the dimension clutter on the 2D drawings. First profile the three circles. Start the **Concentric** constraint. Note that one construction circle is already concentric to the smallest circle. Constrain the other construction circle concentric to the smallest circle. Use the **Tangent** constraint to make circle A tangent to edges D and F. Make circle B tangent to edges C and E. Put one dimension on the smallest circle to fully constrain the profile. Extrude the circle **Join**, **Blind**, and 2.0 units up. The results are shown in **Figure 10-21**.

APPLYING THE **XVALUE** AND **YVALUE** CONSTRAINTS

Name	1st Choice—Object to be Reoriented	2nd Choice—Reference Object
XValue	Line, circle, or arc	Line, circle, arc, or edge
YValue	Line, circle, or arc	Line, circle, arc, or edge

These constraints will make the centers of arcs and circles and/or the endpoints of lines have the same X or Y coordinate. Even though the icons for these constraints show arcs, they can still be applied to lines. Open part Example-10-10.dwg. See **Figure 10-22**. Profile the colored sketch, which should result in 18 required constraints. Show the constraints with the **AMSHOWCON**. They are **Horizontal**, **Vertical**, **Collinear**, and same **Radius**.

Figure 10-22.
The original part with sketch that will be profiled, constrained, and extruded.

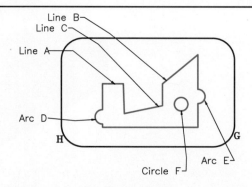

Figure 10-23.
The finished part.
The viewpoint was
changed to present
a better view of the
part.

Use the **Yvalue** constraint and make the center of arc E have the same Y value as the center of arc D. Using the **Yvalue** constraint again, make the center of circle F have the same Y value as arc D. To do this, issue the constraint and pick arc E first and then arc D. Now pick circle F and then arc D again. Using the **Radius** constraint, make circle F have the same radius as arc D.

Using the **Yvalue** constraint, make the lower-left end of line B have the same Y value as line A. Now make line C parallel to line B. This causes line C to cross the base of the sketch.

Press [Esc] to clear the command line. Click on line C to activate the grips, which are the small blue squares. Click and drag the grip at the midpoint of the line and dynamically move the line up. This changes the sketch. Now make arc D have the same X value as fillet radius H. Extrude the part **Join**, **Blind**, and 2.0 units up. The results are shown in **Figure 10-23.**

APPLYING THE **MIRROR** CONSTRAINT

Name	1ˢᵗ Choice—Object to be Reoriented	2ⁿᵈ Choice—Reference Object
Mirror	Line	Line
	Circle	Circle
	Arc	Arc

The **Mirror** constraint is very useful for quickly creating symmetrical features at the sketch level. It can be applied to pairs of lines, arcs, or circles. The second choice must be the same type of object as the first choice—for example, line to line. The drawback is that it requires a work axis or a construction line to use as a centerline for the mirror.

Open part Example-10-11.dwg as shown in **Figure 10-24.** The construction line was drawn from the midpoint of edge D to the midpoint of B. Using the AutoCAD **MIRROR** command, mirror the colored polyline about the construction line. Profile both polylines and the construction line, using edges A and C to close the sketch. This sketch turns into four lines, two arcs, and one construction line. It will require eight dimensions, or constraints.

Note that there are four tangent, four project (the ends of the lines to the edge of the part and the symbol is J), two parallel, one radius, and one horizontal constraints applied by MDT. Use the **AMDELCON** command to remove both parallel constraints on the lines.

Using the **Mirror** constraint, pick the construction line as the mirror axis. Do *not* press [Enter]. The line will turn blue and can be used for any number of pairs of objects. Now pick the arc on the right and mirror it to the arc on the left. Next, pick the line on the right and the corresponding line on the left, and then pick the other

Figure 10-24.
The original part
with ear that will be
profiled,
constrained,
mirrored, and
extruded.

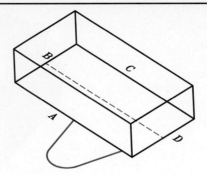

pair of lines. The combination of the project and mirror constraints locks the construction line to the Y-axis center of the base. The constraints applied so far are shown in **Figure 10-25.**

Apply an equal length constraint on the original pair of lines. Apply a 0.75 dimension to the original radius, the one on the left. Apply a 70° angle dimension between one of the lines on the left and the edge A.

Pick a grip on the left arc and you can move both ears back and forth and in and out to the approximate position desired. Two more dimensions will fully constrain the ears. Extrude the sketch **Join**, **Blind**, and 1.0 units up. The results are shown in **Figure 10-26.**

Figure 10-25.
The part with
constraints
displayed.

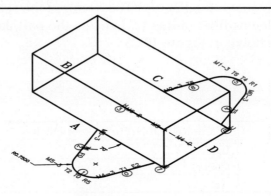

Figure 10-26.
The finished part.

❑ Open the drawing Practice-10-02.dwg. First the red polyline is to be constrained using construction geometry so that one dimension will control the offset from the edges of the base. The same results could be obtained with four dimensions, but the objective is to practice and understand constraints.

    ❑ Profile the four construction circles, the two construction lines, and the red polyline. This should require 18 dimensions, or constraints.

    ❑ Apply the mirror constraint using the vertical construction line E as the mirror axis and relating circle B to A and D to C.

    ❑ Apply the mirror constraint using the horizontal construction line F as the mirror axis and relating circle C to A. All four circles are now the same size.

    ❑ Using the **Xvalue** constraint, relate the left side of the rectangle to circle A and the right side to circle B.

    ❑ Using the **Yvalue** constraint, relate the top red line to circle B and the bottom red line to circle C.

    ❑ Make circles A and D tangent to the two edges of the part in the corner. This locks the construction lines to the center of the base.

    ❑ Pick the grip on the right end of the lower line in the rectangle and dynamically move the rectangle.

    ❑ Apply a 0.5 unit vertical dimension between the top red line and the top edge of the base.

    ❑ Extrude the fully-constrained sketch **Join**, **Blind**, and 1.5 units up.

    ❑ Change any dimension of the base to confirm that the feature stays at a constant offset.

    ❑ Put a 0.25 fillet around the intersection of the extrusion and the base.

❑ Edit the ExtrusionBlind2 to **Cut**, **Through**, and downward.

APPLYING THE PROJECT CONSTRAINT

Name	1st Choice—Object to be Reoriented	2nd Choice—Reference Object
ProJect	Endpoint of a line or center of a circle or arc	Line, circle, arc, and any edge or work object that appears as those in the sketch view

The **Project** constraint is the only constraint that requires an osnap on the first object. You can pick the endpoint of a line or arc, or pick the center of a circle or an arc. The second object can be a line, arc, circle, edge, or any work object that looks like lines in the sketch view.

Open the drawing Example-10-12.dwg. See **Figure 10-27.** Profile the two construction lines, the polyline, and the two circles. MDT applies horizontal, vertical, parallel, and radius constraints. The sketch will require 15 dimensions, or constraints. The technique used here is to project the vertices 1, 2, 3, and 4 to construction line A and 5, 6, and 7 to construction line B. This will make the *steps* all the same height and width.

Use the **Project** constraint and select the end point at the corner labeled 1. Then select construction line A as the second object. Repeat this process for corners 2, 3, and 4. Also, repeat the process for corners 5, 6, 7, and construction line B.

Place a 1.50 horizontal dimension between line 4-7 and circle C. Then project the center of circle C to line 7-3 that will locate the circle vertically. Now, put a 1.50 vertical dimension between line 5-1 and circle D and project the center of circle D to line 5-2.

Four more dimensions are required to fully constrain the sketch. These are shown in **Figure 10-28.** The length of the construction line has no effect on the geometry of the sketch, but a dimension is required to fully constrain it.

Figure 10-27.
The original part
with constraints
displayed.

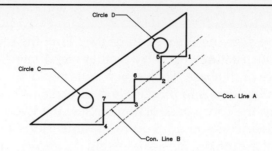

Figure 10-28.
The part with
dimensions
displayed.

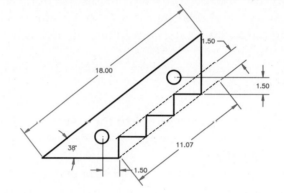

PROFESSIONAL TIP

There are often several ways to constrain a sketch to achieve the same results. Which method you select will depend on the design intent, the constraints and dimensions you think are the simplest, or the constraints and dimensions you are the most comfortable using.

PRACTICE 10-3

❑ Open the drawing Practice-10-03.dwg. First the four circles are going to be constrained to the four corners of the base and then extruded **Cut**, **Through**, and down. The resulting part is shown here.

   ❑ Profile the four circles. They will have radius constraints and require additional nine constraints.

   ❑ Use the **Project** constraint and pick the center of circle A and then line B. This sets the Y position of the circle.

   ❑ Project the center of circle A to line C that sets the X position.

   ❑ Repeat the process for the other three circles locating them at the other three corners (vertices) of the top of the block.

   ❑ Place a 1.5 unit diameter dimension on circle A to fully constrain the sketch.

   ❑ Extrude the sketch **Cut**, **Through**, and down.

❑ Close the drawing without saving.

PRACTICE 10-4

❑ Open the drawing Practice-10-04.dwg. The **Project** constraint can also be applied to work points that can then locate a hole feature. The sketch plane is the inclined face labeled A.
❑ Place a work point on the sketch plane.
❑ Project the end point of the work point to the red work plane.
❑ Locate the other coordinate with a dimension of 1.0 from the upper edge.
❑ Drill a 1.0 diameter hole 1.5 units deep on the work point.

Chapter Test

Answer the following questions on a separate sheet of paper.

1. There are _____ geometric constraints that can be applied to profiled sketches and also used to locate work axis and work points.
2. *True or False?* There are five commands that will temporarily display the constraints that were automatically applied to a sketch.
3. What does the **AMADDCON** command do?
4. What does the **AMPARDIM** command do?
5. What does the **MNU_SHOW_COM** command do?
6. What command is used to increase the display size of the constraints?
7. What is the only limitation of using the silhouette edge to close a profile?
8. Describe how the **Xvalue** and **Yvalue** constraints are used.
9. What is the **Radius** constraint used for?
10. How could you achieve the same results in *Practice 10-2* with only two construction circles?

Chapter Exercises

Exercise 10-1. Start a new drawing. Create the base block as shown. Use two lines, one equal length constraint, and 2 dimensions for the sketch for the cuts on the corner.

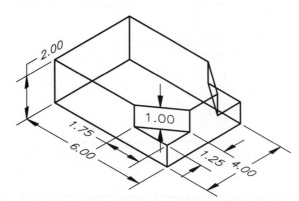

Exercise 10-2. Start a new drawing. Construct the part shown. Use a construction circle and two construction lines to locate the 0.5 diameter features.

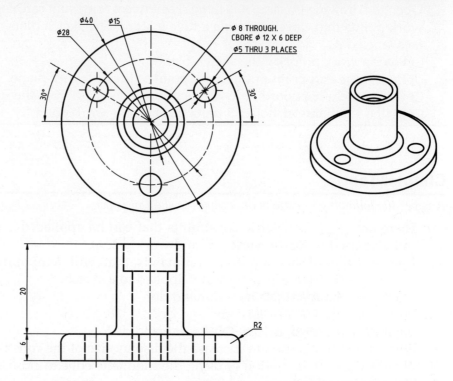

Exercise 10-3. Start a new drawing. Construct the part as shown using the mirror constraint to constrain the ears. Edit the sketch and dynamically move the ears.

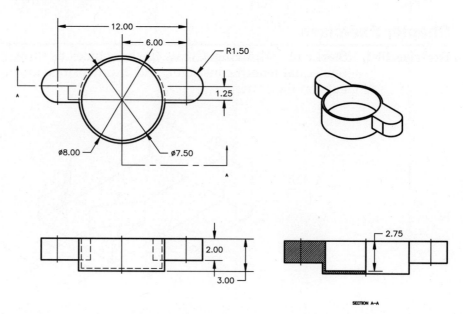

Exercise 10-4. Start a new drawing and construct the part as shown.

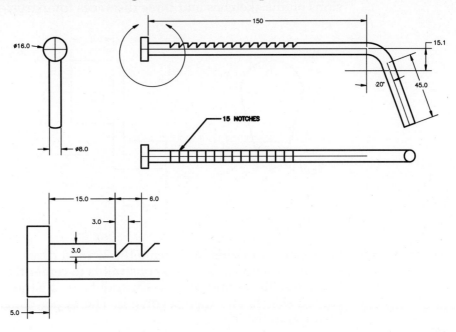

Exercise 10-5. Open the drawing Exercise-10-5.dwg. Project the ends of each construction line to the two edges on the corners—a total of eight constraints. Extrude the circle **Cut**, **Blind**, and 1.0 unit down. Is the circle always at the center of the base?

Exercise 10-6. Open the drawing Exercise-10-06.dwg. Put a work point on the current sketch plane. Project its end point to the white work plane. Put a 2.0 unit vertical dimension to fully constrain the point. Drill a 0.5 diameter hole through, on the point.

Exercise 10-7. Open the drawing Exercise-10-07.dwg. Constrain this sketch with mirror, project, and as few dimensions as possible. Extrude **Join**, **Blind**, and 3.0 units as shown.

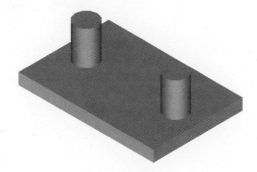

Exercise 10-8. Start a new drawing. Create the crank shown. Use only three dimensions on the sketches and three distances for extrusions.

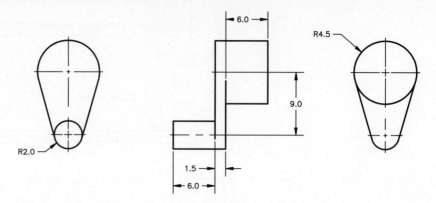

Exercise 10-9. Start a new drawing and create the ³⁄₈ to ¹⁄₄ drive adapter as shown. Use construction circles and constraints to center the square features. Use the collinear and project constraints to constrain the 0.500 radius curved sketch, and then revolve it. The upper cut curved feature is identical to the lower.

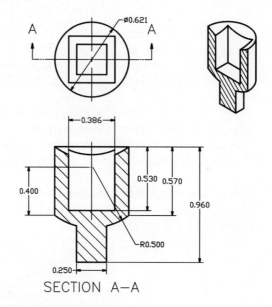

SECTION A—A

CHAPTER 11

Editing and Annotating Drawings

Objectives

After completing this chapter, you will be able to:

* Relocate parametric dimensions in the drawing view.
* Hide parametric dimensions.
* Add MDT reference dimensions.
* Move dimensions to a different view.
* Edit the format and value of a dimension.
* Align, join, insert, and break dimensions.
* Place and edit hole notes.
* Place leaders and annotations on the drawing.
* Place and edit surface finish symbols.
* Put GD&T information on the drawing.
* Use automatic dimensioning on drawings and sketches.

Drawings Needed

*The following is a list of files that you will need to work through this chapter. These files can be found on the **User's Drawings** CD included with this text.*

Examples		Exercises
Example-11-01.dwg	Example-11-08.dwg	Exercise-11-01.dwg
Example-11-02.dwg	Example-11-09.dwg	Exercise-11-02.dwg
Example-11-03.dwg	Example-11-10.dwg	Exercise-11-03.dwg
Example-11-04.dwg		Exercise-11-04.dwg
Example-11-05.dwg	**Practice**	Exercise-11-05.dwg
Example-11-06.dwg	Practice-11-01.dwg	Exercise-11-06.dwg
Example-11-07.dwg		

EDITING AND ANNOTATING

As you saw in *Chapter 4*, the green dimensions, automatically applied to a drawing view when it's created, are the parametric dimensions that were used to constrain the sketches. As was stated in *Chapter 4*, do *not* delete these green dimensions on the drawing. Instead we are going to hide unwanted parametric dimensions on the drawing views and this is one of the first steps we're going to look at.

Several types of objects will not be automatically dimensioned on the drawing. These objects include fillets, chamfers, and holes. If you create a hole by extruding a fully-constrained circle, it will have a parametric dimension in one of the views. If you create one using the **AMHOLE** command, it will not and will require a *hole note*.

MOVING PARAMETRIC DIMENSIONS

Open part Example-11-01.dwg. Use **AMREPLAY** to review the construction and get a better understanding of how it was created. Now go to Layout1 by clicking on the tab at the bottom of the drawing area or by clicking **Drawing** tab at the top of the **Browser**. Zoom in on the front view as shown in **Figure 11-1**. There are four linear and two diameter dimensions. The rounded corners are not dimensioned, because they were created as fillets. The bored hole in the center requires a hole note for a dimension.

The first step in correcting this drawing, so that it meets standards, is to move some dimensions. For example, the 6.00 dimension needs to move down below the section line A. The easiest way to do this is with grips. Click on the dimension and click again on the square in the middle of the 6.00 to turn it red. Now you can drag the dimension to a new position. If **ORTHO** is on, the dimension will move either horizontally or vertically. If **ORTHO** is off, it will move both directions. You can even position the value outside the dimension lines.

Select all three vertical dimensions. Then hold down the [Shift] key and click on the grip in the value in each of them and make them red. Then release the [Shift] key, click on one of the red grips, and move all three dimensions at once.

Note that the upper extension line, for the 4.00 and 1.68 dimensions, does not touch the part. This is because when the sketch was created the corner was square. There is a point called the *defpoint* just to the right of the end of the extension line. Defpoints are used to relate the dimension to the geometry and are on a layer called DefPoints. In the layout you will see these points, but they will not be in the *plot*.

When you pick the dimension, the defpoint has a grip on it. Make it red and snap it to the left endpoint of the line across the top of the part. It is a good idea to have **ORTHO** on when moving extension lines or you may cause the value of the dimension to change. The change will not affect the size of the part and the value will correct at the next view update.

Unfortunately if the value of any parametric dimension is changed, say the 4.00 is changed to 5.00 and the part and drawing are updated, then the end of the extension line for the 1.68 dimension will move back to its original position. To make this change permanent, the **Reattach** option in the **AMMOVEDIM** command is used. This option and the other options in **AMMOVEDIM** command are discussed in the next section.

Figure 11-1.
The front view of the part used to illustrate moving parametric dimensions.

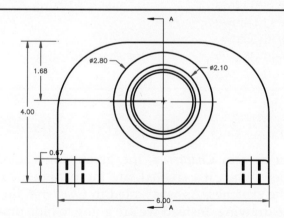

Figure 11-2.
The part with the
two diameter
dimensions and the
1.68 dimension
moved.

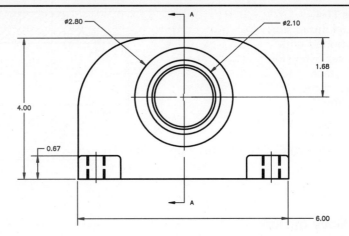

Now move the two diameter dimensions one at a time above the part using grips. It is a good idea to have **ORTHO** off when moving diameters. Move the 1.68 dimension from the left side to the right side of view. Also move its dimension line. All these changes are shown in **Figure 11-2.**

Using the AMMOVEDIM Command

AMMOVEDIM

A**n**notate
→ **E**dit Dimension
→ **M**ove
Dimension

Drawing Layout
toolbar

Move Dimension

To move dimensions, the **AMMOVEDIM** command is used. To access this command, enter AMMOVEDIM at the Command: prompt, select **M**ove **Dimension** from the **E**dit **Dimension** cascading menu in the **A**nnotate pull-down menu, or pick the **Move Dimension** button on the **Drawing Layout** toolbar. This command has three options. These options are **Flip**, **Reattach**, and **Move**.

The Flip option

The **Flip** option moves the text of the dimension in relationship to the dimension lines. Flipping the 0.67 dimension will move the text between the lines. Another flip will move it below the lines.

The Reattach option

The **Reattach** option will attach the end of the extension line to the opposite end of the object line or centerline. You do not need to use grips or osnaps, just click on the extension line and then on the object line near its end. In **Figure 11-3,** the top extension line of the 1.68 dimension was reattached to the right end of the top of the part and the bottom extension line to the end of the centerline.

Figure 11-3.
The part with the
top extension line of
the 1.68 dimension
reattached to the
right end of the top
of the part and the
bottom extension
line to the end of the
centerline.

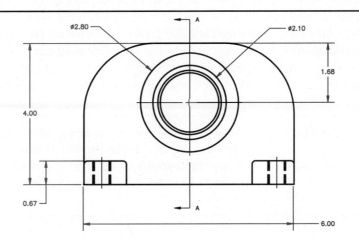

Figure 11-4.
The 1.68 dimension repositioned with the lower extension line reattached to the end of the centerline.

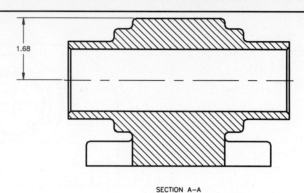

SECTION A–A

The Move option

The **Move** option moves a dimension from one view to another. For example, zoom out to see all three views, issue the **AMMOVEDIM** command and select the **Move** option. Click on the 1.68 dimension and then on the section view. Position the dimension and then use the **Reattach** option to attach the lower extension line to the end of the centerline as shown in **Figure 11-4.**

HIDING PARAMETRIC DIMENSIONS AND ADDING REFERENCE DIMENSIONS

Open part Example-11-02.dwg. This is the same part that was used in the previous example. We want a dimension from the center of the bored hole to the machined surface on the bottom of the part. We have two choices. The first choice is to go back to the model, edit the sketch, erase the 1.68 dimension, add the required dimension, and update the part and the views. The second choice, the one that we will do now, is to hide the 1.68 parametric dimension and add a reference dimension.

NOTE The difference between parametric dimensions and reference dimensions was covered in *Chapter 4*. If you are not clear on the concept, please review that chapter now.

AMDT_DRAWING_
VISIBILITY

Drawing
↦ Drawing
Visibility

Drawing Layout
toolbar

Hide Picked
Drawing Objects

First, do not erase parametric dimensions. The 1.68 dimension is related to dimension d4 by the equation (d4*0.6). Erasing it would eliminate that information.

Now, hide the 1.68 parametric dimension using the **AMDT_DRAWING_VISIBILITY** command. To access this command, enter AMDT_DRAWING_VISIBILITY at the Command: prompt, or select **Drawing Visibility** from the **Drawing** pull-down menu. This accesses the **Desktop Visibility** dialog box. Make sure **Drawing** tab is showing and the **Hide** radio button is picked. Now, pick the **Select** button and you are prompted to select the drawing objects to hide. Pick the 1.68 dimension, press [Enter], and press **OK**.

NOTE You can also initiate the **MNU_DRAWING_VISIBILITY** command by picking the **Hide Picked Drawing Objects** button on the **Drawing Layout** toolbar. You are prompted to select the drawing objects to hide. The dialog box is eliminated if this method is used.

One of two commands can be used to add a reference dimension. You can use either the **AMREFDIM** command or the **AMPOWERDIM** command.

The **AMREFDIM** command adds the dimension in the current dimension style. To access the reference dimension command, enter AMREFDIM at the Command: prompt, select **Reference Dimension** from the **Annotate** pull-down menu, or pick the **Reference Dimension** button on the **Drawing Layout** toolbar.

To access the power dimension command, enter AMPOWERDIM at the Command: prompt, select **Power Dimension** from the **Annotate** pull-down menu, or pick the **Power Dimension** button on the **Drawing Layout** toolbar. With this command you can add a dimension, change the number of places after the decimal, add and change the tolerances, add and change the fits, and add text.

With the power dimension command accessed and the **Osnaps** on and set to endpoint, pick the left end of both the centerline and bottom line of the view. As you move the dimension placement left and right, it will turn red when it is at the correct distance from the object lines. Click to place the dimension at this point. This accesses the **Power Dimensioning** dialog box. Change the precision to 3, and click on the **Add Tolerances** button. This expands the dialog box as shown in **Figure 11-5A.**

Now click on the large image button in the lower-right corner. This accesses the **Select Tolerance Type** dialog box, which gives you eight choices for the tolerance format. See **Figure 11-5B.** Pick the default, which is shown by the red square in the lower-left of the image button, and pick **OK**. You can type in the values for the deviations or click on the button with the three dots and pick them from the list. Set the upper to +0.001 and the lower to –.001. Click the **OK** button to see the effect of the change. The result is shown in **Figure 11-6.**

AMREFDIM
Annotate ↳ _Reference Dimension
Drawing Layout toolbar
Reference Dimension

AMPOWERDIM
Annotate ↳ _Power Dimension
Drawing Layout toolbar
Power Dimension

> **NOTE**
>
> The **Power Dimensioning** dialog box is accessible at any time by double-clicking on any dimension. You may change dimensions using this method.

Figure 11-5.
A—The **Power Dimensioning** dialog box. B—The **Select Tolerance Type** dialog box offers eight choices for the tolerance format.

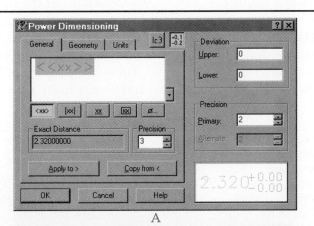

A

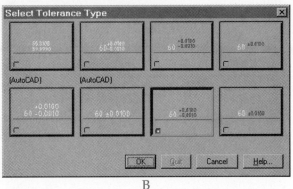

B

Figure 11-6.
The part with
tolerance in place.

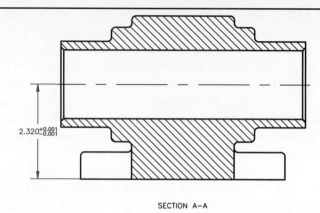

2.320±0.001

SECTION A–A

Changing Dimensions with the **Power Dimensioning** Dialog Box

Double-clicking on either a reference or parametric dimension will access the **Power Dimensioning** dialog box and the characteristic of the dimension can be changed—characteristics such as tolerances, fits, and number of decimal places. The only difference between the two types of dimensions is that the value of a reference dimension cannot be changed directly—the **Exact Distance** field in the dialog box is grayed out. Of course the value of the dimension will automatically change if the part is modified and the drawings updated. The value of a parametric dimension can be changed in the **Exact Distance** field and this will change both the model and the drawings when the drawings are updated.

Modifying Dimensions with **AMDIMFORMAT**

Another command used to modify the format of an individual dimension is **AMDIMFORMAT**. We are still working in Example-11-02.dwg. Zoom in on the front view. To access this command, enter AMDIMFORMAT at the Command: prompt.

AMDIMFORMAT

You are prompted to select a dimension entity. Pick the 2.10 diameter dimension. This displays the **Dimension Formatter** dialog box. See **Figure 11-7**. In this case for a diameter, the dialog box has five tabs. These tabs let you change the appearance of the dimension, but not its value. There are properties here that are not changeable with power edit, such as the gap and center mark size.

Figure 11-7.
The **Dimension Formatter** dialog box can be used to change the appearance of the dimension, but not its value.

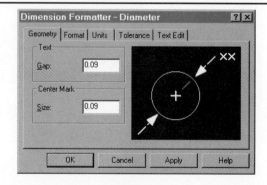

PROFESSIONAL TIP

The menu choices on the **Edit Dimensions** cascading menu are also available on a context-sensitive pop-up menu for controlling views and layouts. To access this menu, right-click anywhere on the drawing and the menu appears. Select **Annotate Menu** at the bottom and you get the cascading menu with editing choices. See **Figure 11-8.** The next time you access this pop-up menu, **Drawing Menu** is at the bottom and **Repeat Annotate Menu** is at the top.

Figure 11-8.
The cascading menu
with editing
choices.

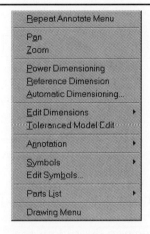

DIMENSION MODIFYING COMMANDS

MDT will also automatically align and arrange parametric dimensions when a new view is created. To see this happen, open Example-11-03.dwg. First we want to see how the dimensions were placed on the sketch. Right-click on ExtrusionBlind1 in the **Browser** and edit the sketch. Enter 9 to change the view. Notice that the dimensions along the top are not in line and those along the sides overlap. Update the part and move to Layout1. Now we'll create a new view and see how MDT positions the dimensions. Use **AMDWGVIEW** to create a base view of the active part at a scale of one. Select the top face of the part and then the top front edge to set the viewpoint and then locate the view at the center of the drawing. Notice how MDT arranges and aligns the dimensions.

To get the diameter dimension off the part, use **AMDIMFORMAT** command. In the **Format** tab, select the **Arrows Inside Only** button, which is the one on the right side. Then move the dimension with grips. Results are shown in **Figure 11-9.**

Open part Example-11-04.dwg. See **Figure 11-10.** There are four dimension-modifying commands available on the **Annotate** pull-down menu and on the drawing toolbar. These are as follows:

- **AMDIMBREAK.** Breaks either a dimension or extension line. It works similar to AutoCAD's **BREAK** command. You pick both the object and the first break point with the first pick and then pick the second break point. Issue the **AMDIMBREAK** command and pick the dimension line, for the 2.73 diameter dimension, just below the edge of the part. For the second point, pick just above the edge of the part. The gap will remain until the dimension is moved twice or when the drawing is updated after a parametric dimension was changed that moved the hole or the part edge.
- **AMDIMJOIN.** Joins two or more parametric or reference dimensions and replaces them with one reference dimension. The parametrics are hidden and not

Figure 11-9.
The part after the
diameter dimension
is moved and
formatted.

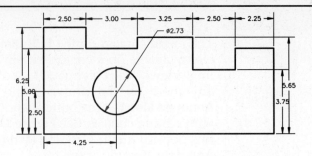

Figure 11-10.
The part used to
demonstrate the
four dimension
modifying
commands.

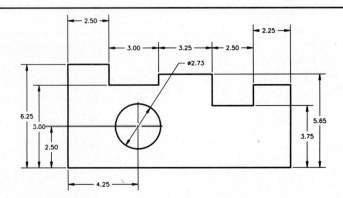

erased and they can be unhidden. Issue the **AMDIMJOIN** command and pick the three vertical dimensions on the left end of the part. Start with the 6.25 as the base dimension. Now join the 3.00 and 3.25 horizontal dimensions at the top of the part.

- **AMDIMINSERT.** Turns one parametric or reference dimension into two reference dimensions. Issue the command and pick the 6.25 reference dimension as the base dimension. Now pick the endpoint of the left end of the centerline of the hole.
- **AMDIMALIGN.** Aligns two or more linear parametric or reference dimensions. Pick the 6.25 reference dimension as the base dimension and then the 2.50 horizontal dimension that is to the left and above, and then 2.25 dimension to the right. Press [Enter] and the three dimensions will line up.

PRACTICE 11-1

❑ Open the drawing Practice-11-01.dwg.
❑ Use the four dimension modifying commands of break, join, insert, and align to arrange the dimensions as shown here.

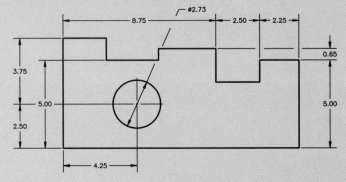

❑ Note there will be two extension lines through the 5.00 dimension text and you will have to break twice.

PLACE AND EDIT HOLE NOTES

Open part Example-11-05.dwg. See **Figure 11-11.** The four holes shown in this view were put in the model with the **AMHOLE** command. These holes do not get automatic dimensions and therefore require *hole notes*. To place a new note or edit an existing one, use the **AMHOLENOTE** command.

To access this command, enter AMHOLENOTE at the Command: prompt, select **Hole Note** from the **Annotation** cascading menu in the **Annotate** pull-down menu, or pick the **Hole Note** button on the **Drawing Layout** toolbar.

Select New and pick the hole in the upper-left corner. This accesses the **Create Holenote** dialog box. See **Figure 11-12.** MDT knows that this is a through hole so it displays the names of the templates for the text for through holes. The name of the template may not accurately describe the template text. Click on the **Edit Template** button that opens the **Multiline Text Editor** dialog box. Change the word THROUGH to THRU and add the text 2 PLACES. Press **OK** and this results in a one time only, temporary template called *EDITED*. Press **OK**, select a location for the note, and press [Enter] twice to place the note. See **Figure 11-13.**

AMHOLENOTE

A**n**notate
➡ **A**nnotation
➡ **H**ole Note

Drawing Layout toolbar

Hole Note

Figure 11-11.
The part used to place and edit hole notes.

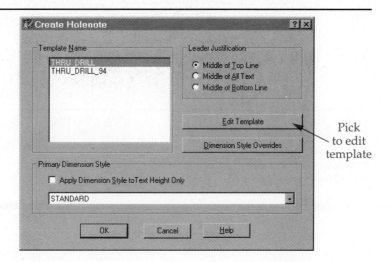

Figure 11-12.
The **Create Holenote** dialog box.

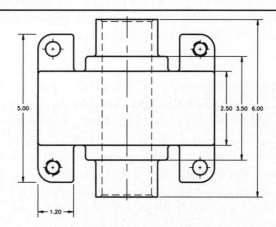

Pick to edit template

Figure 11-13.
The part with the
hole note in place.

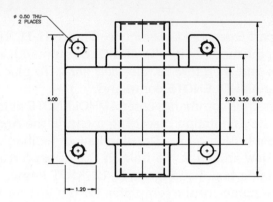

AMTEMPLATE

Annotate
➥ **A**nnotation
➥ Hole **N**ote
Template

Drawing Layout
toolbar

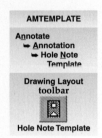

Hole Note Template

Edit and Create Permanent Templates

To edit and create permanent templates, use the **AMTEMPLATE** command. To access this command, enter AMTEMPLATE at the Command: prompt, select **Hole Note Template** from the **Annotation** cascading menu in the **Annotate** pull-down menu, or pick the **Hole Note Template** button on the **Drawing Layout** toolbar.

This accesses the **Hole Templates** dialog box. The process is to rename and edit an existing template to form a new one. From the **Templates:** drop-down list, select THRU_DRILL and double-click on the **Name:** text box and change the name to DRILLTAP_13. See **Figure 11-14.** Click the **Save** button to create DRILLTAP_13 from THRU_DRILL. Now click the **Edit Template** button to access the **Multiline Text Editor** dialog box. Delete the %%c that puts in the diameter symbol, leave the %%AMDIA that puts in the hole size, and change the word THROUGH to – 13 UNC – 2B THRU. Click the **OK** button twice to save the changes in the template.

Issue the **AMHOLENOTE** command, select New, and pick the tapped hole in the lower-left corner. Now select the DRILLTAP_13 template, click **OK**, and place the note. This gives the diameter of the inside of the threads.

Press [Enter] to issue the command again and select **Edit**. Select the note just placed to access the **Edit Holenote** dialog box. Pick the **Edit Template** button and change the note to read .50 –13 UNC–2B THRU 2 PLACES. Put a return after the word THRU to make it a two-line note. Press **OK** twice to view the edited note in place. See **Figure 11-15.** These hole notes are automatically attached to the view. If you move the view or the position of the hole, they will move with it.

Figure 11-14.
The **Hole Template**
dialog box is used
to create permanent
templates.

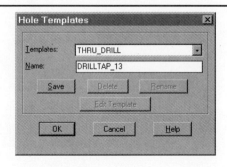

Figure 11-15.
The part with the new hole note in place.

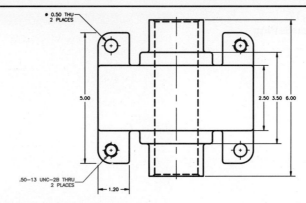

LEADERS, ANNOTATION, AND ATTACHMENT

QLEADER

Annotate
➥ Annotation
➥ Leader

Drawing Layout
toolbar

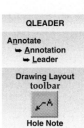

Hole Note

Open part Example-11-06.dwg. See **Figure 11-16.** Leaders are created with the **QLEADER** command. To access this command, enter QLEADER at the Command: prompt, select **Leader** from the **Annotation** cascading menu in the **Annotate** pull-down menu. The **Leader** button on the **Drawing Layout** toolbar accesses the **LEADER** command—not the **QLEADER** command.

 NOTE The **Annotation** menu can also be accessed from the pop-up menu by right-clicking anywhere on the screen.

You can pick the first leader point or by pressing [Enter] you can modify the leader settings using the **Leader Settings** dialog box. See **Figure 11-17.** On the **Leader Line & Arrow** tab you can select any of twenty types of arrowheads including none, have straight or curved (spline) leader lines, have as many segments as you want, and set the angles for the first two segments. The **Attachment** tab controls the position of the text in relationship to the end of the leader line. The **Annotation** tab gives you several choices for the type of annotation including **Mtext** and **Tolerance**, lets you justify the text or put a box around it, and reuse existing annotations.

Using this dialog box, change the settings to frame text, spline leader line, no arrowhead, and the text on right side should be at top of the top line. Place a leader with the text as shown in **Figure 11-18.** You are limited to three points for the spline and the fourth pick sets the width of the text box.

Figure 11-16.
The part used to demonstrate variations of leaders.

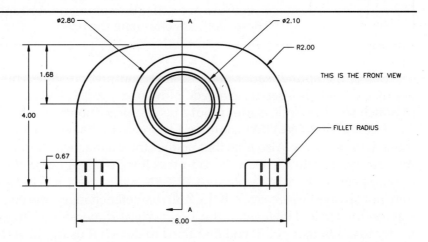

Figure 11-17.
The **Leader Setting** dialog box.

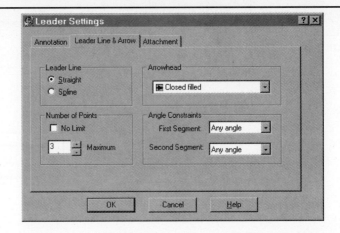

Figure 11-18.
The part with leader and text in place.

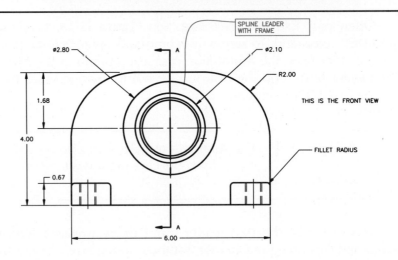

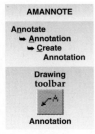

AMANNOTE

A̲nnotate
↳ A̲nnotation
  ↳ C̲reate
     Annotation

Drawing
toolbar

Annotation

The leader you just created, the existing ones with the text FILLET RADIUS and THIS IS THE FRONT VIEW note, will *not* move with the view when it is repositioned. They must be attached to the view with the create-annotation command **AMANNOTE**. To access this command, enter AMANNOTE at the Command: prompt, or pick the **Annotation** button on the **Drawing Layout** toolbar. When prompted to enter an option, enter C for create.

> **NOTE**
>
> The **AMANNOTE** command can be also be accessed by selecting **C̲reate Annotation** from the **A̲nnotation** cascading menu in the **A̲nnotate** pull-down menu or by right-clicking anywhere on screen and accessing the command through the pop-up menu. These methods eliminate the need to enter the **Create** option.

Select the spline leader and the text you just created. This requires two picks, because the text is a separate object. Now press [Enter] and pick the point in the view to attach to—the circle center would be a logical choice.

Use the **AMMOVEVIEW** command to move the view. Note that the leader and text move with the view. Also note that neither the unattached leader on the fillet radius nor the text move with the view. If you press [Esc] while moving the view, the leader text may appear in the wrong place. A **REGEN** will correct it. This annotation attachment may not always be *parametric*. If the 2.80 diameter changes, the end of the leader will not stay on the circle. However, if the 1.68 vertical dimension changes, it will. To attach a leader to a different point, you first need to detach it using the **AMANNOTE** command.

SURFACE FINISH SYMBOLS

Open part Example-11-07.dwg. Symbols can be added to the edges of the part, both linear and circular. We are going to put five symbols, labeled A through E, on this drawing. See **Figure 11-19.** The symbols can have leaders or not, arrowheads or not, and a variety of numerical and symbolic information. For ease of viewing in this example, all the symbols are scaled up by a factor of 2 using the regular AutoCAD **SCALE** command. The first symbol to be inserted is shown at A. To do this, use the **AMSURFSYM** command. To access this command, enter AMSURFSYM at the Command: prompt, select **Surface Texture** from the **Symbols** cascading menu in the **Annotate** pull-down menu, or pick the **Surface Texture** button on the **Drawing Layout** toolbar.

AMSURFSYM

A__notate
➡ **Symbols**
➡ **Surface**
Texture

Drawing Layout
toolbar

Surface Texture

>
> **NOTE**   The **Annotation** menu can also be accessed from the pop-up menu by right-clicking anywhere on the screen.

Select the object you wish to attach the symbol to—use the top line across the part. You are prompted to pick the *start point*. This is the start of the leader, or in this case where we do not want a leader, and the location of the symbol on the line. Pick a location on the line. It will probably be necessary to turn off **Osnap**. Press [Enter] for the *next point* and then pick above the line to select the side on which the symbol will be placed. This accesses the **Surface Texture** dialog box. See **Figure 11-20A.**

There are three surface type symbols—the first is a basic symbol, the second adds a machining required designation, and the third adds a machining prohibited designation. The **Miscellaneous** button adds a horizontal line called a *tail* to the symbol. There are several data fields (text boxes) in the dialog box and a different selection is available for each of the surface type symbols. Pick the **Basic** symbol and leave the data fields blank.

> **PROFESSIONAL TIP**   The **Properties...** button in the lower-left corner controls the surface texture properties for ANSI. See **Figure 11-20B.**

Figure 11-19.
Part with a number of surface finish symbols in place.

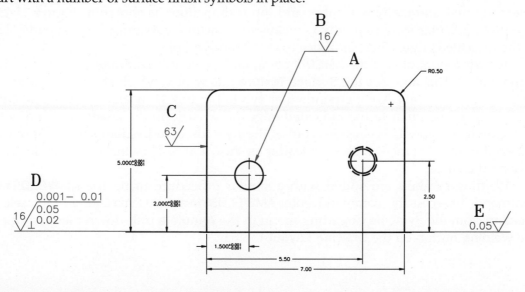

Figure 11-20.
A—The **Surface Texture** dialog box. B—The **Standard Properties** dialog box is used to make changes to surface finish symbols.

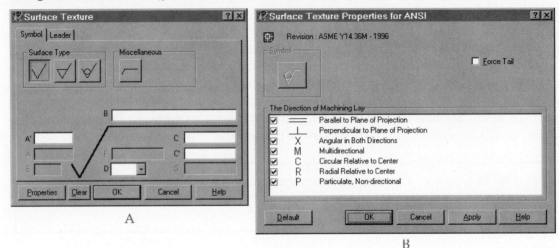

A

B

Now put a symbol with a leader on the hole as shown at B. First pick the circle as the object to attach and then pick it again as the start point for the leader. Then pick two more points for the end points of the two legs of the leader and touch [Enter]. Put a 16 (micro inch) value for **A'** in the **Surface Texture** dialog box. Click **OK** twice to draw the symbol.

Now put another **Basic** symbol with a leader on the line as shown at C. First pick the vertical line as the object to attach and then pick it again as the start point for the leader. Then pick one more point for the end point of the straight leader and touch [Enter]. Put a 63 (micro inch) value for **A'** in the **Surface Texture** dialog box. Click **OK** twice to draw the symbol.

For the machining required symbol shown at E, pick the line across the bottom of the part as the object to attach. Then pick the right end point as the start point for the leader. Pick one more point for the end point of the horizontal leader and touch [Enter]. Select the second **Surface Type** symbol and put a 0.05 value in text box E. Click on the **Leader** tab, pick **Arrowhead Type** and select the button without an arrowhead. Click **OK** twice to draw the symbol.

For the **Standard** symbol shown at D, pick the line across the bottom of the part as the object to attach. Then pick the left end point as the start point for the leader. Then pick one more point for the end point of the horizontal leader and touch [Enter]. Select the first **Surface Type** symbol and input the values as shown in **Figure 11-21**. The perpendicular symbol in box D means the machine cuts must be perpendicular to the indicated edge. Click **OK** twice to draw the symbol.

To edit a symbol, use the **AMEDIT** command. Issue this command and select the symbol at D. This accesses the **Surface Texture** dialog box, which shows the location of the numerical data. The perpendicular symbol in box D means the machine cuts must be perpendicular to the indicated edge. Now click on the **Leader** tab and note the arrowhead type. This is where you set the style for an individual leader. Input the material removal symbol without a leader as shown at E. The start of the leader will stay on the circle.

Welding symbols are added with a similar procedure using the **AMWELDSYM** command. To access this command, enter AMWELDSYM at the Command: prompt, select **Welding** from the **Symbols** cascading menu in the **Annotate** pull-down menu, or pick the **Welding** button on the **Drawing Layout** toolbar.

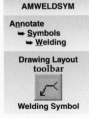

AMWELDSYM

A**n**notate
➥ **Symbols**
  ➥ **Welding**

Drawing Layout toolbar

Welding Symbol

Figure 11-21.
The **Surface Texture** dialog box showing setting for symbol at D.

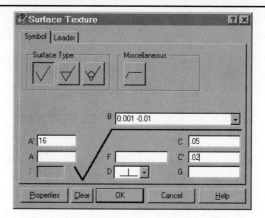

> **NOTE**
>
> The **Annotation** menu can also be accessed from the pop-up menu by right-clicking anywhere on the screen.

GD&T SYMBOLS

Open part Example-11-08.dwg. This section is in no way meant to be a discussion of the definitions or meanings of the geometric dimensioning and tolerancing methods. It is about how you create the symbols and place them on the drawing views. Generally, you first need to identify the datum surfaces and then add feature control frames. There is one datum identifier on the top view labeled "A." We want to add two more datum identifiers and then one feature control frame. See **Figure 11-22.**

Figure 11-22.
The part with datum and feature control frame.

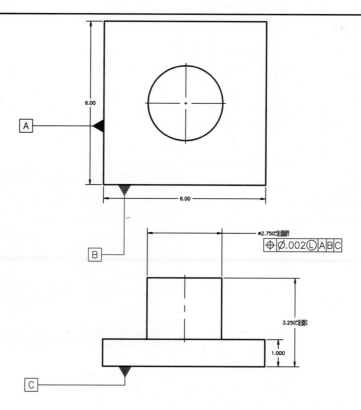

Adding a Datum Identifier

AMDATUMID

Annotate
➡ Symbols
  ➡ Datum
    Identifier

Drawing Layout
toolbar

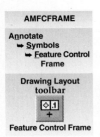

Datum Identifier

The **AMDATUMID** command is used to add a datum identifier. To access this command, enter AMDATUMID at the Command: prompt, select **Datum Identifier** from the **Symbols** cascading menu in the **Annotate** pull-down menu, or pick the **Datum Identifier** button on the **Drawing Layout** toolbar.

NOTE	The **Annotation** menu can also be accessed from the pop-up menu by right-clicking anywhere on the screen.

Now, pick the bottom edge of the top view. This is the object where it will attach. For the start point, pick a point somewhere on the bottom edge. Move the cursor down to get the leader and pick the second point. Now move it to the left and pick the third point. This accesses the **Datum Identifier** dialog box. Enter a B for the letter and pick **OK**. Scale the symbol up by a factor of 2 to emphasize the symbols in this drawing. Now add symbol C to the front view and also scale it up by a factor of 2.

Adding a Feature Control Frame

AMFCFRAME

Annotate
➡ Symbols
  ➡ Feature Control
    Frame

Drawing Layout
toolbar

Feature Control Frame

The **AMFCFRAME** command is used to add a feature control frame. To access this command, enter AMFCFRAME at the Command: prompt, select **Feature Control Frame** from the **Symbols** cascading menu in the **Annotate** pull-down menu, or pick the **Feature Control Frame** button on the **Drawing Layout** toolbar.

NOTE	The **Annotation** menu can also be accessed from the pop-up menu by right-clicking anywhere on the screen.

Pick an object to attach the frames and then pick the points for the leader. This accesses the **Feature Control Frame** dialog box. Using the small "keypad" buttons in the dialog box enter the symbol, tolerance, and datum as shown in **Figure 11-23**. Press **OK** and then scale this frame up by a factor of 2. To edit either of these symbols use power edit by double-clicking on the symbol.

Figure 11-23.
The **Feature Control Frame** dialog box showing settings.

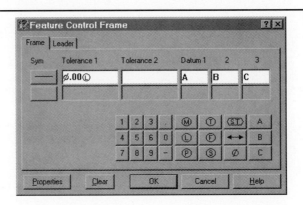

AUTOMATIC DIMENSIONING

Open part Example-11-09.dwg. See **Figure 11-24.** To get the parametric dimensions automatically placed on the drawing views to standards, a large amount of editing is sometimes required. In these cases it may be faster to hide most of the parametric dimensions and apply automatic reference dimensions. These dimensions are applied to all linear edges and centerlines, so you may want to keep diameter dimensions as parametric.

Now set up the top view for automatic dimensions. First select **Drawing Visibility...** from the **Drawing** pull-down menu or pick the **Drawing Visibility** button from the **Drawing Layout** toolbar. This accesses the **Drawing Visibility** dialog box. Click the **Select** button and select all the linear parametric dimensions, leaving only the one diameter and one radius. Next, since we don't want dimensions to hidden lines, right-click on the TOP view in the **Browser**, pick **Edit**, and click off the **Display Hidden Lines** check box.

PROFESSIONAL TIP
    In more complex parts with many dimensions, it is easier to hide all the parametric dimensions by clicking **Parametric Dims** in **Desktop Visibility** dialog box. Then reissue the command and individually unhide the few dimensions you want visible.

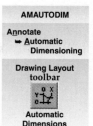

AMAUTODIM

A**n**notate
➡ **A**utomatic
   Dimensioning

Drawing Layout toolbar

Automatic
Dimensions

To add automatic dimensions, enter AMAUTODIM at the Command: prompt, select **Automatic Dimensioning** from the **A**nnotate pull-down menu, or pick the **Automatic Dimension** button on the **Drawing Layout** toolbar.

NOTE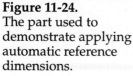
    The **Annotation** menu can also be accessed from the pop-up menu by right-clicking anywhere on the screen.

This displays the **Automatic Dimensioning** dialog box. See **Figure 11-25.** In this dialog box, there are the **Parallel, Ordinate,** and **Shaft/Symmetric** tabs for dimensions. On the **Parallel** tab, you may select **Baseline** or **Chain** from the **Type** drop-down list. Select the **Baseline** option. Select the **Both Axes** check box, because both vertical and horizontal dimensions are desired. If you select the **Display Power Dimensioning Dialog** check box, as each dimension is placed the dialog box is displayed and you can change decimals, tolerances, etc. For our purposes, leave this unchecked.

Figure 11-24.
The part used to demonstrate applying automatic reference dimensions.

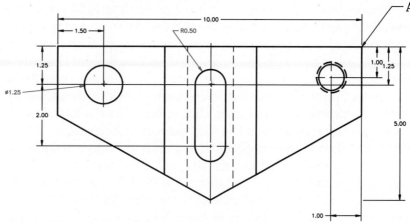

Figure 11-25.
The **Automatic Dimensioning** dialog box.

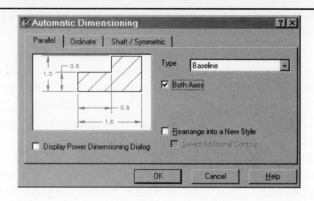

Figure 11-26.
Part with parametric dimensions removed and automatic reference dimensions in place.

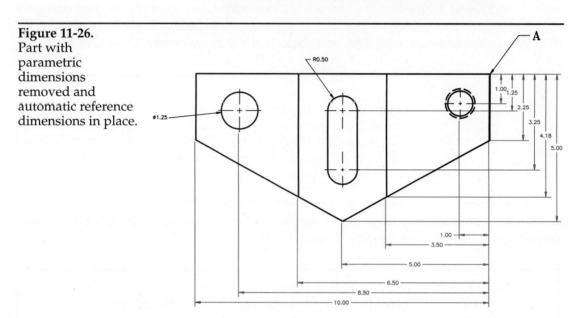

Click **OK** and put a window around the top view to select all the edges and press [Enter]. Use endpoint to pick the point labeled A as the first extension line origin. Press [Enter] and move the cursor to the right and place the vertical dimensions. When the dimensions turn red, they are at the correct ANSI standard distance from the edge. Press [Enter] and locate the horizontal dimensions. Erase the two horizontal dimensions on the slot. See **Figure 11-26.** The properties of any dimension can be edited by double-clicking on that dimension.

Automatic Dimensioning Sketches

The same automatic dimensioning process can be used on sketches. A sketch containing a spline is one of the most difficult to fully constrain. As an example, open part Example-11-10.dwg. Using the **SPLINE** command, draw a spline through the upper endpoints of the construction lines—starting at A and ending at B. For the starting tangent, select the endpoint on point C and then point D for the ending tangent. Profile the three lines and the spline resulting in 19 dimensions required. See **Figure 11-27.**

Now apply the automatic dimensions. Pick the **Baseline** and **Both Axes** check box in the **Automatic Dimensioning** dialog box. Now select all four objects in the sketch. Pick the endpoint at A as the first extension line origin. Place the dimensions to the right and then above the sketch as shown in **Figure 11-28.**

Two more dimensions, the angles of the tangents at the ends of the spline, are required to almost fully constrain the sketch. Zoom in on the corner of the sketch showing both D and B. Select **New Dimension (AMPARDIM)**, the tangent line, (a dashed line) will appear. As shown in **Figure 11-29,** place angular dimensions between the

Figure 11-27.
Part with spline in place.

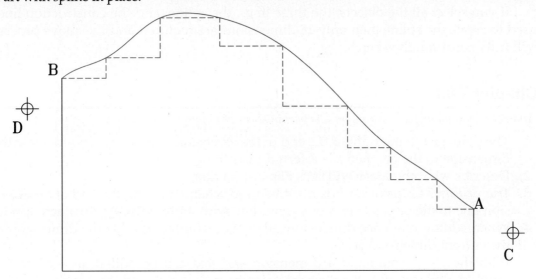

Figure 11-28.
Part with automatic reference dimensions in place.

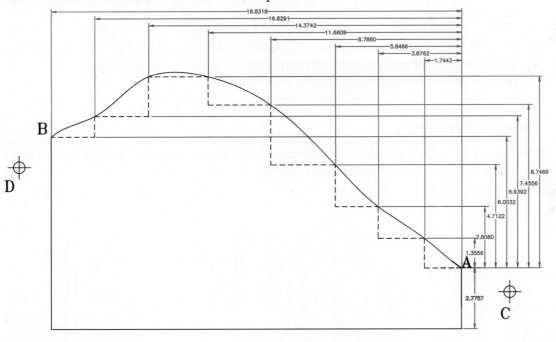

Figure 11-29.
A and B show the two dimensions, the angles of the tangents at the ends of the spline, required to fully constrain the sketch.

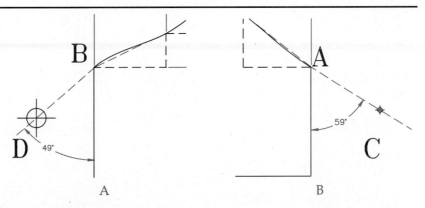

tangent line and the left edge of the sketch. You may have to enter an "N" to get an angle dimension. Repeat the process adding another angle dimension at A and C.

If you select all the objects: the three lines, the spline, and the construction lines used to create the spline then only 18 dimensions are required and the above process will fully constrain the sketch.

Chapter Test

Answer the following questions on a separate sheet of paper.

1. The point just to the right of the end of the extension line that is used to relate the dimension to the geometry is referred to as the _____.
2. Describe what the **AMMOVEDIM**'s **Flip** option does.
3. *True or False?* Grips or osnaps must be used when attaching the end of an extension line to the opposite end of a centerline with **AMMOVEDIM**'s **Reattach** option.
4. When adding reference dimensions, the _____ command adds the dimension in the current dimension style.
5. With the _____ command a dimension and text can be added, as well as the number of places after the decimal, tolerances changed, and the fits changed.
6. *True or False?* When using the **Dimension Formatter** dialog box to change dimensions, the appearance of the dimension can only be changed.
7. The _____ command is used to align two or more linear parametric or reference dimensions.
8. The _____ command is used to edit and create permanent templates.
9. Describe the difference between the **AMDATUMID** and **AMFCFRAME** commands.
10. What command can be used to eliminate the problem of getting automatically-placed, parametric dimensions to standards?

Chapter Exercises

Exercise 11-1. Open the drawing Exercise-11-01.dwg. Use grips to relocate the parametric dimensions as shown.

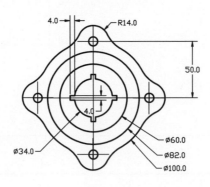
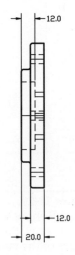

Exercise 11-2. Open the drawing Exercise-11-02.dwg. Use **AMDWGVIEW** to create the front view of the part. Hide all the parametric dimensions and apply reference dimensions as shown.

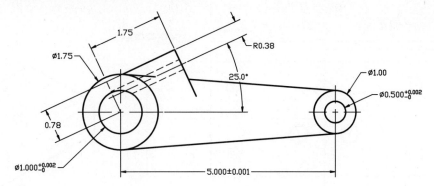

Exercise 11-3. Open the drawing Exercise-11-03.dwg. Move the parametric dimensions to different views and add reference dimensions. Edit the values to look like the figure shown.

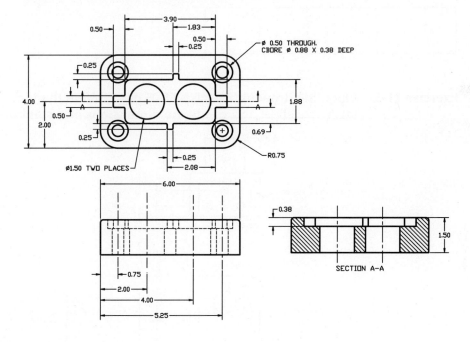

Exercise 11-4. Open the drawing Exercise-11-04.dwg. Add hole notes, leaders, and notes as shown in the figure. Attach the leaders and notes and move the views as shown.

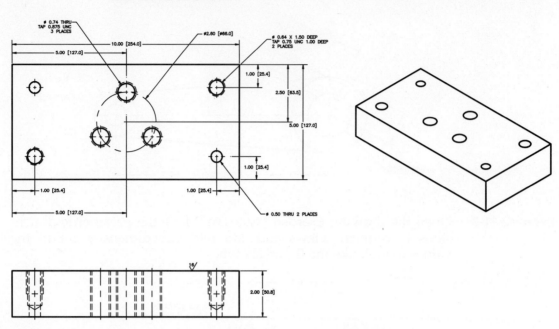

Exercise 11-5. Open the drawing Exercise-11-05.dwg. Create the views as shown on an A-size sheet and create the dimensions and symbols.

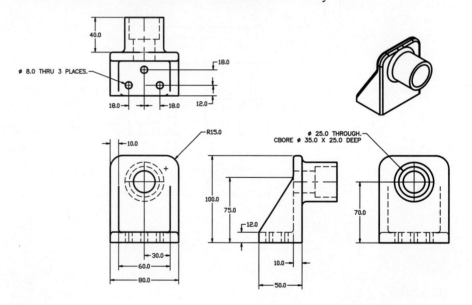

CHAPTER 12

More Features

Objectives

After completing this chapter, you will be able to:

* Understand the concept of draft angles.
* Create parts with draft angles.
* Change the draft angle on a face.
* Understand the concept of a shadow draft angle.
* Divide a face into two parts.
* Shell a part excluding faces.
* Activate a part in a multipart drawing.
* Split a part into two parts.
* Combine two parts into one.
* Bend a part with the **AMBEND** command.
* Use thin extrusions to make a web.

Drawings Needed

*The following is a list of files that you will need to work through this chapter. These files can be found on the **User's Drawings** CD included with this text.*

Examples		Exercises
Example-12-01.dwg	Example-12-08.dwg	Exercise-12-01.dwg
Example-12-02.dwg	Example-12-09.dwg	Exercise-12-02.dwg
Example-12-03.dwg	Example-12-10.dwg	Exercise-12-03.dwg
Example-12-04.dwg	Example-12-11.dwg	Exercise-12-04.dwg
Example-12-05.dwg		Exercise-12-05.dwg
Example-12-06.dwg	**Practices**	Exercise-12-06.dwg
Example-12-06-Constrained.dwg	Practice-12-01.dwg	
	Practice-12-02.dwg	
Example-12-07.dwg	Practice-12-03.dwg	

DRAFT ANGLES

Some manufacturing processes, such as die-casting and injection molding, require the sides of the part to have a draft angle. A *draft angle* is usually less than 3° and is used to facilitate part removal from a mold. A draft angle can be applied to extruded profiles.

Larger draft angles can be used to easily create shapes that would be difficult with the techniques previously discussed. In this chapter, we will discuss a command that allows you to put a draft angle on any face of a part. You will also learn to make a part hollow with a specified wall thickness and to make a given face open as in an open box.

EXTRUDING WITH A DRAFT ANGLE

The draft angle for an extrusion is input in the **Extrusion** dialog box. See **Figure 12-1.** The angle must be between a –89.999° and +89.999°. If an unacceptable value is input, the number will turn red.

To see how different values affect the part, open Example-12-01.dwg as shown in **Figure 12-2.** Part A was extruded upward with a positive draft angle of 15°. Positive angles make the part larger in the extrusion direction. Part B with a negative draft angle gets smaller in the direction of extrusion. Part D was extruded mid-plane with a negative draft angle and the part gets smaller in both directions.

Part C was extruded upwards a distance of High and has the maximum allowable draft angle. Exceeding the maximum angle would create a part like E where the draft passes through itself, causing the modeler to fail. Part E cannot be extruded in one operation. The angle that would extrude part C to a knife-edge at the top is calculated from the formula: –Atan (High/(Wide/2)). The value of this angle is –26.5651 and this will also cause the modeler to fail. The angle has to be at least 0.0001 units less or –26.5610. If you are interested, zoom in on the top several times and measure the width of the flat.

Figure 12-1.
The draft angle is input in the **Draft angle:** text box in the **Extrusion** dialog box.

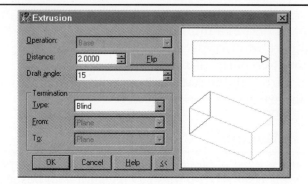

Figure 12-2.
Various examples of draft angles applied to parts.

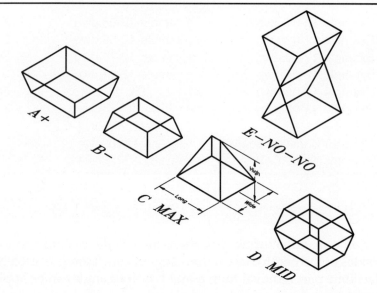

Remember that draft angles are applied to all edges of the profile equally in extrusion. Draft angles can also be applied to 2D sweeps (covered in *Chapter 13*), but not to any of the 3D paths—**Edge**, **Helix**, **Pipe**, or **Spline** (also covered in *Chapter 13*).

PRACTICE 12-1

❏ Open the drawing Practice-12-01.dwg.
❏ Profile the sketch and extrude 3.0 units up with a 10° draft angle.
❏ Edit the extrusion and find the maximum allowable draft angle. This will be when the slot on the right closes.
❏ Edit the extrusion to a draft angle of –10°. Then find the maximum negative allowable draft angle. This will be when the slot on the left touches the left edge.

PLACING A DRAFT ANGLE ON EXTRUDED FEATURES

Draft angles can also be placed on extruded features. Open Example-12-02.dwg as shown in **Figure 12-3A.** Extrude the profiled sketch **Cut**, **Blind**, 4 units with a 14° draft angle. See **Figure 12-3B.** Note that as the cut moves up, it gets larger. This is because of the positive draft angle.

Figure 12-3.
A—The part used to demonstrate a draft angle. B—The part with resulting draft angle.

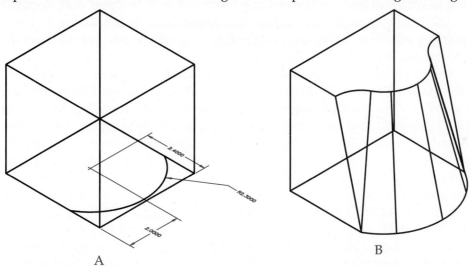

A

B

PUTTING A DRAFT ANGLE ON INDIVIDUAL FACES

A unique draft angle can be placed on one or more faces of a model. As an example, open Example-12-03.dwg as shown in **Figure 12-4.** To place a unique draft angle, use the **AMFACEDRAFT** command. To access this command, enter AMFACEDRAFT at the Command: prompt, select **Face Draft** from the **Placed Features** cascading menu in the **Part** pull-down menu, or pick the **Face Draft** button on the **Part Modeling** toolbar.

This command accesses the **Face Draft** dialog box. See **Figure 12-5.** The angle in this case is measured from a plane called the *draft plane*. The first step is to select the draft plane by picking the **Draft Plane** button and selecting the bottom of the part as the planar face. A small symbol appears on the screen that looks somewhat like a badminton birdie with an arrow. If the arrow points down, the face will rotate about the bottom edge moving out of the part and increasing its volume. If it is flipped upward,

AMFACEDRAFT

Part
➥ Placed Features
➥ Face Draft

Part Modeling
toolbar

Face Draft

Figure 12-4.
The part used to
demonstrate a face
draft.

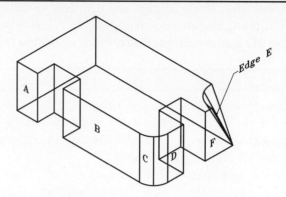

Figure 12-5.
The values for face
drafts are input in
the **Face Draft** dialog
box.

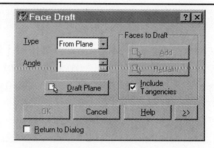

the face will rotate about the bottom edge moving into the part and decreasing its volume. The arrow is required, because the **Angle** field in the dialog box will only accept positive numbers. In either case, the face to draft always rotates about the edge common with the draft plane.

Flip the arrow up, accept it, and input an angle of 10° in the dialog box. Click the **Add** button in the **Faces to Draft** area and select the face labeled B. Since the corner has a radius and the **Include Tangencies** box is checked, the selection acts like the **AMFILLET** command and the three faces B, C, and D are selected. Press [Enter] and pick **OK**, and the face is drafted. The maximum angle on this face is about 18°, which is where the radius of the filleted corner goes to zero at the top edge. Now, delete the face draft just created.

If the **Include Tangencies** box is *not* checked, only face B will be selected for the face to draft. Issue the **AMFACEDRAFT** command, input an angle of 12°, pick the bottom of the part as the draft plane, and flip the arrow so it points up. Uncheck the **Include Tangencies** check box, pick the **Add** button, and select face B. Press [Enter] and pick **OK**. The fillet conforms to the new angle for face B, but does not change the radius.

To rotate a face about an edge that is not part of the draft plane, select From Edge in the **Type** drop-down box as shown in **Figure 12-6**. For example, input a 12° draft angle, select the bottom of the part as the draft plane, and point the arrow downward. Select face F as the face to draft and edge E as the *fixed edge*. The results of these face drafts are shown in **Figure 12-7**.

Figure 12-6.
Settings for a face
draft from edge E at
a 12° angle.

Figure 12-7.
The part showing the results of the face drafts.

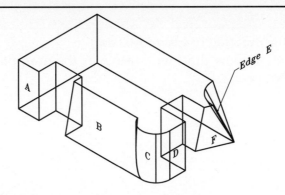

THE SHADOW DRAFT

Shadow is the third option in the **Type** drop-down box. To understand what this means open Example-12-04.dwg as shown in **Figure 12-8.** This part has a flat face A and a curved face B. A light shines on the part at a 50° angle to the vertical and casts a shadow from B onto A. To see the shadow in the isometric view, enter RENDER at the Command: prompt and click the **Render** button in the **Render** dialog box.

The **Shadow** option will fill this shadow with material, increasing the size of the part. Perform a **REGEN** and press [Enter]. Issue the **AMFACEDRAFT** command and select Shadow and input 50°. See **Figure 12-9.** Select face A as the draft plane and make sure the arrow points upward to add material. Select B as the face to draft and click on **OK**.

Figure 12-8.
The part used to demonstrate a shadow draft.

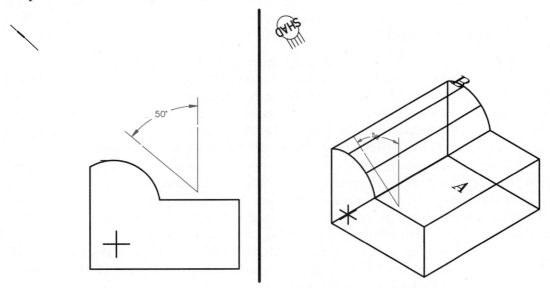

Figure 12-9.
The **Shadow** option is found in the **Type** drop-down list in the **Face Draft** dialog box.

❑ Open the drawing Practice-12-02.dwg.
❑ Put a 60° shadow draft with face A as the draft plane and B as the face to draft as shown here.

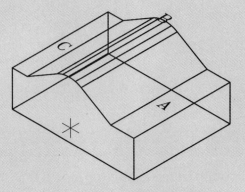

❑ Note the shadow goes both ways and since faces A and C are collinear they both get the same draft.

DIVIDING A FACE INTO TWO PARTS

Dividing a face into two parts works well with the **AMFACESPLIT** command, because a different angle can be applied to each of the new faces with the **AMFACEDRAFT** command. The face can be divided by the *planar* method, which requires a work plane as the dividing object, or the *project* method, which requires a predefined and profiled *split line*.

The Planar Method

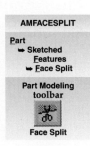

AMFACESPLIT

Part
↳ Sketched
  Features
↳ Face Split

Part Modeling
toolbar

Face Split

Open Example-12-05.dwg as shown in **Figure 12-10.** Face A is to be divided into two parts and the work plane shown in color will divide them. The work plane was drawn through the corners of face A using the three vertices option. To split the face, use the **AMFACESPLIT** command. To access this command, enter AMFACESPLIT at the Command: prompt, select **Face Split** from the **Sketched Features** cascading menu in the **Part** pull-down menu, or pick the **Face Split** button on the **Part Modeling** toolbar.

AMFACESPLIT uses the command line rather than a dialog box. At the Enter facesplit type [Planar/pRoject] <pRoject>: prompt and enter a P for Planar. Select face A as the face to split and then press [Enter] to accept the face. Press [Enter] again to indicate the end of the face select phase. Select the work plane as the "work plane for split." A line appears on face A. This is the division of the two new faces.

Use the **From Edge** option of the **AMFACEDRAFT** command to place a 15° draft angle on both of the new faces. To get the results shown in **Figure 12-11,** the procedure must be applied twice—once for each face.

Figure 12-10.
The part used to demonstrate the planar method of dividing a face into two parts.

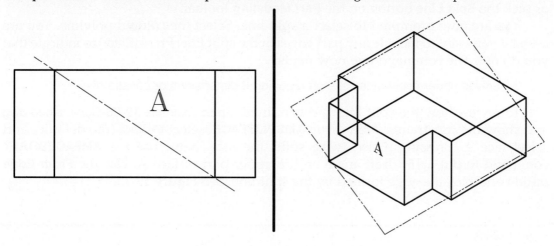

Figure 12-11.
The result of dividing face A and adding different draft angles to the two new parts.

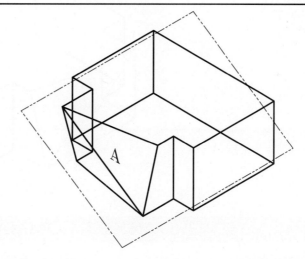

The Project Method

The **Project** option of the **AMFACESPLIT** command projects a *split line* onto a face and divides the face along this line. The split line is a line or polyline of the Continuous line type that is profiled with the **AMSPLITLINE** command. Open Example-12-06.dwg as shown in **Figure 12-12.** Face A is the sketch plane and a polyline was drawn on this face in red. The color is immaterial—red is used for emphasis only.

Figure 12-12.
The part used to demonstrate the project method of dividing a face into two parts. The colored line is the split line.

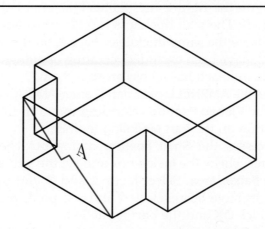

AMSPLITLINE

Part
↳ Sketch Solving
↳ Split Line

Part Modeling
toolbar

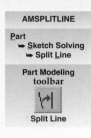

Split Line

To issue the **AMSPLITLINE** command, enter AMSPLITLINE at the Command: prompt, select **Split Line** from the **Sketch Solving** cascading menu in the **Part** pull-down menu, or pick the **Split Line** button on the **Part Modeling** toolbar.

You are now prompted to select a split line. Select the colored polyline. You are asked if you want to include any part edges in the split line. Press [Enter] to indicate that you do not. The command line now reads:

Solved under constrained sketch requiring 8 dimensions or constraints.

To review how the split line is constrained, open Example-12-06-Constrained.dwg and show the constraints. Now issue **AMSPLITFACE**, select **Project** (the default), and select face A. Pressing [Enter] twice splits the face. Now, use the **AMFACEDRAFT** command to put a 15° draft angle on the upper part of face A. Use the **From Edge** option and pick the edge formed by the split line. See **Figure 12-13**.

Figure 12-13.
The result of dividing face A at the split line and adding different draft angles to the two new faces.

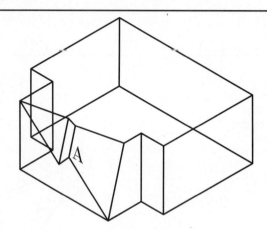

THE AMSHELL COMMAND

The **AMSHELL** command turns solid parts into boxes with a specified wall thickness. The thickness can be added to the inside of the part, outside, or mid-plane. Different faces can have different thicknesses. Faces can also be *excluded* to create open boxes. This command must be used judiciously as it can only be used once on a part.

Open part Example-12-07.dwg as shown in **Figure 12-14.** This is the first drawing we have used that contains two parts. The part on the left named UNSHELLED is the active part. The *active part* is the one you can work on. This is indicated by the dark green icon in front of its name in the **Browser**. This part is a solid block with four small holes in the bottom and a boss on the front.

The part on the right is an identical part named SHELLED with the shell operation added to it. The wall thickness is 0.125 units added to the inside of the part and all the faces have the same thickness. Face A on the connector boss and face B on the top of the box have been excluded, creating openings. The four holes do not go all the way through and each has a boss around them.

AMSHELL

Part
↳ Placed Feature
↳ Shell

Part Modeling
toolbar

Shell

To access the **AMSHELL** command, enter AMSHELL at the Command: prompt, select **Shell** from the **Placed Feature** cascading menu in the **Part** pull-down menu, or pick the **Shell** button on the **Part Modeling** toolbar.

This accesses the **Shell Feature** dialog box. See **Figure 12-15**. In the **Default Thickness** area, check the **Inside** box and set the value to 0.125. Pick the **Add** button in the **Exclude Faces** area. Select faces A and B and press [Enter] to accept selections. Once the faces have been selected and accepted, press [Enter] again to return to the dialog box. Pick **OK** and the part is shelled.

Figure 12-14.
A—The part used to demonstrate the **AMSHELL** command. B—The shelled part.

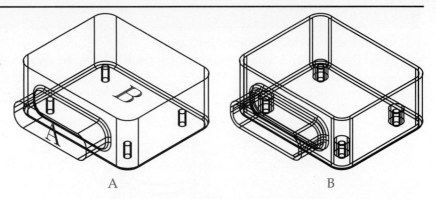

A

B

Figure 12-15.
The **Shell Feature** dialog box showing the settings for the shelling of the part.

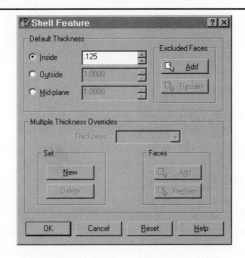

Now we will give one of the faces a different thickness. Right-click on Shell1 in the **Browser** and select **Edit** from the pop-up menu. This accesses the **Shell Feature** dialog box. In the **Multiple Thickness Overrides** area, pick the **New** button and enter 0.25 in the **Thickness:** field. In the **Faces** area, click on **Add** and pick the small hole just to the left of the letter B. Make sure only the hole is highlighted and not the entire part. Press [Enter] twice, click **OK**, and update the part. The results are shown in **Figure 12-16.**

Figure 12-16.
The result of the shelling and editing of the wall thickness for one boss.

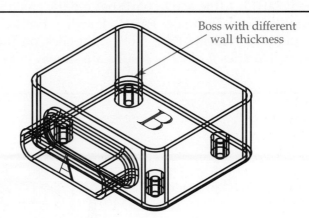

Boss with different wall thickness

❑ Open the drawing Practice-12-03.dwg.
❑ Shell the part with a 0.05 wall mid-plane. Exclude the face on the top. Results are shown here.

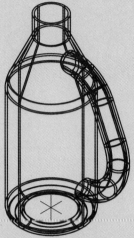

SPLITTING A PART

You can cut, or split, a part into two parts with the **AMPARTSPLIT** command. One part keeps the original name and the other gets a new name. The part can be split along a face, a work plane, or a split line. That is the same type of split line just used to split a face of a part. As an example open the part Example-12-08.dwg. See **Figure 12-17**. This is a shelled part with a work plane through the center.

To issue the **AMPARTSPLIT** command, enter AMPARTSPLIT at the Command: prompt, select **Part Split** from the **Placed Feature** cascading menu in the **Part** pull-down menu, or pick the **Part Split** button on the **Part Modeling** toolbar.

You are prompted to select either a planar face, a work plane, or a split line. Select the work plane. Flip the arrow to define the right side as the new part and press [Enter]. The default name for the new part is PART2, but you may give it any name you wish. Accept the default by pressing [Enter]. The part name appears on the **Browser** as PART2_1.

Compare the icons in front of the part names in the **Browser**. As previously mentioned, PART1_1 (the active part) has a dark green icon. PART2_1 is inactive and has an unfilled icon. To make PART2_1 active, right-click on the part name in the **Browser** and pick **Activate Part**. Now right-click on the inactive PART1_1 and pick **Visible**, which toggles it off. The results are shown in **Figure 12-18**.

AMPARTSPLIT

P̲art
↪ P̲laced Feature
  ↪ Pa̲rt Split

Part Modeling
toolbar

Part Split

Figure 12-17.
The part used to demonstrate the **AMPARTSPLIT** command.

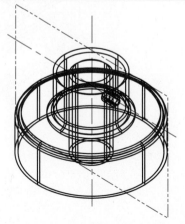

Figure 12-18.
The result of
splitting the part.

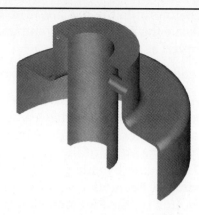

COMBINING PARTS

You can combine two parts together into one with the **AMCOMBINE** command. The part you are combining with must be the active part. The part being combined is an inactive part and is referred to as the *toolbody*. It can be combined in one of three different ways—cut, join, or intersect. Open the part Example-12-09.dwg. See **Figure 12-19.** This drawing consists of the base (the active part) and two housings, which will be used as the toolbodies. These three parts are positioned with respect to each other with 3D constraints, which you can see listed in the **Browser.** The 3D constraints are covered in detail in *Chapter 15*.

To issue the **AMCOMBINE** command, enter AMCOMBINE at the Command: prompt, or select **Combine** from the **Placed Feature** cascading menu in the **Part** pull-down menu, or pick the **Combine** button on the **Part Modeling** toolbar.

You are prompted for the combining method. Enter J for join or right-click on screen and pick **Join** from the pop-up menu. Select one of the housings to join to the base. Then repeat the command and join the other housing. Notice that the housings have become a part of the BASE in the **Browser.** They now are listed as Combine1 and Combine2. The **AMCOMBINE** command is covered in greater detail in *Chapter 14*.

AMCOMBINE

Part
↳ Placed Feature
  ↳ Combine

Part Modeling
toolbar

Combine

Figure 12-19.
The part used to
demonstrate the
AMCOMBINE
command.

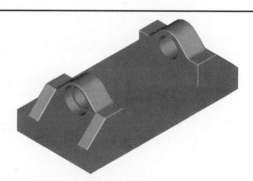

THE **AMBEND** COMMAND

The **AMBEND** command is used to bend a part a specified angle about a straight line at a specified bend radius. The line has to be profiled as an open profile. This is not an exact sheet metal bend where the neutral axis moves, but **AMBEND** is useful because it makes part construction easier.

Open the part Example-12-10.dwg. In this example, the colored line (red on screen and green in figure) is the bend line and all the faces it touches (ears A and B) will be bent. Since the line does not touch ear C, it will not be bent.

The first step is to profile the red line. The line can only be one straight line, no polylines or radii. Issue the **AMPROFILE** command and select the line and press [Enter]. You are then prompted as follows:

Select part edge to close the profile <open profile>:

Press [Enter] to accept the default, which is an open profile. Open profiles are not allowed for the base sketch, however they can be used for features such as bends and ribs. The sketch has a horizontal constraint on it but needs to be constrained further. Add a dimension with the value of the design variable OFFS. See **Figure 12-20**. Now issue the **AMBEND** command by entering AMBEND at the Command: prompt, selecting **Bend** from the **Sketched Features** cascading menu in the **Part** pull-down menu, or picking the **Bend** button on the **Part Modeling** toolbar.

This displays the **Bend** dialog box. See **Figure 12-21A**. There are three options availabe in the **Combination:** drop down list. They are **Angle+Radius**, **Radius+ArcLen**, and **ArcLen+Angle**. The two data fields change as you change combinations. See **Figures 12-21B** and **C**.

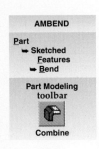

AMBEND

Part
➥ Sketched
    Features
➥ Bend

Part Modeling
toolbar

Combine

Figure 12-20.
The part used to demonstrate the **AMBEND** command.

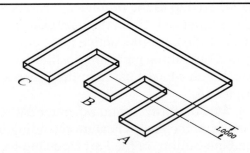

Figure 12-21.
A—The **Bend** dialog box showing the settings for the bending ears A and B.
B—The dialog box showing the **Radius+ArcLen** option.
C—The dialog box showing the **ArcLen+Angle** option.

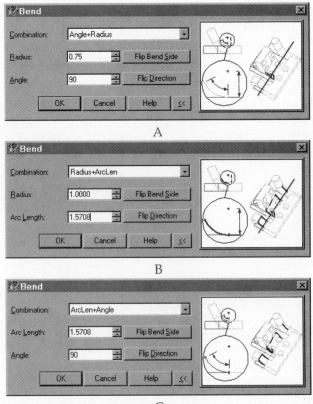

Figure 12-22.
A—The result of the **AMBEND** command. B—Illustrations showing the relationship of the bend radius and the part.

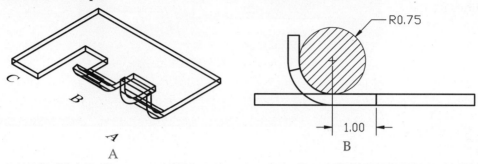

Select **Angle+Radius** and input 0.75 for the radius and 90° for angle. The button **Flip Bend Side** sets which side of the part moves as shown by the linear arrow. The **Flip Direction** button sets which way it is going to bend as shown by the circular arrow on the part. Click **OK** and the two ears A and B will be bent up. See **Figure 12-22A**. The part was bent about an imaginary 0.75 radius cylinder whose center is aligned horizontally with the bend line and is 0.75 above it. See **Figure 12-22B**. Change the value of the design variable OFFS from 1.0 to 2.0 and then to 0.0 to see the effects on the bend.

CREATING THIN EXTRUSIONS FROM AN OPEN PROFILE

Open the part Example-12-11.dwg as shown in **Figure 12-23**. This is a shelled part with two bosses, with holes. A colored polyline (red on screen and green in figure) has been constructed through the centers of the bosses with a leg on the left end at an angle. This procedure will convert this pline into a web of specified width and extend the web until it meets the faces at the bottom and sides of the inside shell. The colored pline could have many segments and could contain lines and arcs.

The pline sketch first has to be profiled as an open profile. Issue the **AMPROFILE** command, select the pline, and press [Enter] twice creating an open profile. The sketch can be constrained to the center of one boss and the angle dimensioned but in this example we will skip that step.

Issue the **AMEXTRUDE** command. In the **Extrusion** dialog box, select the **Extend** check box. This should set the termination type to **Next** and gray it out. If it is not grayed out, uncheck **Extend**, change the termination type to **Next**, and recheck **Extend**. Now set the type to **Midplane**, the operation to **Join**, and the thickness to 0.20. See **Figure 12-24**.

Move the dialog box so that you can see the part. Make sure that the arrow is pointing downward as shown in **Figure 12-25A**. Note that this also shows the thickness of this web. Click on **OK** and the web will be extended downward until it meets

Figure 12-23.
The part used to demonstrate adding a thin extrusion as a web.

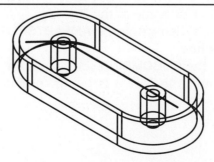

Figure 12-24.
The **Extrusion**
dialog box showing
the settings for the
web.

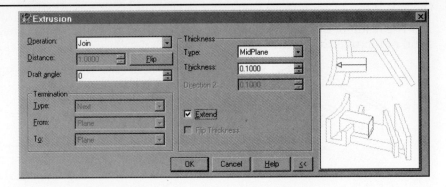

Figure 12-25.
A—Part showing
the thickness of the
web and direction
of the extrusion.
B—The result of the
extrusion.

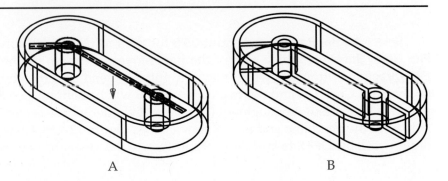

A   B

the first face, which is the bottom. This is called the "Next" face. The web is also extended along the ends of the lines until it meets a "Next" face on the wall of the part. See **Figure 12-25B**. Note that the web was not created through the inside of the holes.

Chapter Test

Answer the following questions on a separate sheet of paper.

1. A(n) _____ _____ is usually less than 3° and is used to facilitate part removal from a mold.
2. *True or False?* The draft angle for extrusion is input in the **Extrusion** dialog box and must be between a –89.999° and +89.999°.
3. A unique draft angle can be placed on one or more faces of a model using the _____ command.
4. Dividing a face into two parts can be done well with the _____ command.
5. Describe the difference between the *planar* and *project* methods of dividing a face into two parts.
6. What is a *split line*?
7. *True or False?* With the **AMSHELL** command, the specified wall thickness can be applied to the inside or the outside of the part only.
8. The _____ part is the one that can be worked on and shown with a dark green icon in front of its name in the **Browser**.
9. *True or False?* When using the **AMPARTSPLIT** command, a part can only be split along a split line.
10. When using the **AMCOMBINE** command, the part being combined to must be the active part and the part it is being combined with is referred to as the _____ _____.

Chapter Exercises

Exercise 12-1. Open the drawing Exercise-12-01.dwg. The top face of the decorative plate is the current sketch plane. Profile the face and extrude upward 2.5 units with a –74° draft angle as shown.

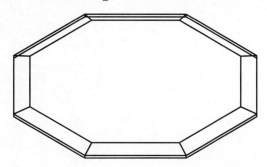

Exercise 12-2. Open the drawing Exercise-12-02.dwg. Extrude the profiled sketch 6.0 units mid-plane with a –4° draft angle. Fillet the top and bottom faces with a 0.25 radius as shown.

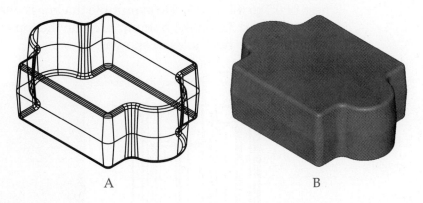

A   B

Exercise 12-3. Open the drawing Exercise-12-03.dwg. Use the **Face Draft** command to place 20° draft angles on the faces as shown.

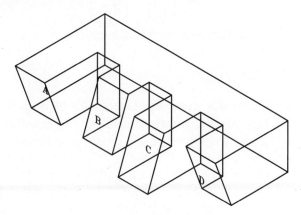

Exercise 12-4. Open the drawing Exercise-12-04.dwg. Place a 30° shadow draft on each side of the part from the spherical faces as shown.

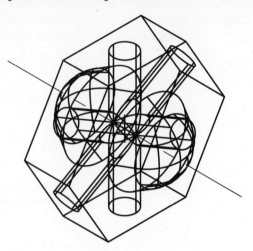

Exercise 12-5. Open the drawing Exercise-12-05.dwg. Shell the part with a wall thickness of 0.5 to the inside. Do a multiple thickness override of 1.0 on the two holes and 3.0 on the bottom. Exclude the top face of the part and the tops of the two holes as shown.

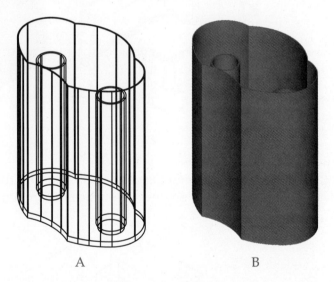

A   B

Exercise 12-6. Open the drawing Exercise-12-06.dwg. Split the circular face A into two parts through the center and use the face draft command from a plane to create the 45° chisel point as shown.

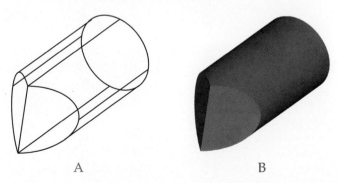

A   B

CHAPTER 13

Threads, Springs, and Sweeps

Objectives

After completing this chapter, you will be able to:

* Create helical sweeps such as screw threads or springs.
* Edit the values controlling the helical path, thereby changing the part.
* Create parametric sweeps along splines.
* Create parametric sweeps using an existing part edge.
* Create parametric sweeps along a pipe centerline.
* Edit the values controlling the sweep path, thereby changing the part.
* Learn the difference between these types and when to use them.

Drawings Needed

*The following is a list of files that you will need to work through this chapter. These files can be found on the **User's Drawings** CD included with this text.*

Examples	Exercises
Example-13-01.dwg	Exercise-13-01.dwg
Example-13-02.dwg	Exercise-13-02.dwg
Example-13-03.dwg	Exercise-13-03.dwg
Example-13-04.dwg	Exercise-13-04.dwg
Example-13-05.dwg	Exercise-13-05.dwg
	Exercise-13-06.dwg

SWEEPING FEATURES

All of the operations in this chapter consist of a process called *sweeping*. In this you define a path and a cross section. The cross section is then swept along the path resulting in a 3D parametric shape. This shape can be helical, as in a spring or the threads on a bolt. The shape can also be free-form, as that derived from a spline drawn in 3D space. The fundamental process for creating a swept feature of a part follows a consistent, straightforward procedure. The basic steps are as follows:

1. In some cases, you will *Create* the geometry to define the path. This path consists of either a spline, or a series of lines in 3D space.
2. *Create* a Mechanical Desktop feature of this path with **AM3DPATH**. It will ask you to select the path and the start point. Furthermore, you may opt to place a work plane and a work point at the start of the path.

3. *Create* a cross section at the start of the path (this is 0,0,0). Use any of the normal AutoCAD objects used to create the basis for a sketch profile. You cannot use the new MDT R4 multiple-loop type of sketch profiles for sweeps.
4. *Profile* the cross section using **AMPROFILE**. Constrain it until it is fully constrained. If you want to tie it to the start of the point, you could try dimensioning it to the work point using dimensions of a value of zero.
5. Sweep the fully-constrained profile along the path using **AMSWEEP**. You have a few options to choose from in the dialog box that pops up depending on what type of sweep you are doing. You then should have a finished shape. Please keep in mind that this feature is fully parametric—you can change the dimensions/parameters of the path or the dimensions/constraints that control the cross section.

THREADS AND SPRINGS

For our first part, we will cut threads into a ½-13UNC Hexagonal Head bolt. You will start out with a bolt without threads. The finished part is shown in **Figure 13-1A**.

To start this part, open Example-13-01.dwg. See **Figure 13-1B**. Use the cylindrical shaft of the bolt as a basis for the helical 3D path. First the 3D path must be created using the **AM3DPATH** command. To access this command, enter AM3DPATH at the Command: prompt, select **3D Helix Path...** from the **Sketch Solving** cascading menu in the **Part** pull-down menu, or click the **3D Path** button underneath the **Profile a Sketch** flyout on the **Part Modeling** toolbar.

AM3DPATH

Part
➥ Sketch Solving
  ➥ 3D Helix Path

Part Modeling
toolbar

3D Path

NOTE

There is only one icon for **AM3DPATH** on the toolbar, called **3D Path**, and it requires that you select which type of path you want by entering a letter at the command line. In the **Sketch Solving** cascading menu in the **Part** pull-down menu there are four choices. These are **3D Edge Path**, **3D Helix Path**, **3D Pipe Path**, and **3D Spline Path**. When you select one of these pull-down menus, it issues the **AM3DPATH** command and automatically makes the first selection on the command line.

Once the command is issued, you receive the following prompts:

> Enter path type [Helical/Spline/Edge/Pipe] <Helical>: *(enter H for Helical)*
> Select work axis, circular edge, or circular face for helical center: *(Pick the bottom cylindrical edge of the bolt's shaft)*

This accesses the **Helix** dialog box. See **Figure 13-2**. Fill in the various fields with the data as shown. Set the **Profile Plane** to Center Axis/Path and there is no **Flip** involved. Set the diameter to .4219, the revolutions to 17, and the pitch to 0.0769, which is the pitch of the 13TPI UNC thread. When you have entered all of the data, pick the **OK** button.

Figure 13-1.
A—The finished hexagonal head bolt.
B—The drawing used to demonstrate creating threads.

A   B

Learning Mechanical Desktop R5

Figure 13-2.
The **Helix** dialog box with needed values.

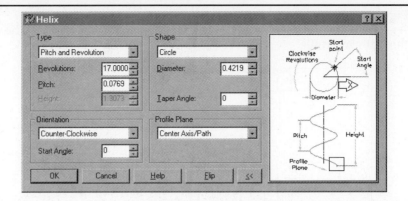

At the Select edge to align X axis or [Z-flip/Rotate] <Accept>: prompt, left-click the mouse once to rotate the coordinate system around the Z-axis so that the X-axis is pointing toward the top of the screen. If you have changed the view, the X-axis should be pointing away from and to the right of the bolt. Now press [Enter] to accept.

The *sketch plane* is oriented parallel to the long axis of the bolt. See **Figure 13-3.** It is also coincident with the start of the helical path, which is also the origin (0,0,0) of the new coordinate system. Please notice that a work plane, a work point, and a work axis have all been automatically added to your model. Later the work point will be used to constrain the threadform sketch profile. The work point is at the new 0,0,0.

The helical path will show up in a blue color and should be progressing into the bolt, not away from it. If this is not the case, undo and try again.

Figure 13-3.
The bolt with helical 3D path, work point, and work plane.

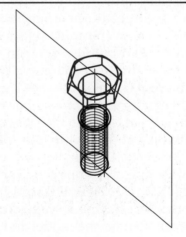

Profiling and Constraining the Thread Form

Our next step is to profile and constrain the thread form. Turn on the Profile layer and you will find that there is a small polygonal shape at 0,0,0. Orient your display to the plan view of the current *sketch plane*. Entering 9 on your keyboard most easily does this. You may also access this command by using the **Sketch** option of the **AMVIEW** command, or click the **Sketch View** button underneath the **New Sketch Plane** flyout on the **Part Modeling** toolbar. Once the command is issued, you should have the display showing the bolt in a vertical orientation with the hexagonal head on top. See **Figure 13-4A.** Zoom in on the threadform shape at the bottom of the bolt. Profile the four lines that make up the small polygon. There should be six dimensions or constraints needed. The dimension values to use in constraining this profile are shown in **Figure 13-4B**.

Figure 13-4.
A—The bolt vertically oriented is the sketch plane plan view. B—The dimensions for the threadform profile.

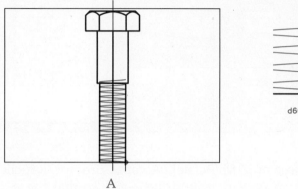

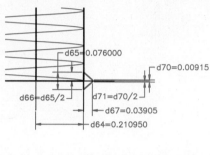

A

B

> **NOTE**
>
> Your dimension label numbering scheme may be different— d64 may be d52 or d53 in your exercise. MDT retires a dimension label number if that dimension has been erased.

Please note that d64 is dimensioned to the work axis and d66, as well as d71, are dimensioned to the work point. The rest of the dimensions are straightforward.

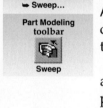

AMSWEEP

Part
➥ **Sketched Features**
➥ **Sweep...**

Part Modeling toolbar

Sweep

Once the sketch profile is fully constrained, move onto the next step, which is quite simple because the software does everything for you. We will use the **AMSWEEP** command to sweep the threadform around the helix. To access this command, enter AMSWEEP at the Command: prompt, select **Sweep...** from the **Sketched Features** cascading menu in the **Part** pull-down menu, or click the **Sweep** button underneath the **Sketched Features-Extrude** flyout on the **Part Modeling** toolbar.

This accesses the **Sweep** dialog box. See **Figure 13-5.** You are not prompted to select anything before this dialog box appears because there is one path and there is one profile—MDT knows what to do. In the **Operation:** and **Termination:** drop-down lists, select **Join** and **Path-Only**, respectively. Use **Join** in this case because material is being added to the shaft of the bolt. Then pick **OK**.

Depending on the speed of your computer, the computing operation may take a minute or two for it to finish. Helical threads are typically very complex features for the software to compute. Imagine if we had done a 10-32 UNF thread. Zoom out to view your handiwork. Render the model and use the **Dynamic 3D Orbit** to admire it. Save the model, go into Explorer, and notice how large the drawing (.dwg) file is.

Figure 13-5.
The **Sweep** dialog box with needed values.

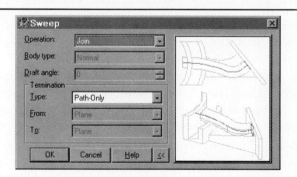

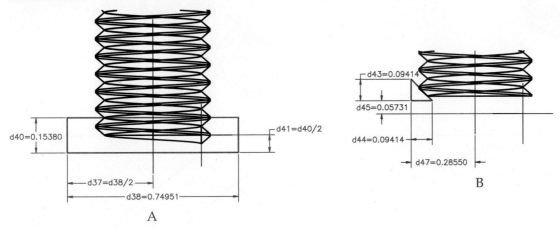

Finishing the Thread

If you want to finish the thread so it looks like a real bolt, there are two more profiles to be used to *machine* off the ragged end. They are on layers Profile2 and Profile3, respectively. Do each operation in order.

First, turn on the Profile2 layer and profile and constrain the shape that you see at the bottom of the bolt. This is a simple rectangle and will be used to cut off the bottom of thread. Refer to **Figure 13-6A** for the dimensioning scheme. Notice that some of the dimensions use the work axis or work point. Extrude the constrained profile **Cut** and **Mid-Through**.

Now, turn on the Profile3 layer and profile and constrain the shape that you see to the left of the thread. This is a simple triangle and will be used to trim the thread further to create the chamfer effect on the end of a bolt. Refer to **Figure 13-6B** for the dimensioning scheme. Again, we use the work axis and the work point to base some of the dimensions on. Use **AMREVOLVE** to revolve the constrained profile. Use the work axis as the axis for the rotation. Use the **Cut** option with an angle of 360° in the **Revolution** dialog box. You should have a nicely finished hexagonal head bolt with real threads.

CAUTION	You would rarely use this detailed fastener in an assembly. If you had ten or twelve of these in an assembly along with all of the other parts, the size and complexity of the model would be too large for most computers to handle. Please save your work under a different name.

CREATING SPRINGS

Creating a spring shape is really very similar to what was done for the thread. We will define a helical path, draw a circle, profile it, constrain it, and then sweep it along the helical path. The finished part is shown in **Figure 13-7**.

Open Example-13-02.dwg as shown in **Figure 13-8**. Now, define the helical path using the **AM3DPATH** command. At the Select work axis, circular edge, or circular face for helical center: prompt, select the green work axis. When the **Helix** dialog box appears, fill in the data fields as shown in **Figure 13-9**. Please note that the **Flip** option must be invoked.

Figure 13-7.
The finished spring.

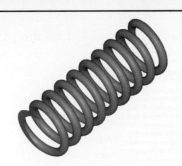

Figure 13-8.
The work planes
and work axis used
to create the spring.

Work axis

Figure 13-9.
The **Helix** dialog box
with needed values.

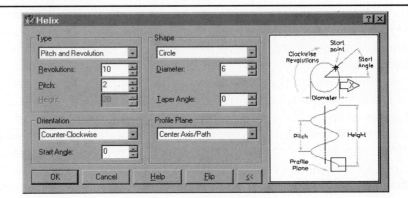

PROFESSIONAL TIP  If you move and drag the **Helix** dialog box to the upper-left hand corner of the screen, you can see the helical path being dynamically defined as you change the values in the dialog box.

When you have entered all the required data, pick **OK**. At the Select edge to align X axis or [Z-flip/Rotate] <Accept>: prompt, press [Enter] to accept the default orientation for the sketch plane. You should have a blue helical path progressing toward the right side of the screen. Next, draw the cross section, profile, and constrain it. Use the **CIRCLE** command to draw a circle with a center point of 0,0 and a radius of .5. Profile the circle you just drew. Dimension the circle with a radius of 0.5. See **Figure 13-10.**

Now use **AMSWEEP** to sweep the circle along the helical path. Use the default of Path-Only in the **Sweep** dialog box and pick **OK**. This also should take some time to compute, but you should be left with a spring. It may not look like much in wireframe mode, but if you do want to increase the wireframe lines used to display the spring, change the **ISOLINES** setting in AutoCAD to 8. Now render and view it with the **Dynamic 3D Orbit**. See **Figure 13-11.**

Figure 13-10.
The helical 3D path with profiled and constrained circle.

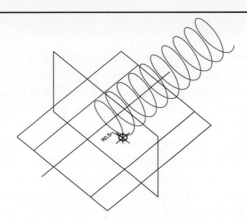

Figure 13-11.
The spring with ends not finished.

Machining the Ends of the Spring

Most springs are ground down on their ends so that they can fit into a mechanical cavity. This step will detail the steps required to create a rectangle, profile it, constrain it, and extrude so both ends of the spring are trimmed off. *This step is optional.*

Using the **RECTANG** command and create a rectangle. Use 1,–1 as its first corner point and the relative point of @18,8 for its other corner point. Use **AMPROFILE** to profile this rectangle. Notice that there are four dimensions needed to constrain the rectangle.

Dimension the length and the width of the rectangle—do not change the size of the rectangle. Also, dimension the lower-left hand corner of the rectangle to the work point. Again, accept the defaults for the dimension values. See **Figure 13-12.**

Now, extrude the rectangle **Intersect** and **Mid-Through**. The ends of the spring are now *machined* off and your spring should look like **Figure 13-7.** For practice, double-click on SweepPathOnly1 in the **Browser** and change the values in the **Helix** dialog box to change the parameters controlling the helical path. Try changing the **Shape** to an Ellipse or changing the value of the **Taper Angle** to something other than zero. There is no option for variable pitch springs. Please save your work under a different name.

Figure 13-12.
A rectangle is constrained and extruded to finish the ends of the spring.

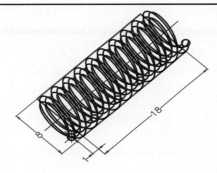

USING AN EDGE OF A PART FOR A SWEEP

We will now investigate a problem found in practically all injection-molded parts. Usually the two halves of the plastic part fit together using a tongue-and-groove arrangement along their parting lines. With the next model, you will define the edge of a part as a path for a 3D sweep of the profile you will create. The finished part is shown in **Figure 13-13**.

Open Example-13-3.dwg as shown in **Figure 13-14A**. The 3D path must be created using the **AM3DPATH** command. Use the **Edge** option and begin selecting segments that define the inner edge of the shelled feature. Please pick the segments in a sequential order starting with the first as shown in **Figure 13-14B**. There are 24 in all. The first segment to be picked as well as the start point of the edge path is shown. You will be asked for the start point later on.

Figure 13-13.
The finished part showing the recessed edge.

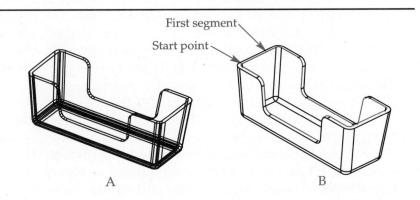

Figure 13-14.
A—The part used to demonstrate sweeping an edge cut. B—The edge to be used is shown in color. The first segment and start point are shown.

First segment

Start point

A

B

PROFESSIONAL TIP

If you should happen to pick near two or more *visually* overlapping lines, then you may see the mouse icon cursor. There is a subtle indicator warning you this is about to happen when you hover over a line. It is a little red cross that flashes as you hover. Selecting that point will trigger the cycling mouse cursor and command interaction. You will have to left-click until you cycle through the possible candidates for this pick point. There may be more than two. Keep clicking until you see the edge you want highlighted. Once it is, press [Enter].

Continue selecting model edges until done. Press [Enter] at the Select model edges (to add): prompt when you have selected all of them back to the first one you picked. There are 24 of them in all.

Next, you will be prompted for the *start point* of the new edge path. Please do not use an object snap. A work plane is created normal (perpendicular) to the first path segment of the path if you answer Yes to the Create workplane? [Yes/No] <Yes>: prompt. On this work plane, a new sketch plane and a new work point will be created automatically. It is advisable that you do this. What follows is classic sketch plane X axis orientation. Left-click to rotate the coordinate system around the Z axis. Usually the X axis is to be pointing toward the right-hand side of the screen. If the X axis is properly displayed, press [Enter] at the Select edge to align X axis or [Z-flip/Rotate] <Accept>: prompt. You could enter R (Rotate) or Z (Z-flip) to manipulate the coordinate system.

You now have a work plane and a work point on your new sketch plane. See **Figure 13-15.** The new work point should be coincident with the specified start point of the path. You will also notice that the sketch plane passes through the work point and is coplanar with the work plane.

There is a 2D shape drawn on layer Profile. Turn the layer on. Enter 9 to orient your display to the plan view of the current sketch plane. Zoom in on the 2D shape. Now profile and fully constrain the shape as shown in **Figure 13-16.** Please note that the .13 dimension is to the work point. The top edge of the profile is constrained to be collinear with the top edge of the part. Use any top edge of the part—there are eight of them, pick one. Geometric constraints work through 3D space, not just in the same plane as the sketch profile. There are five dimensions/constraints needed—four dimensions and the **Collinear** constraint.

Figure 13-15.
After edge is selected, a work point and work plane are shown.

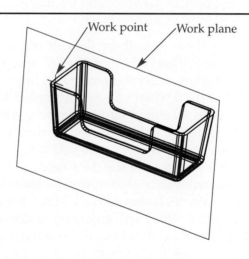

Figure 13-16.
The constrained profile used for the edge sweep.

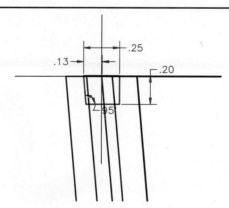

After the sketch is profiled and constrained, it is simple to issue the **AMSWEEP** command and finish the sweep. Remember, MDT does not ask you to select anything before accessing the **Sweep** dialog box. There is one path and there is one profile—it knows what to do. Use the defaults in the **Sweep** dialog box and pick **OK**.

There should be a nice groove running along the entire inner edge of the part. This will mate with the top piece of the model, which is not shown. The difficult part of this operation is defining the edge, because MDT handles the sweep. For practice, try changing the length or width of the Base Feature of the part. Notice that the new swept feature updates very nicely. Please save your work under a different name.

USING A PIPE CENTERLINE FOR A SWEEP

Now we will create pipes, tubes, or wiring. You will start with a blank screen and construct a centerline. Use **AM3DPATH** to define it as a pipe centerline, add radii at the vertices, draw a cross section, profile it, constrain it, and sweep the profile along the centerline. Finally, you will *shell* the whole pipe so as to make it hollow. Please refer to **Figure 13-17** for a finished product.

Open Example-13-04.dwg and draw the centerline as a series of lines in 3D space. Use the following prompts to create the centerline:

    Command: **LINE**↵
    Specify first point: **0,0**↵
    Specify next point or [Undo]: **@5,0**↵
    Specify next point or [Undo]: **@0,0,5**↵
    Specify next point or [Close/Undo]: **@0,5**↵
    Specify next point or [Close/Undo]: **@5,0**↵
    Specify next point or [Close/Undo]:↵

Next, we want these lines to be the centerline of the pipe. Issue the **AM3DPATH** command and enter P (Pipe) when prompted for path type. At the Select polyline path source: prompt, window select the group of 4 lines *and press* [Enter]. Pick the first line near its endpoint at 0,0,0 when you are prompted for the start point.

This accesses the **3D Pipe Path** dialog box, which looks like a spreadsheet. See **Figure 13-18.** It has all the values for the vertices that make up the centerline. Move the dialog box out of the way so that you can watch the centerline update on the screen as you enter numbers in the data fields of this *spreadsheet*. If you click on the cell labeled 2, you will notice that vertex #2 shows up red on the screen. Notice its value for its X coordinate. It should read 5.00. Change it to 7.00. Try using the *spinner* arrows next to the field—they will adjust up or down by one. Try right-clicking in one of the fields. The centerline should update on the screen. Practice changing the values for the coordinates for each of the vertices. Finally, add the radii values to the vertices as indicated, thereby creating 3D fillets. Once all the values have been input, press **OK**. The results are shown in **Figure 13-19.**

Figure 13-17.
The finished pipe.

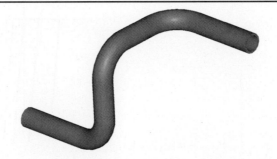

Figure 13-18.
The **3D Pipe Path** dialog box with data for the vertices of the pipe.

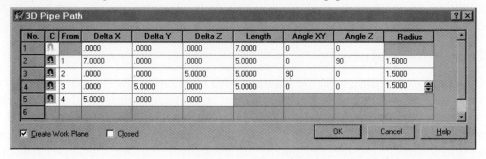

No.	C	From	Delta X	Delta Y	Delta Z	Length	Angle XY	Angle Z	Radius
1			.0000	.0000	.0000	7.0000	0	0	
2		1	7.0000	.0000	.0000	5.0000	0	90	1.5000
3		2	.0000	.0000	5.0000	5.0000	90	0	1.5000
4		3	.0000	5.0000	.0000	5.0000	0	0	1.5000
5		4	5.0000	.0000	.0000				
6									

☑ Create Work Plane   ☐ Closed   OK   Cancel   Help

Figure 13-19.
The 3D path used
for the pipe.

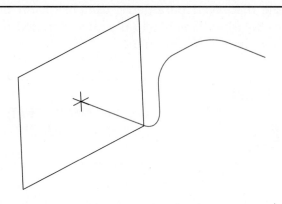

Drawing the Geometry

We now need to draw the geometry for the cross section. This will be a pipe, so we use a circle. MDT does allow sketch profiles with inner geometry (multiple loops), but **AMSWEEP** does not support them. That would have been a good idea to try it with this pipe—as in sweeping two concentric circles as one sketch profile to form the hollowed out pipe in one step.

Use the **CIRCLE** command to create a circle with a center of 0,0 and a radius of .5. Profile the circle and use **AMPARDIM** to parametrically dimension it. You should receive a prompt stating you have a fully-constrained sketch.

Now, we want to sweep the circle along the centerline using the most common termination for this operation. Invoke the **AMSWEEP** command, choose the Path-Only option in the **Sweep** dialog box, and pick **OK.**

Finally, we want to hollow out the pipe, so it really becomes a pipe. To do this, use the **AMSHELL** command by entering AMSHELL at the Command: prompt. In the **Shell Feature** dialog box, enter an inside wall thickness of 0.1—this means there is an inside offset.

Then, we want to clear the end caps so we will be able to see inside the pipe. Remember, if a face is *shelled* you cannot see through it because it has a wall thickness. To do this, pick the **Add** button in the **Exclude Faces** area of the dialog box. You will be back in the drawing now. Pick the flat circular faces at the start and the end of the pipe run. Remember, you may need to cycle through the faces so the correct face is selected. If you do pick and select the wrong face, you can do an **Undo.**

PROFESSIONAL TIP

If you are lost at this point, press [Esc] to get you back to the **Shell Feature** dialog box. If something else goes wrong, you can use the **Reclaim Faces** button in the **Shell Feature** dialog box to go out and *reclaim* faces that had been excluded. You are basically adding them back to those that will be shelled.

When you have selected the two circular end faces and nothing else, press [Enter] to get back to the **Shell Feature** dialog box. Pick **OK** to start the operation. It should not take long. Take time to look your work over—**Render** and **3D Orbit** to view the object.

Editing Pipe or Tubing

Creating and editing pipe or tubing in MDT is easy. Try double-clicking on the SweepPathOnly1 feature in the **Browser.** There are actually three methods of editing a pipe. You can change the values in the **3D Pipe Path** dialog box and experiment while the model updates on the screen. Pressing **OK** accesses the **Sweep** dialog box and allows you to change the termination of the sweep. Remember, there are three options and the most common (Path-Only) is the one we used. Finally, you can change the value of the displayed parametric dimension on the circle. This will redefine the diameter of the pipe. This is actually an automated form of **AMEDITFEAT.** Please save your work under a different name.

<table>
<tr><td>PROFESSIONAL TIP </td><td>Try to use the AMEDITFEAT from the command line. You have to force an AMUPDATE to see the revision after the changes.</td></tr>
</table>

FREE FORM SPLINE-BASED SWEEP

Of course, there are times when you will need a freeform curve to construct tubing, or hosing—as the case may be. To demonstrate this we will construct one side of a pair of handlebars. The finished version is shown in **Figure 13-20.**

Open Example-13-05.dwg and you will see that a spline has already been constructed for you. After working with this one, you can create your own spline. This is a nice shape for the left side of the handlebars. Convert it to an MDT spline-based path. To do this, enter the **AM3DPATH** command, and enter S (Spline) when prompted for a path type. Pick the left-hand endpoint of the spline for the start point, and press **OK** in the dialog box. Press [Enter] at the Select edge to align X axis or [Z-flip/Rotate] <Accept>: prompt. Draw a circle with a radius of 0.25 at the coordinate system origin (0,0,0). Profile, constrain, and sweep the circle using the **Path-Only** option and pick **OK.** See **Figure 13-21.**

If you **3D Orbit** this tubing, you will notice it is basically all in one plane—there is no curvature in 3D space. We saved that for the editing. Double-click on SweepPathOnly1 feature in the **Browser.** This is similar to what we did when we edited the previous pipe. We are going to change some values to give the handlebars a real curve. Notice that you can change many items that control the shape of this spline. We will be changing the **Delta Z** values for vertices 1 through 4 to give a flowing shape to this handlebar. Refer to **Figure 13-22A** for the specific values.

We will not be changing the **i j k** information or the **Weight.** The **i j k** information is local rotational data for each vertex. The **Weight** value can cause the spline segment after this vertex to have more influence over the eventual shape of that segment. Normally, the weight is evenly distributed between the two vertices that border a segment.

Please notice how the spline updates on the screen as you make value changes. Pick **OK** when you are ready. Pick **OK** in the **Sweep** dialog box (remember this one allows you to select a different termination). Finally, at the Select object: prompt press [Enter]. Remember, this last one allows you to change the diameter of the circular cross section. The part should update automatically. See **Figure 13-22B.**

Figure 13-20.
The finished set of handlebars.

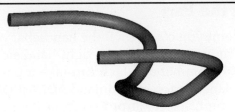

Figure 13-21.
The spline sweep of the profiled circle.

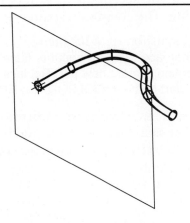

Figure 13-22.
A—The **3D Spline Path** dialog box with data for the vertices of the handlebar. B—The resulting handlebar.

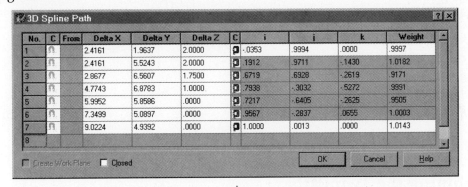

No.	C	From	Delta X	Delta Y	Delta Z	C	i	j	k	Weight
1			2.4161	1.9637	2.0000		-.0353	.9994	.0000	.9997
2			2.4161	5.5243	2.0000		.1912	.9711	-.1430	1.0182
3			2.8677	6.5607	1.7500		.6719	.6928	-.2619	.9171
4			4.7743	6.8783	1.0000		.7938	-.3032	-.5272	.9991
5			5.9952	5.8586	.0000		.7217	-.6405	-.2625	.9505
6			7.3499	5.0897	.0000		.9567	-.2837	.0655	1.0003
7			9.0224	4.9392	.0000		1.0000	.0013	.0000	1.0143
8										

A

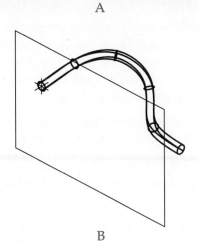

B

Mirroring the Part

At this point you could shell the handlebar out and make it hollow. We will skip this step and move on to create the right-hand side of these handlebars. To do this use the **AMMIRROR** command, which can take a part and create its mirror image about any defined plane. To access command **AMMIRROR** command, enter AMMIRROR at the Command: prompt, select **Mirror Part** from the **Part** cascading menu in the **Part** pull-down menu, or click the **Mirror Part** button underneath the **New Part** flyout on the **Part Modeling** toolbar. If you accept the default option of the command, this creates a new part in the drawing. The command sequence is as follows:

> Select part to mirror: *(pick the tubing)*
> Select planar face to mirror about or [Line]: *(pick the end circular face, press* [Enter]*)*
> Enter an option [Create new part/Replace instances] <Create new part>:⏎
> Enter new part name <TOOLBODY1>: **RIGHTSIDE.**⏎

You now should have two parts—mirror images of one another and connected at where the handlebar stem (gooseneck) would be. See **Figure 13-23**. In the **Browser**, notice that the toolbody RIGHTSIDE_1 is the active part. MDT can have only one part active at any given time. Double-click on the first part in the **Browser**. Notice the change in color of the icon.

Figure 13-23.
The mirrored and combined handlebar resulting into a set of handlebars.

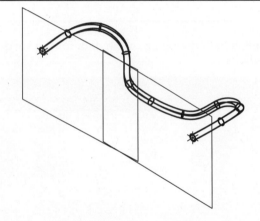

Combining the Two Parts

The final step is to combine the two parts using the **AMCOMBINE** command. To access this command, enter AMCOMBINE at the Command: prompt, select **Combine** from the **Placed Features** cascading menu in the **Part** pull-down menu, or click the **Combine** button underneath the **Hole** flyout on the **Part Modeling** toolbar. Once this command is accessed, the command sequence is as follows:

> Enter parametric boolean operation [Cut/Intersect/Join] <Cut>: **J** *or* **JOIN.**⏎
> Select part (toolbody) to be joined: *(select the second mirrored part)*

MDT takes a few moments to figure everything out and then gives you a finished set of handlebars. Now all you need is the rest of the bicycle. Please save your work under a different name.

Chapter Test

Answer the following questions on a separate sheet of paper.

1. Describe the fundamental process for creating any swept feature.
2. Name the four types of features that can be used to define a sweep path. This is even before entering the dialog box portion of the command.
3. Can a window be used to select the edge segments when defining a part's edge to be used as a path for a sweep?
4. Can a sketch profile be dimensioned to the work point placed at the start of a 3D sweep path?
5. Can a concentric constraint be used to tie a circular cross section to the start of a 3D sweep path?
6. If elbows are needed in a 3D pipe, is it necessary to draw the arcs in the pipe centerline before you turn it into an MDT sweep path? Why or why not?
7. When creating a spline-based sweep path, what is the difference between constraining a spline's vertex to a work point or to the path itself? Where can this be changed?
8. Why was it necessary to *dress* the ends of the threads modeled in this chapter with additional *cutting* features?
9. When creating a 3D path with MDT, there are several operations done automatically at the *start* of the 3D path. The automatic creation of a work point is one, what else is done automatically?
10. What is the best way to edit a 3D sweep path's properties, such as the number of revolutions of a helix or the XYZ lengths at a vertex on a spline?

Chapter Exercises

Exercise 13-1. *Tapped Hole.* Open Exercise-13-01.dwg and use **AMREPLAY** to understand the steps necessary to construct an inner thread in a tapped hole. Using this information, make your own tapped hole.

Exercise 13-2. *Threaded Shaft.* Open Exercise-13-02.dwg and use **AMREPLAY** to reveal the methods used to create the threaded shaft. See if you can edit the sketch profile defining the thread shape and update the part to see the results. Can you change the pitch or the number of revolutions?

Exercise 13-3. *Coil Spring.* Open Exercise-13-03.dwg and create a helical 3D path using the rod seen in the drawing. The values for the dialog box are **Pitch and Revolution**, 1.00 for the pitch, 8 revolutions, a counterclockwise helix, and use a circle for the helical shape with a diameter of 2.0. The **Profile Plane** is set to Center Axis/Path.

After MDT has created the helical path, create a circle located at 0,0,0 with a radius of 0.125. Profile and constrain it to the work point using the **Concentric** constraint. Use **AMSWEEP** to sweep it along the helical path. Examine the resulting spring. After that, double-click on the 3DhelixPath1 feature in the **Browser** and modify the values to change the spring. You could even change the shape from a **Circle** to an **Ellipse**, also with a taper.

Exercise 13-4. *Wood Molding.* Open Exercise-13-04.dwg and in the **Browser** find Profile2. Right-click on Profile2 and pick **Sweep** from the pop-up menu. Use the defaults in the dialog box and pick **OK**. The profile is then swept along the edge of the part. How would you copy this feature over to the opposite side to sweep and cut that edge? Do so now.

Exercise 13-5. *Plaque's Edge.* Open Exercise-13-05.dwg and in the **Browser** find Profile2. Right-click on Profile2 and pick **Sweep** from the pop-up menu. Use the defaults in the dialog box and pick **OK**. The profile is then swept along the part's edge. Try to experiment with creating your own cross section to be swept along the part's edge.

Exercise 13-6. *Spline Sweep Constrained by Work Points.* Open Exercise-13-06.dwg and notice there is a red spline passing through a few of MDT's work points. Use the **AM3DPATH** command with the **Spline** option to create a **3D Path** along this spline. Notice how it is constrained to the work points that it passes through automatically. You can toggle this constraint off in the dialog box by right-clicking on the little work point icon in the dialog box for each vertex. Leave all the vertices constrained to their respective work points and pick **OK**. Draw a 0.25 radius circle at 0,0,0 and constrain it to the work point that is there (use the **Concentric** constraint). Go ahead and sweep it using the spline as a 3D path, making certain to choose **Join**. To change the shape of the spline sweep, you have to edit the 3DhelixPath1 feature in the **Browser** by double-clicking on it. Pick **OK** to exit the dialog box and then you can edit the dimensions controlling the locations of the work points. Changing the location of the work points results in changing the spline sweep shape. Some of the work points are located on offset work planes. Edit the offset value for the work plane—thereby changing the location of the work point. This results in the changing of the shape of the spline sweep.

CHAPTER 14

Multiple Parts in One File

Objectives

After completing this chapter, you will be able to:

* Understand the difference between a part file and an assembly file.
* Identify the ways MDT helps you with feedback from the user interface.
* Understand the various operations possible or not possible in a part file.
* Use **Combine** operations in a part file—**Join, Cut**, or **Intersect**.
* Set up an Excel spreadsheet to do table-driven parts.
* Do some simple assembly constraints.

Drawings Needed

*The following is a list of files that you will need to work through this chapter. These files can be found on the **User's Drawings** CD included with this text.*

Examples	Exercises
Bottle.dwg	Exercise 14-01.dwg
Bottle.xls	Exercise 14-02.dwg
	Exercise 14-03.dwg

THE DIFFERENCE BETWEEN AN ASSEMBLY FILE AND A PART FILE

First, we need to define some terms and make sure that everybody is on the same page, so to speak. When you have more than one part in a file, that drawing is referred to as an *assembly drawing*. In fact, when you create a new drawing in MDT with the standard AutoCAD **NEW** command you are actually creating an assembly drawing. Try it right now. Use the **NEW** command to create a drawing and give it any name you would like. Please notice that the **Browser** has the **Model, Scene**, and **Drawing** tabs. Also, the **Part Modeling, Assembly Modeling, Scene**, and **Drawing Layout** buttons are found on the **Desktop Main** toolbar. See **Figure 14-1.**

To create a part drawing, select **New Part File** from the **File** pull-down menu. This is a special MDT menu selection and there is no command line equivalent. This is very similar to the **NEW** command. Notice that the **Browser** now contains the **Model**

Figure 14-1.
The **Desktop Main**
toolbar in its
floating state.

Figure 14-2.
The **Browser** after
creating a part
drawing. Notice
that the icon next to
the part name is
different.

and **Drawing** tabs only. Also, the **Assembly** pull-down menu has been replaced with **Toolbody** and that there are only three icons on the **Desktop Main** toolbar, instead of the usual four. There are also some subtle clues in the **Browser** that indicate that you are in a part file as opposed to an assembly file. See **Figure 14-2**. Notice that the name of the drawing is displayed at the top of the **Browser**, but the icon next to it differs.

STARTING A NEW PART

Up to this point, you probably have been starting your part drawings with the **NEW** command, not knowing that you were actually starting an assembly drawing. When working in an assembly drawing, you must realize that you have to *initialize* a new part when you want to start a new part. Otherwise, all your new sketch profiles, and hence, their features, will be subordinate to the existing part. There is only one part active at any time in MDT. If you do want to begin a new part in an assembly drawing, use the **AMNEW** command. To access this command, enter AMNEW at the Command: prompt, select **New Part** from the **Part** cascading menu from the **Part** pull-down menu, or pick the **New Part** button on the **Part Modeling** toolbar.

AMNEW

Part
↳ Part
  ↳ New Part

Part Modeling
toolbar

New Part

Depending on how you initiate the command, the prompts differ. If you enter the command at the command line, you are prompted to enter an option. Enter P for part. Notice the **AMNEW** command is also used to create new instances of a part, a new scene, or a new subassembly. Subassemblies are covered in greater detail in *Chapter 15*.

No matter which way you initiate the command, at this point you are asked to enter a name for the new part. Please give the part a meaningful name, and do not accept the default name of PART1. After you begin a new part, please keep in mind that all of the subsequently created sketch profiles, and hence new features, will be for that new part.

RULES FOR PART FILES

If you begin a part file, there are certain restrictions inherent in this mode of MDT. For example, if you create a new part file there can be only one part in that file. Any additional parts in the file are called *toolbodies*. If you create a new part using the previous method, you will notice that you are actually creating a toolbody. This is a part that will interact with the main part in a Boolean operation. Later in this chapter, toolbodies that interact with the main part will be covered. In MDT, the rule is that the only additional parts that can reside in a part file are toolbodies. Try to copy

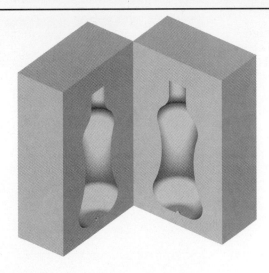

Figure 14-3.
The finished bottle mold.

the main part and you will get an error message stating that the main grounded part cannot be copied. Try to use **AMMIRROR** on the main part and elect to create a new part. You will notice that you are only creating a toolbody. This is simply true to the rule for a part file.

A collection of parts in a drawing is known as a *local assembly*. The parts it is made up of all reside in the assembly file. This differs from an *external assembly*, where the parts that make it up are all externally referenced from part files on the hard disk. You may also have any combination of locally defined parts and externally referenced parts in one assembly file. This is covered in more detail in *Chapter 16*.

In this chapter we will be using a somewhat lengthy exercise to illustrate all of these principles and other related operations. We will start with a blank part file, define our main part, bring in a toolbody, make the toolbody table-driven, combine it with the main part, split the main part into two, and assemble the two resultant halves (optional). It is a moldbase for a blow-molded plastic bottle. Many of your favorite soft drinks come in these blow-molded plastic bottles, as well as milk or liquid detergent. An example of the finished assembly is shown in **Figure 14-3**.

CREATING THE MOLDBASE

To start please begin a new part file and name the file MOLDBASE. Please notice how the name of the part in the **Browser** updates to reflect the name of the drawing file. Draw a rectangle beginning at 0,0 and make it 8″ per side. Before you profile it, three design variables need to be made. Initiate the **AMVARS** command and pick the **Global** tab in the **Design Variable** dialog box. Use the **New** button to create three new variables. See **Figure 14-4**.

 CAUTION It is very important that new variables be created in the **Global** tab in the **Design Variable** dialog box.

The first one is called LENGTH and it has a value of 8.00, the second one is called WIDTH and it too has a value of 8.00, the last one is called HEIGHT with a value of 12.00. Notice that comments have been added to describe the variables. This is very important because you may have to come back to this model later and will want to know what these variables control. After you have done that, exit the dialog box.

Figure 14-4.
The three design variables are shown on the **Global** tab.

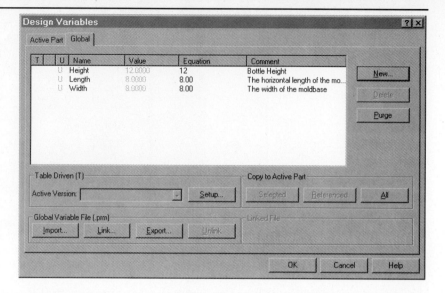

Now, profile the rectangle and dimension it with two dimensions. Do the horizontal dimension first so that it has the dimension label D0. Instead of a numeric value type in LENGTH. The vertical dimension has a value of WIDTH. Both dimensions equate to 8.00, but they refer to the design variables previously created. We will use these later on when constraining the bottle to lie centered in the moldbase. Now extrude the profile with an extrusion height of HEIGHT+2.00. You should now have a 3D box on your screen. This is the moldbase; a simplified one at that. Enter 8 to orient your display so that it shows the front-right isometric view.

CAUTION

It is very important that you orient your display so that it shows the front-right isometric view, because many of the steps that follow refer to this orientation.

AMCATALOG

Toolbody
➧ Catalog...

Part Modeling
toolbar

Part Catalog

Importing in the Toolbody

Now we want to import our toolbody using the **AMCATALOG** command. To access this command, enter AMCATALOG at the Command: prompt, select **Catalog...** from the **Toolbody** pull-down menu, or pick the **Part Catalog** button on the **Toolbody Modeling** toolbar.

PROFESSIONAL TIP

To access the **AMCATALOG** command, you can also right-click in the graphics area and select **Catalog...** or pick the **Catalog** icon at the bottom of the **Browser.**

This accesses the **Part Catalog** dialog box, which is used to manage parts in the drawing. See **Figure 14-5.** At this point we will be using it to import a part file. This is identical to the way AutoCAD XREFs in a file. As a matter of fact, external part files that are imported into MDT are XREFs. You can also edit them *in-place* the same way AutoCAD 2000 lets you edit XREFs. **AMCATALOG** has many other uses and they will be covered in later chapters.

In the **External** tab right-click in the right **Directories** window and choose **Browse & Attach** from the pop-up menu. Navigate the directory structure to the location of

Figure 14-5.
The **Part Catalog** dialog box is used to manage parts in a drawing.

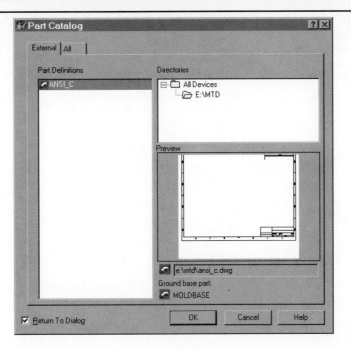

the exercise files and select the BOTTLE.DWG file. This is a drawing file of a bottle. Pick **OPEN** to exit the dialog box and you will be prompted to specify a new insertion point for the new part. Pick a point near the moldbase and then press [Enter].

You will then be back in the **Part Catalog** dialog box. The bottle is now externally referenced into this part file and we want to make it local to this file. In the **All** tab you will see BOTTLE is listed in the **External Toolbody Definitions** area. See **Figure 14-6.** Right-click on it and choose **Localize** in the pop-up menu. BOTTLE should move and be listed in the **Local Toolbody Definitions** area. The bottle now has become part of this file. This is similar to the AutoCAD **INSERT** command. You could externalize it by right-clicking on it again in the **Local Toolbody Definitions** area. However, we want to keep it **Local** because we are going to drive it parametrically with an Excel spreadsheet—changing sizes on-the-fly. Pick **OK** to exit—we are done for now.

Figure 14-6.
A—The **All** tab showing the BOTTLE as an external assembly definition. B—The **All** tab now showing the BOTTLE as a local toolbody definition.

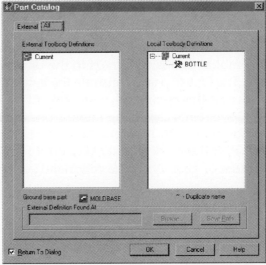

A   B

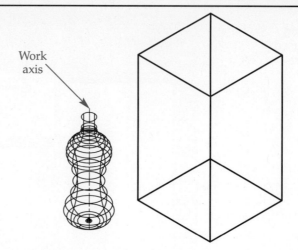

Figure 14-7.
The bottle is placed near the moldbase and a work axis is added to the bottle.

Work axis

To prepare for the next step, we will need to place a work axis in the bottle. Please make sure the bottle is the active part by double-clicking on it in the **Browser**. Place a work axis through the bottle by picking its cylindrical top face when prompted for the placement of the work axis. See **Figure 14-7.** We will use this axis to constrain the bottle to the base.

Constraining the Bottle

The next step requires constraining the bottle to be in the center of the moldbase. We could manually place the bottle in the moldbase, but it would not be constrained to *always* lie in the moldbase. In MDT the more things are constrained the better. We will need to use the *3D assembly constraints*. There are four of them and they are used to constrain two parts together. You will be learning more about them in other chapters as well. We will use one of them to constrain the bottle in three dimensions, so we will use the command three times. The assembly constraint to use at this time is **AMMATE**.

If you look at the **3D Toolbody Constraints** toolbar, notice the **Flush, Angle,** and **Insert** buttons next to the **Mate** button. We will be using them later in this exercise. Together these four are all that is needed to constrain assemblies together. **Mate** is definitely the workhorse of the group, because it does many jobs. As you will see in the next few steps, it can be used to constrain a point to a line, a point to a plane, a line to a line, a line to a plane, or a plane to a plane. When constraining parts together, they do not necessarily have to touch physically. There can be an offset between the mating features. This will be shown in the next few steps.

AMMATE

Tool**b**ody
  ↳ 3D Co**n**straints
    ↳ **Mate**

3D Toolbody
Constraints
toolbar

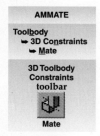

Mate

Before going further, update the assembly by picking the **Update Assembly** icon, which is the last icon on the right underneath the **Browser**. This is for updating assemblies and is right next to the **Update Part** icon.

The first step is to constrain the bottle in the X direction with the **AMMATE** command. To access this command, enter AMMATE at the Command: prompt, select **Mate** from the **3D Co**n**straints** cascading menu from the **Tool**b**ody** pull-down menu, or pick the **Mate** button on the **3D Toolbody Constraints** toolbar.

Pick the work axis sticking out of the bottle when you are prompted to select the first set of geometry. The command sequence is as follows:

> Select first set of geometry: *(pick the work axis sticking out of the bottle)*
> First set = Axis, (line)
> Select first set or [Clear] <accEpt>: ↵
> Select second set of geometry: *(pick the left-rear vertical face of the moldbase)*
> Second set = Plane
> Enter an option [Clear/Next/Flip/cYcle] <accEpt>:

If you get the wrong face, left-click again to cycle through until the proper side is highlighted. At this point we will want to flip the direction of the arrow. The offset which you will use works off this direction; we want it to be pointing inward. See **Figure 14-8.** Command sequence continues:

```
Enter an option [Clear/Next/Flip/cYcle] <accEpt>: F↵
Second set = Plane
Enter an option [Clear/Next/Flip/cYcle] <accEpt>: ↵
Enter offset <0.0000>: LENGTH/2↵
```

This will effectively center the bottle in the X direction. See **Figure 14-9.**

For the Y direction, use the same procedure. First, enter the **AMMATE** command and pick the work axis for the first set of geometry. Then, pick the left-front vertical face of the moldbase for the second set of geometry. Flip the direction of the arrow again so it points inward and press [Enter] to accept. Enter WIDTH/2 for the offset. This will effectively center the bottle in the Y direction. See **Figure 14-10.** If you want to check on this, look at the top view of the part by entering 5. The bottle should be centered in the moldbase.

Figure 14-8.
The **AMMATE** command is used to constrain the bottle in the X direction. The work axis and the left-rear vertical face of the moldbase are used to do this.

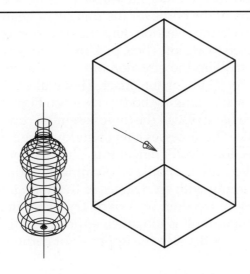

Figure 14-9.
The bottle constrained in the X direction.

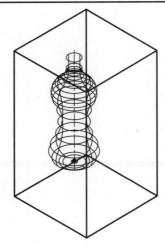

Figure 14-10.
A—The bottle constrained in the Y direction.
B—The top view showing the bottle is constrained to the middle of the moldbase.

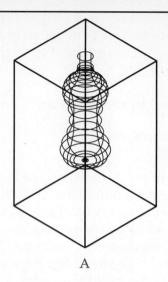

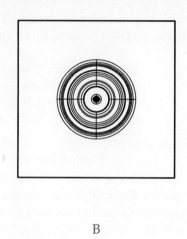

A   B

Now you need to constrain the bottle in the Z direction. Enter **8** to get the right isometric view again. Enter the **AMMATE** command and pick the top cylindrical flat face on the top of the bottle for the first set of geometry. Remember, you may have to left-click to cycle the selection. Flip the direction of the arrow so it points downward and press [Enter] to accept. Then, pick the bottom plane of the moldbase for the second set of geometry. Flip the direction of the arrow so it points upward and press [Enter] to accept. Enter **12.00** for the offset. This will ensure the bottle stays in the box later on when we start changing bottle sizes. See **Figure 14-11.**

After you have placed the three assembly constraints, use the **LIST** command and pick the bottle. This accesses a text window. See **Figure 14-12.** Notice that there is one degree of freedom (DOF) remaining. The rotational DOF about the Y-axis cannot be constrained because the bottle has symmetry about the Y-axis. All parts, except for the base part, start out with six degrees of freedom. It is not necessary to eliminate all six of the degrees of freedom on a part. Try to eliminate four or five at most, with four being the minimum. It is too much effort to eliminate all of them and MDT does not need it. Now pick the X in the upper-right corner of the text window to close it.

Figure 14-11.
The bottle constrained in the Z direction.

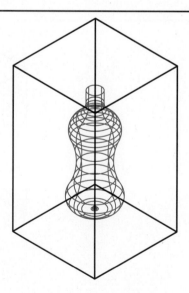

Figure 14-12.
The text window showing that there is one degree of freedom (DOF) remaining.

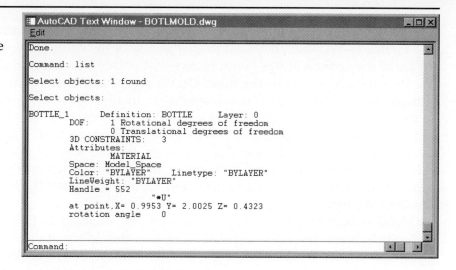

```
AutoCAD Text Window - BOTLMOLD.dwg                              _ □ X
Edit

Done.

Command: list

Select objects: 1 found

Select objects:

BOTTLE_1      Definition: BOTTLE      Layer: 0
     DOF:      1 Rotational degrees of freedom
               0 Translational degrees of freedom
     3D CONSTRAINTS:    3
     Attributes:
                MATERIAL
     Space: Model_Space
     Color: "BYLAYER"      Linetype: "BYLAYER"
     LineWeight: "BYLAYER"
     Handle = 552
                    "*U"
     at point,X= 0.9953 Y= 2.0025 Z= 0.4323
     rotation angle    0

Command:
```

Driving a Part's Dimensions with a Spreadsheet

Now that the bottle is properly assembled to the moldbase, we are going to create a series of bottle shapes with different sizes using Microsoft Excel. We will create a spreadsheet with the various bottle sizes and then link it to the bottle. This is only possible because we assigned design variables to the bottle's dimensions.

CAUTION   You *must* have Microsoft Excel on your computer to complete the following section.

Please make certain the bottle is the active part by double-clicking on it in the **Browser**. Use **AMVARS** to access the **Design Variables** dialog box again. Notice that there is now one tab for **Active Part** and one tab for **Global**. The **Active Part** design variables should look like those in **Figure 14-13**. You could change them here in this dialog box by double-clicking on them and typing in a new value, but we want to drive them all together with a spreadsheet.

Figure 14-13.
The design variables for the bottle.

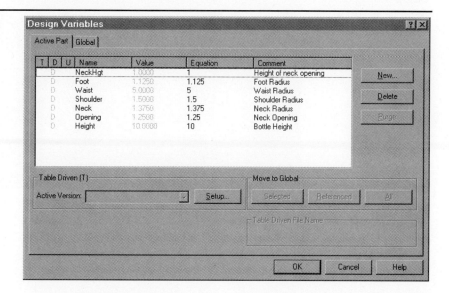

T	D	U	Name	Value	Equation	Comment	
	D		NeckHgt	1.0000	1	Height of neck opening	New...
	D		Foot	1.1250	1.125	Foot Radius	
	D		Waist	5.0000	5	Waist Radius	Delete
	D		Shoulder	1.5000	1.5	Shoulder Radius	
	D		Neck	1.3750	1.375	Neck Radius	Purge
	D		Opening	1.2500	1.25	Neck Opening	
	D		Height	10.0000	10	Bottle Height	

Table Driven (T)   Move to Global
Active Version: [   ] [Setup...]   [Selected] [Referenced] [All]

                                    Table Driven File Name

[OK]   [Cancel]   [Help]

Setting up a spreadsheet

Now click on the **Setup...** button. This accesses the **Table Driven Setup** dialog box. See **Figure 14-14.** We do *not* want to change any of the values here, but you should be aware of the options at this point. Notice that you could change the beginning cell in the spreadsheet or rename the sheet. On the right-hand side, notice that spreadsheets can also be used to drive feature suppression, in addition to the use that we are going to put it to. On the far right side of the dialog box, you will find the **Create** and **Link** buttons. The **Create** button is used to begin a new spreadsheet. You would have to fill in the values by yourself, and this is tedious. Instead we will use the **Link** button to select a prepared spreadsheet. Select the **Link...** button to access the **Link Table** dialog box. Navigate the directory structure and highlight the BOTTLE.XLS file. Now press the **Open** button.

Notice that three new buttons have become enabled in the **Table Driven Setup** dialog box. When linking to a spreadsheet, you should always choose the **Update Link** button once to make sure everything goes smoothly. You could also choose the **Edit...** button to edit the spreadsheet or the **Unlink** button to unlink from the spreadsheet. Editing the spreadsheet would entail Microsoft Excel popping up and loading the spreadsheet in question. We do not want to do this just yet. There is a great deal involved in setting up a model so that the numeric values you use for the different design variables result in a logical solid model. It is very easy to construct a corrupt version of the bottle because of the way the dimensions work together to form a cohesive model. For example, you could make the shoulder radius less than the neck radius thereby resulting in a corrupt version. The other versions would work fine, just the corrupt one would result in an error. If you have not already, exit the dialog boxes by picking **OK** twice until you get back to the MDT drawing screen.

Notice in the Browser that a series of bottle versions have become present. See **Figure 14-15.** To see the different versions of the bottle, simply double-click on a specific version in the Browser under the Table feature. You may have to expand the feature list by clicking on the small plus sign next to each feature. Notice how each version behaves nicely and stays in the moldbase. You now have a table-driven part. Before we go on, please make the 16-ounce bottle the current version. Also, this would be a good time to save your work.

Figure 14-14.
The **Table Driven Setup** dialog box is used to set up a spreadsheet.

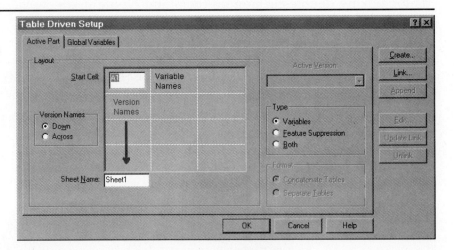

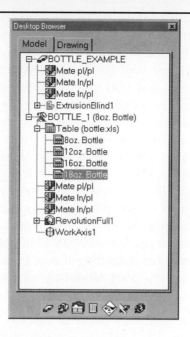

Hollowing Out the Moldbase

Now that we have a table-driven bottle that is parametrically assembled to the moldbase, we will hollow out the moldbase by combining it with the bottle. Basically, we will be hollowing out the moldbase with the shape of the bottle. First, make the moldbase the active part. Then we will use the **AMCOMBINE** command, which provides MDT with Boolean capability between disparate parts. Typical Boolean operations in solid modeling are union, subtract, and intersect. MDT calls two of these operations by different names—join for union and cut for subtract, but intersect is the same. To access this command, enter AMCOMBINE at the Command: prompt, select **Combine** from the **Placed Features** cascading menu from the **Part** pull-down menu, or pick the **Combine** button on the **Part Modeling** toolbar. Since the bottle is going to remove material from the moldbase, we will use the **Cut** option. Once the command is issued, the command sequence is as follows:

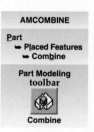

Enter parametric boolean operation [Cut/Intersect/Join] <Cut>: C↵
Select part (toolbody) to use for cutting: *(pick the bottle)*
Computing...

The bottle should have become part of the moldbase. The color of the bottle should now be the same color as the moldbase. After the **Combine**, try choosing a different version of the bottle in the **Browser**—you may have to expand the feature list to get to the bottle versions under the Table feature. Notice that everything behaves normally. Before you proceed onto the next step, make sure that the 16-ounce version is current.

Splitting the Part

The next step is to split the moldbase into two equal halves. This will create two parts that are mirror images of each other. We will do this parametrically with a work plane using a design variable you set up earlier. Using the **AMWORKPLN** command, place a work plane **Planar Parallel** and **Offset**. The workplane will lie in the same direction as the WCS's X-axis. You will want to make the workplane parallel to the left-front vertical face of the moldbase. Use the formula of WIDTH/2 for the offset value. Remember, the design variable WIDTH controlled the width (vertical dimension) of the base sketch profile. Also, you will need to flip the default placement of the workplane. Once the placement is flipped, the work plane should be placed exactly in the middle of the moldbase and it should be oriented vertically. Make sure **Create Sketch Plane** check box is checked. Accept this default orientation of the new sketch plane. See **Figure 14-16** for the proper orientation.

To split a part, we will use the **AMPARTSPLIT** command. This command was covered in *Chapter 12* and you may want to review that section. This command will split a part based on an existing planar face on the part, an existing work plane, or a previously constrained split line. The *split line* is constructed similarly to a sketch profile except it is not a closed shape—instead it is open-ended. The nice thing about a split line is that it can have breaks in it, thereby reflecting those angles in the split parts based on it. Also, since it is parametric if you change the dimensions controlling the shape of the split line, the resulting shape will be reflected in the parts based on it; very powerful indeed for the plastic injection molding industry.

We, of course, will be using the work plane we just placed. To access the **AMPARTSPLIT** command, enter AMPARTSPLIT at the Command: prompt, select **Part Split** from the **Placed Features** cascading menu from the **Part** pull-down menu, or pick the **Part Split** button on the **Part Modeling** toolbar. Once the command is issued, the command sequence is as follows:

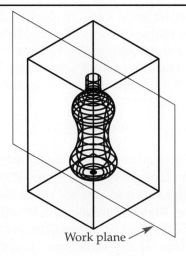

Select planar face, work plane, or split line for split: *(Select the workplane)*
Define side for new solid [Flip/Accept] <Accept>: **F**↵
Define side for new solid [Flip/Accept] <Accept>: ↵
Enter name of the new part <TOOLBODY1>: **MOLDTOP**↵
Computing…

At first, it might be hard to figure out what happened. The two parts are in place next to one another. Use the **MOVE** command and move one part away from the other. Now for the fun part—turn on rendering and orient your display for a good view of the two mold halves. You may also want to hide the display of the workplanes to simplify your display. To do this, make a part active, find the workplane in the **Browser**, right-click on it, and pick **Visible** from the pop-up menu. This toggles the workplane invisible. Do the same procedure to the other part. See **Figure 14-17.**

In the **Browser**, choose a different version of the bottle and watch as MDT updates your model with a new bottle shape in each of the halves. Imagine how much time this saves in industry—one file supporting multiple moldbases. Extend this parametric power to all the tooling that would accompany this design. It could all be linked to this model and updated effortlessly.

Figure 14-16.
The work plane is
used as a split line.

Work plane

Figure 14-17.
The mold after it
has been split.

ASSEMBLING THE TWO MOLD HALVES

We will now use the *3D assembly constraints* to assemble the two halves of the moldbase. Orient your display so that you can get a good view of the two halves and pull them apart so that there is some space between them.

Using **AMMATE**, we want to pick the left-hand vertical edge on one of the mold halves and mate it to the mirrored edge on the other half. Remember, this command can line up two edges as if they were lines. To get this command to cycle through the possibilities, left-click after selecting the vertical edge on the mold half. At first you may be selecting a plane—be careful and make sure to select the vertical edge. You can use the **Clear** option to start over without exiting the command. The two edges needed are shown in **Figure 14-18**. After selecting both edges, use an offset value of 0.00.

The two halves may or may not be correctly lined up—they should not be, at least not yet. We need to make the two top faces of the halves flush with one another. Use the assembly constraint **AMFLUSH**, pick the top face of each half in turn, and use an offset value of 0.00.

Now, things should be looking better, but there is one last step. We want to make an angular constraint between the two hollowed-out faces of the mold halves. Use the assembly constraint **AMANGLE** and pick each of the hollowed-out faces in turn. You may have to move them apart to pick the proper faces. Use a value of 90 for a pleasing arrangement. See **Figure 14-19**. The angle you specify is actually the *angle between the face vectors, not the faces themselves*. Remember those blue arrows that showed up when you were selecting the faces? Those are known as the *face vectors*. It is a perpendicular line coming out from the face.

It should be noted that you can move these parts away from each other, but they are constrained to one another now. To get them to move back together into their constrained assembly, use the **Update Assembly** icon underneath the **Browser**. It is the farthest one on the right and it looks like an assembly with a lightning bolt on it. You always use this icon to update an assembly.

Figure 14-18.
The two edges for
AMMATE line to line.

Figure 14-19.
The mold parts put
together at 90°.

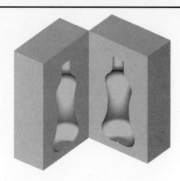

Changing the Value of an Assembly Constraint

Now that we have applied the three assembly constraints to the mold halves, you can change the value of an assembly constraint and watch the assembly update on the screen. In the **Browser**, find the Angle Vec/Vec constraint and double-click on it. This accesses the **Edit 3D Constraint** dialog box, which allows you to change the angle value. See **Figure 14-20.** Type in a new value and press [Enter]. The dialog box stays and the model updates on the screen. This allows you the opportunity to try different values and get a good idea of what is happening to the model as you input different values. You can also use the **Update Constraint** button (button with the lightning bolt) instead of pressing [Enter]. Once you have decided on the value, pick the **OK** button to make the change. Now try choosing a different version of the bottle again and see the mold halves update.

Figure 14-20.
The **Edit 3D Constraint** dialog box is used to change the angle between part faces.

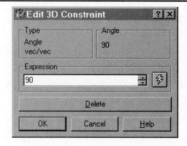

Chapter Test

Answer the following questions on a separate sheet of paper.

1. Describe the difference between a part file and an assembly file.
2. How do you start a part file?
3. How do you start an assembly file?
4. What is essential that you set up and apply to the part's dimensions if you are going to use a spreadsheet to drive a part?
5. What is *feature suppression*? Can you use a spreadsheet to drive feature suppression?
6. You can add more versions to a table-driven part by editing the linked spreadsheet and creating a new entry below the ones already there. What is critical to remember if you want the model to update and not result in an error?
7. Name the four 3D assembly constraints and list their uses.
8. Why does an animated cursor pop up sometimes when you are trying to select a face for assembly constraining? How does it work?
9. Describe how the **AMFLUSH** assembly constraint is used.
10. When using the **AMANGLE** assembly constraint, the angle you specify is not between the selected faces themselves. What does the specified angle indicate?

Chapter Exercises

Exercise 14-1. *Moldbase Part Splitting.* Open Exercise-14-01.dwg and notice that it is an assembly file. Refer to the figure here and split the part as shown on the left. Use the following names for the new parts: CORE for the bottom part split off at the obvious setback. Use the name CAVITY for the top part, split off at the obvious setback. Finally, split the original solid at the centerline into two parts, naming one half SLIDE1 and the other half SLIDE2. Move them apart to get a view as shown on the right.

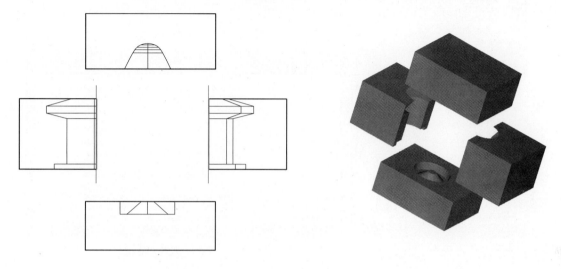

Exercise 14-2. *Table-Driven Part and Feature Suppression.* Open Exercise-14-02.dwg and examine it. Learn the names of the features and use **AMREPLAY** to see the order of construction. Set up some new design variables that will drive the part's shape for the new MS Excel spreadsheet that will be made. Make LENGTH with a value of 10.00, WIDTH with a value of 7.00, SLOTLENGTH with a value of 4.00, and BOSSDIA with a value of 2.00.

Use **AMMODDIM** to apply LENGTH and WIDTH to the base feature's dimensions. Apply SLOTLENGTH to the dimension controlling the slot's length. Apply BOSSDIA to the diameter of the circular boss. Using **AMVARS**, you can change the shape of the model by changing the values for the variables in the dialog box. Set up a spreadsheet to drive these design variables. Include feature suppression in this model—make sure to check **Both** under the **Type** of variable to be driven.

In MS Excel, create different versions of the part and use feature suppression to hide the fillets or holes in some of the versions. Please refer to sample spreadsheet shown on the next page.

Go back to the model and test out the different versions. Edit the table and make changes to the resulting parts. Practice suppressing features using the spreadsheet. Notice how the **Browser** reflects a suppressed feature.

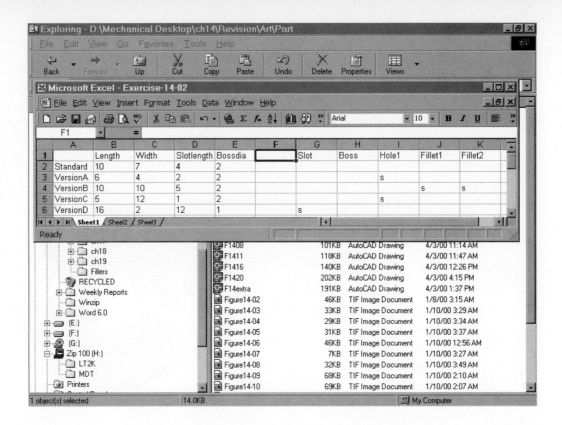

Exercise 14-3.   *Types of Combines.* Open Exercise-14-03.dwg and use **AMCOMBINE** to do a Boolean operation on the parts that you see there. Examine the **Browser** and notice that there are assembly constraints on some of the parts. It should be noted that it is not necessary to have assembly constraints on parts before you do a combine. The CYLINDER has the REVOLVE part subtracted from it. In addition, it has the CUTTER part subtracted from it also. Make sure you make the CYLINDER part active first. The HOLDER is joined with the CLAMP part. Finally, the HEXHEAD intersects with the CHAMFER part. Make certain the HEXHEAD part is active. Examine the resulting parts; you now have a graphic example of the types of combines done in MDT. Please notice how a combine feature appears in the **Browser**.

CHAPTER 15

Building Assemblies

Objectives

After completing this chapter, you will be able to:

* Understand the difference between a part definition and a part instance.
* Use the 3D assembly constraints to connect the parts in your assembly model.
* Understand degrees of freedom and how to eliminate them.
* Edit assembly constraints and update your assembly model.
* Edit an externally referenced part in the assembly.
* Use global design variables and their power to revise an assembly.

Drawings Needed

*The following is a list of files that you will need to work through this chapter. These files can be found on the **User's Drawings** CD included with this text.*

Examples

Example-15-01.dwg
Axle.dwg
Base.dwg
Bolt.dwg
Bushing.dwg
Spin.lsp
Tire.dwg

Wheel.dwg
WheelAssy.xls
Yoke.dwg

Practice

Practice-15-01.dwg

Exercises

Exercise-15-01.dwg
Exercise-15-02.dwg
Exercise-15-03.dwg
Exercise-15-04.dwg

PART DEFINITIONS OR PART INSTANCES

When you create a part in an assembly drawing, you have actually created a *part definition*. You cannot see the part definition on the screen—what you do see are *part instances*. You can have many part instances, only one part definition. For example, a ¼-20UNC screw might be used in 20 locations in your assembly, but they are all just copies of the one part definition. If you change the part definition, you change all 20 copies present on the screen. This is similar to management of blocks in AutoCAD, where all occurrences of a block will update to reflect changes if the block is redefined. Part definitions are managed by the **AMCATALOG** command. This will be discussed in greater detail later in this chapter.

Methods of Building an Assembly

There are essentially three methods of building an assembly in MDT. These methods consist of top-down, bottom-up, or middle-out.

Top-down

Top-down is the process of building an assembly from existing parts, perhaps from a corporate parts database. You start out with an empty drawing and instance parts into the drawing from an external source—a network drive or CD-ROM. You build the assembly piece by piece. Most engineers working at the concept level would not be doing this. Designers or even draftsmen most often build top-down assemblies.

Bottom-up

Bottom-up is the opposite of top-down and is typically how an engineer would build an assembly. *Bottom-up* is the process of building an assembly with an empty drawing and building a layout of the assembly first. This layout is a rough approximation of the parts correctly aligned with each other—essentially placeholders for the more detailed parts to come. Think about how it is done on a board or even 2D CAD. A conceptual layout starts out with the centerlines of equipment, the work envelopes, and the arrangement of the parts to one another. The thing to keep in mind is that in this process the parts do not exist. They will not exist until the engineer finalizes the assembly. The parts will be *fleshed-out* once the layout is done. This is perhaps the most difficult type of assembly for novices to 3D CAD to do. It requires a huge amount of imagination and work *inside* the mind of the engineer.

Middle-out

Middle-out is typically used when a combination of top-down and bottom-up is possible. You might have an assembly that is partially laid out, some of the finished parts are in place, and some of the subassemblies are known. The rest of the design is finished by basing the new parts on existing parts or distances between those existing parts. This type of assembly requires a high degree of skill, but not as much as the bottom-up method.

BUILDING AN ASSEMBLY

The assembly we will be building is a standard 12" industrial wheel assembly for application in a factory environment. We will be using a top-down method. An empty drawing will be the starting point and we will be bringing in finished parts that were supplied for this chapter on the *User's Drawings* CD. We will constrain them together using the 3D assembly constraints in MDT. Furthermore, we will be adding in a new part and editing the assembly to accommodate the new part. Finally, we will set up a Microsoft Excel spreadsheet to drive the different versions of the assembly. **Figure 15-1** shows the finished example.

Open Example-15-01.dwg and prepare to bring in our parts. To do this, we will be using the **AMCATALOG** command. To access this command, enter AMCATALOG at the Command: prompt, select **Catalog...** from the **Assembly** pull-down menu, or pick the **Assembly Catalog** button on the **Assembly Modeling** toolbar.

AMCATALOG

Assembly
 ↳ Catalog...

Assembly Modeling
toolbar

Assembly Catalog

PROFESSIONAL TIP

To access the **AMCATALOG** command, you can right-click in the graphics area and select **Catalog...** or pick the **Catalog** icon located at the bottom of the **Browser.**

Figure 15-1.
The completed
wheel assembly.

This accesses the **Assembly Catalog** dialog box, which contains the **External** and **All** tabs. Using the **External** tab, attach a directory to the right-hand pane so that it will always be there when you want to go back to it. Right-click in the background of the right-hand pane and choose **Add Directory...** from the pop-up menu. You will have to navigate the directory structure to indicate the location of the files supplied on the *User's Drawings* CD. Click on the directory where the files for this chapter are located and pick **OK**.

In the right-hand pane, you should see your directory listed in the window. In the left-hand pane, you should see a listing of drawing files. See **Figure 15-2.** The highlighted file will be displayed graphically in the **Preview** window. If you click on a few different drawing files, they will be displayed in the window. When you bring a part into your assembly, the process is termed *instancing*. This means you are creating an *instance* of the part in your assembly. You can have many copies of the part in the assembly; they are all called *instances*. As previously mentioned, if you make changes to one instance, they all change. Instances are listed in the **Browser** as you insert them. Instances are not the true part, however. That exists in MDT's database and is stored as a *part definition*. The part definition is the *basis* for all of the instances in the assembly.

We are going to try to build this assembly as you would in reality. We will instance the parts in the same order that we are going to assemble/constrain them. With the proper directory listed in the left-hand pane, right-click on the drawing file BASE and choose **Attach** from the pop-up menu. Pick a point on the screen when you are prompted

Figure 15-2.
The **Assembly Catalog** dialog box showing part drawings that are used for creating the assembly.

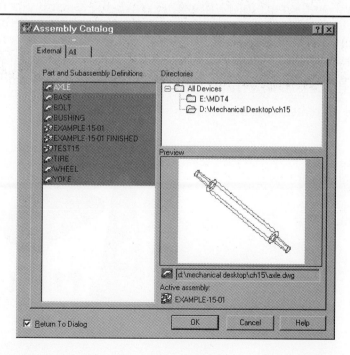

to specify a new insertion point. Now you are prompted for an insertion point of another instance. Since we want to put in only one instance, press [Enter] to continue and then pick **OK**.

Notice that the BASE came in lying flat on the screen from a plan view overhead. That is because all new AutoCAD drawings start out in plan view of the WCS. Also, the part is brought in with its WCS aligned with the WCS of your assembly and the BASE drawing was made with the part laying down on the WCS. To give yourself a good isometric view, orient your display using the **3D Orbit** icon, enter 8 for a front-right isometric, or enter 88 for a front-left isometric view.

Next, use **AMCATALOG** to instance the YOKE. Find it in the listing, right-click, choose **Attach**, and place *two* parts on the screen in and around the BASE. Continue on with the BOLT. You will need four of them. Place them in and around the BASE.

Click **OK** to exit the dialog box. Zoom the display so that it looks like the view shown in **Figure 15-3.** Now, the two side yoke supports need to be connected to the base and bolted in place with the bolts. Notice that there are no threads shown on the bolts—this is typical for assemblies. Unnecessary details will slow computer performance. This is important when there are hundreds of parts in an assembly.

Please take a moment and study the **Browser**. The parts should be listed in the order in which you instanced them. The first part should be BASE, followed by YOKE_1 and YOKE_2, and finally the four bolts listed BOLT_1 through BOLT_4. The **Browser** *only* lists *part instances*, that is, the actual parts you have in the assembly.

The part definitions are listed and managed in the **Assembly Catalog** dialog box. If you wanted to rename *all* of the bolts in this assembly, you would do it from this dialog box. Renaming something in the **Browser** will only affect that part instance's name. Please take note of the names of the two supports—YOKE_1 and YOKE_2 and the names of the four BOLT instances. This *incrementing* of names, or appending increasing numbers, is done so that each part instance has a unique name in the assembly.

As stated before, only one part can be active at a time in MDT. To make a part active, you can double-click on its name in the **Browser**. Make the BASE the active part and notice that it is now highlighted and all the other parts are grayed out. There is also a tiny red padlock on the icon for the part. This indicates that a file lock has been put on that part's drawing file. If you try to open that part drawing in another window of AutoCAD, you will get the *Read Only* message indicating the drawing is available in read-only status. The assembly *owns* it until released because you double-clicked on the part in the **Browser**. When the assembly is made up of external parts, double-clicking them in the **Browser** activates them for editing, but the software has to lock the external disk file so that no one else can edit the file as long as the assembly is open. This is to ensure data integrity in a workgroup project with many people opening the project's drawings and working on them. Notice that you can open the other drawings, such as YOKE or BOLT, in read/write status. This is because they were never activated and the file lock was never applied.

Figure 15-3.
The initial parts used for the assembly are instanced into the drawing.

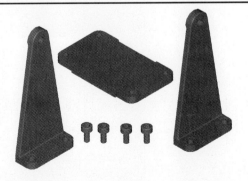

Assembling the Parts

The 3D assembly constraints will be used for the next section. The main purpose of the 3D constraints is to assemble the parts of the assembly so that they are anchored to one another. This anchoring effect traces its way up to the first part instanced into the assembly. This first part, therefore, becomes the *base* part or *grounded* part. If the assembly is all constrained together and the base part is moved, then all other parts will follow after you use the **Assembly Update** icon.

Degrees of freedom

At this point, it is important that you understand the concept of "degrees of freedom" (DOF) and how it relates to building assemblies. Every part that you bring in, except for the base part, has six degrees of freedom. There are three translational degrees of freedom along the three axes (X, Y, and Z), and three rotational degrees of freedom about those same axes. This means that the part is free to move in a linear path along any of the axes—to the left or right, up and down, and front to back. The part also is free to rotate about the X-axis, the Y-axis, or the Z-axis. For example, this would be comparable in an aircraft to latitude, longitude, altitude, roll, pitch, and yaw. Your task is to remove as many of these degrees of freedom as are required to anchor the part to the rest of the assembly.

How many degrees of freedom you need to remove depends upon its shape and the shape of the instance you are constraining it to. For example, in our exercise there will be a wheel constrained to an axle, which in turn will be constrained to the side yoke supports. The wheel can have five degrees of freedom removed with one left, which will be rotational freedom about the axis of the axle—just like an actual wheel. You want a rotational degree of freedom on the wheel so it can spin—we will actually be doing that later on in this exercise.

The side yoke supports will be constrained to the BASE. For each of these, you will need to remove all six degrees of freedom since they must not move with respect to each other. The bolts that hold them in place only require removing five DOF's because they are symmetrical about their axis of insertion.

To see the DOF symbol on your parts, right-click on a part in the **Browser** and choose **DOF Symbol**. At the geometric center of each part you should see a blue and red symbol. See **Figure 15-4A.** If you zoom in on one of them, you will notice the blue arrows indicate translational freedom and the red arc arrows indicate rotational freedom. The red arrow indicates the positive Z-axis direction for this part's local coordinate system. As you apply the assembly constraints in the next steps, watch the DOF symbol and see how it changes as it loses degrees of freedom. This is great visual feedback on what might be a tricky situation.

The **3D Constraints** toolbar is shown in **Figure 15-4B.** Picking the **Launch 3D Constraints Toolbar** button on the **Assembly Modeling** toolbar activates this toolbar. These are the four assembly constraints used in MDT. The first one is Mate (AMMATE) and is the workhorse of the four. You can pick a point, a line, or a face on either of the parts being constrained. This results in six possible combinations that are shown in **Figure 15-5.** Please note the arrow lines coming out of the faces—they are called *face vectors*. You will also be asked to input a distance for an offset to separate the parts. A positive offset separates the parts and a negative causes the parts to intersect. A design variable can be used for the offset value. The Mate assembly constraint is very powerful and one you will be using often.

To the right of the **Mate** button is the **Flush (AMFLUSH)** button. This can only be applied to two faces and it forces the face vectors to point in the same direction. See **Figure 15-6.** For example, the bricks in a wall are all flush with each other but mate face to face with the mortar. The offset can be positive or negative and which way the parts move depends on which one you selected first. Picking the base first, a positive offset moves the block in the direction of the base face vector.

Figure 15-4.
A— The DOF symbol for an unconstrained part. B— The **3D Constraints** toolbar contains the **Mate, Flush, Angle,** and **Insert** constraint buttons.

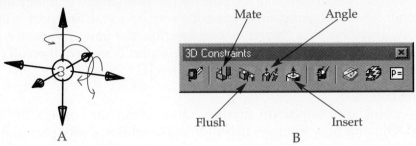

A   B

Figure 15-5.
The six possible combinations of the **Mate** assembly constraint.

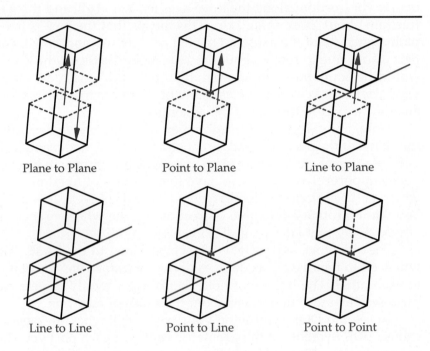

Plane to Plane   Point to Plane   Line to Plane

Line to Line   Point to Line   Point to Point

Figure 15-6.
The **Flush** assembly constraint—the face vectors are always in alignment.

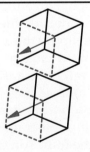

Next to **Flush** is the **Angle** (**AMANGLE**) button. This is used whenever there is an angle between two parts as in the case of two lever arms with a specified angle between them. See **Figure 15-7A**. It is important to note that the angle is between the face vectors and not the faces themselves. **Figure 15-7B** shows the dialog box used to edit the angle constraint. **Figure 15-7C** shows where the angle constraint between the two blocks has been edited. This displays the face vectors and the angle.

The last of the four buttons is the Insert button. This is typically used when you have a bolt going into a hole. This constraint does a coaxial alignment between the centerline of the bolt and the centerline of the hole. It then does a **Mate Face-to-Face** (plane-to-plane) of the underside of the head of the bolt with the mounting face of the hole. All you have to do is pick the proper faces/planes. See **Figure 15-8**.

Figure 15-7.
A— The **Angle** assembly constraint— angle value is applied to the angle between the face vectors.
B—The **Edit 3D Consraint** dialog box.
C—The face vectors and the angle.

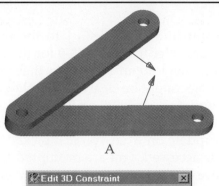

A

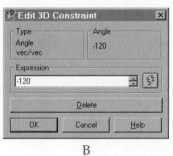

B

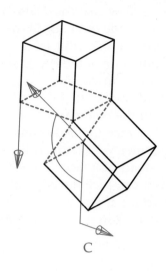

C

Figure 15-8.
The **Insert** assembly constraint is commonly used for bolts.

PRACTICE 15-1

❑ Open the drawing Practice-15-01.dwg.

❑ Try each of the six possible combinations of the Mate assembly constraint between the Block and the Base. In each case, check the DOF symbol of the Block and compare it to this table:

Constraint on Block	Constraint on Base	Translational DOF Removed	Rotational DOF Removed
Plane	Plane	1	2
Point	Plane	1	0
Line	Plane	1	1
Line	Line	2	2
Point	Line	2	0
Point	Point	3	0

❑ Now repeat each of the constraints this time with a 0.50 offset.

Attaching the YOKE

Now use the **Mate** constraint to attach YOKE_1 to the proper location on the underside of the BASE part. Please refer to **Figure 15-9** for the proper faces to pick. If you use **3D Orbit** to rotate your view about 90° up so that you are looking at the underside of the BASE, then things might be clearer. Even though the YOKE_1 part is upside down in relation to where it needs to be on BASE, by using **Mate** you can *flip* it around so that the two faces/planes must face each other. As you do this, always read the command line for an update on what you just picked.

Once the **Mate** constraint is accessed, you are prompted to select the first set of geometry. Pick the indicated face on the BASE part. If you are getting a point or a line instead of a face, left-click to cycle through all the geometry under your pick point. You can pick the edge of a face or you can pick directly inside its edges—picking its surface, if you will. The second method hardly ever works when you are in **Rendered** mode. You now receive the following prompts:

> First set = Plane *(Notice the command line feedback. At times you will see*
> *other data here like* First set = Point *or* First set = Arc*)*
> Enter an option [Clear/Next/Flip/cYcle] <accEpt>: *(Press* [Enter] *when you have the*
> *proper plane)*
> Select second set of geometry: *(pick the plane on the* YOKE *part)*
> Second set = Plane
> Enter an option [Clear/aXis/Point/Next/Flip/cYcle] <accEpt>:⏎
> Enter offset <0.0000>: *(An offset of zero makes the faces/planes touch. A positive*
> *value would place them that distance from each other, but still they would be*
> *facing each other)*

What happened? Why does the YOKE part flip and not the BASE part? The BASE part is the first one in the assembly and is considered *grounded* with no degrees of freedom. Your drawing should look similar to **Figure 15-10.**

Figure 15-9.
The first two faces/planes to be constrained are shown highlighted.

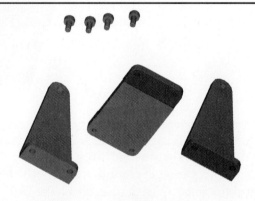

Figure 15-10.
The first two parts assembled together.

Figure 15-11.
The proper holes for
the **Mate Line to Line**
constraint.

For the next step, use the **MOVE** command to move the two parts apart. Remember, they are constrained partially together now, moving them apart will not change their assembly. You can always use the **Update Assembly** icon under the **Browser** and all the *constrained* parts will come together. We will use the **Mate** constraint again for the alignment between the holes. As you apply these constraints, take notice of the DOF symbol and how it changes. When prompted for the first set of geometry, pick the cylindrical edge of a hole on the YOKE part. See **Figure 15-11.** Press [Enter] when you see a blue line passing through the center of the hole. You are now prompted to select the second set of geometry. Pick the cylindrical edge on the corresponding hole on the BASE part. Press [Enter] when you see a blue line passing through the center of the hole. At the Enter offset <0.0000>: prompt, enter a value of zero for coaxial alignment (shared/identical centerlines). Of course, a positive value would send them that distance apart, with their centerlines still parallel.

Notice as you add constraints that they are listed in the **Browser**. Use **Mate** again for the other set of mating holes. How many degrees of freedom are left on YOKE_1? Find out by using the **LIST** command on the part. All of the degrees of freedom are removed. Is that because we used three constraints? In this case yes, but that is not always the case. Sometimes you will use four assembly constraints to remove all degrees of freedom. As previously stated, it is not necessary to remove all six degrees of freedom—four will do for most applications.

Using the 3D Manipulator

The next step will be to attach the second YOKE support to its mating place on the BASE part. We could again use **Mate** to flip the YOKE_2 into proper position, but we will use the **3D Manipulator** this time. First, the part is rotated –180° about its Z-axis with the **AMMANIPULATOR** command, which is used to move/rotate parts in 3D space. To access this command, enter AMMANIPULATOR at the Command: prompt, select **3D Manipulator** from the **Assembly** pull-down menu, or pick the **3D Manipulator** button from the **Assembly Modeling** toolbar.

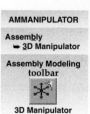

AMMANIPULATOR

Assembly
➥ 3D Manipulator

Assembly Modeling
toolbar

3D Manipulator

Select the YOKE_2 part when you are prompted to select an object. The **3D Manipulator** will appear on the object. See **Figure 15-12A.** You then receive the following prompt:

> Middle handle is FREE MOVE, other handles are ROTATE and MOVE Select handle
> to manipulate (Selecting no handle shows option dialog) or [Undo/Mode (follow)]
> <Exit>: *(click and hold on the –X node on the 3D manipulator)*

Slowly drag it around so that the part is seen rotating around its Z-axis. Rotate it 180° as shown in **Figure 15-12B.** Hitting [Esc] to try again is okay if you get the wrong

Figure 15-12.
A—The **3D Manipulator**. B—YOKE after being rotated 180° around its Z-axis. C—YOKE after being rotated 180° around its Y-axis.

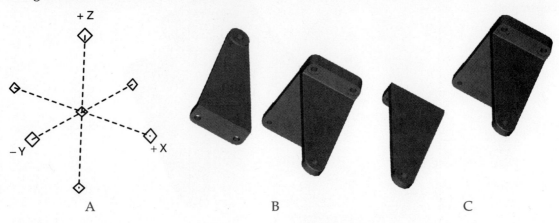

A   B   C

Figure 15-13.
Completed assembly at this point.

rotational movement. It is very subtle and you have to get the right feel for which hand movement creates the desired rotation. After that rotation, do it again. This time around the part's Y-axis—again swinging the part around 180°. Pick the +X node to make it easier. Slowly rotate the part 180° around its Y-axis. See **Figure 15-12C.**

Apply the same 3D assembly constraints to the same features on the BASE and the YOKE_2 part. **Figure 15-13** shows the completed assembly.

Editing a 3D constraint

Now we want to edit one of the **Mate Plane to Plane** constraints on one of the YOKE parts. In the **Browser** find the Mate pl/pl assembly constraint underneath YOKE_1 (you may have to expand the listing of YOKE_1 features). Double-click on the constraint and the **Edit 3D Constraint** dialog box is accessed. The current **Expression** value is zero. Change it to 2.00 and pick **OK**. What happened to the assembly? This is the effect of a positive value on an assembly constraints' offset value—it moves the parts apart. Most offset values used in assembly modeling are zero, but it is good to know how to enter your own values after the fact. Change the value back to zero.

Figure 15-14.
Each BOLT is
inserted with the
Insert assembly
constraint.

Adding the Bolts

Now the bolts need to be added. To do this you will use the **Insert** constraint. It was specifically designed for this application. You may have to use **3D Orbit** so you are looking at the underside of the BASE/YOKE assembly and the bolts. The command sequence is as follows:

> Command: **AMINSERT**↵
> Select first circular edge: *(pick the cylindrical edge of the underside of the head of the BOLT)*
> First set = Plane/Axis
> Enter an option [Clear/Flip] <accEpt>:↵
> Select second circular edge: *(pick the cylindrical edge of the hole)*
> Second set = Plane/Axis
> Enter an option [Clear/Flip] <accEpt>:↵
> Enter offset <0.0000>:↵

That is all the constraints needed on a bolt in its hole. Use the **LIST** command on the bolt and notice that its degrees of freedom have been reduced by five. Repeat the procedure for the remaining bolts. See **Figure 15-14.**

Importing Additional Parts

At this point you need to import some new parts and constrain them as well. Using the **AMCATALOG**, attach two BUSHING, one AXLE, and one WHEEL parts (in that order). Place them in and around the assembly. Use the **Insert** constraint to constrain the two bushings to their respective holes. See **Figure 15-15.** Make sure you pick the larger diameter faces—not the smaller diameter faces at the BUSHING end. You may choose to assemble BUSHING_1 with YOKE_1, but it is not necessary. Again, that is all there is to do for the BUSHING—one constraint and five degrees of freedom removed. Which DOF is remaining?

Figure 15-15.
Each BUSHING
shown inserted into
the respective YOKE.

In reality, the axle would be placed in the wheel first and then both parts would be mounted to the supports together. So that is how you will do it. You can do these two first, because the AXLE need not be constrained to the BUSHING first. It is not necessary to always constrain the parts in the order in which you inserted them. The AXLE and the WHEEL will be constrained into coaxial alignment using **Mate Line to Line**. In addition, the AXLE needs to be constrained along the axis. We will center it in the WHEEL between its hub faces, preventing it from sliding out. First, apply the **Mate** to the centerline of the AXLE and the centerline of the WHEEL using an offset of zero. Why did the WHEEL flip 90° to line up with the AXLE? Why not the other way around? The next constraint is to align the 3D center of the AXLE to the 3D center of the WHEEL, known as a ***coincident alignment***. Orient your display as shown in **Figure 15-16** and refer to the points shown on the AXLE. Initiate the **AMMATE** command and you get the following command sequence:

> Select first set of geometry: *(pick any cylindrical edge on the* AXLE*)*
> First set = Axis, (arc)
> Select first set or [Clear/fAce/Point/cYcle] <accEpt>: *(left-click to cycle to the point)*
> First set = Point, (arc) *(a point should have shown up at the center of the arc you picked)*
> Select first set or [Clear/aXis/fAce/cYcle] <accEpt>: *(do not press [Enter])*

Hover over the corresponding mirrored cylindrical edge at the other end of the AXLE. Notice you get a pick box instead of the object-cycling animated cursor; pick the cylindrical edge. Another point should have shown up at this end of the AXLE, you now should have two and a centerline showing up.

> First set = Axis, (arc)
> Select first set or [Clear/Midpoint] <accEpt>: **M** *or* **MIDPOINT.**⏎
> First set = Point, (arc) *(a point midway between our two outside points is located)*
> Select first set or [Clear/aXis] <accEpt>:⏎

Do the same method on the WHEEL, using the large cylindrical edges of the hub faces. Pick one of the edges of the hub faces.

> Select second set of geometry:
> Second set = Axis, (arc)
> Select second set or [Clear/fAce/Point/cYcle] <accEpt>: *(left-click until the edge of the hub face is selected and press [Enter])*
> Second set = Point, (arc)
> Select second set or [Clear/aXis/fAce/cYcle] <accEpt>: *(pick the mirrored edge on the other hub face and press [Enter])*
> Second set = Axis, (arc)
> Select second set or [Clear/Midpoint] <accEpt>: **M** *or* **Midpoint.**⏎
> Second set = Point, (arc)
> Select second set or [Clear/aXis] <accEpt>:⏎
> Enter offset <0.0000>: ⏎

Figure 15-16.
The WHEEL rotates
90° to align with
the AXLE.

Figure 15-17.
The AXLE and
WHEEL are
assembled.

The WHEEL and the AXLE are assembled together. See **Figure 15-17.** That is how you solve the tricky problem of centering two parts on their geometric centers. Is there one more DOF left on the WHEEL?

Assembling the **AXLE** to the **BUSHING**

To assemble the AXLE to the BUSHING, make it easier by hiding the display of the WHEEL. In the **Browser**, right-click on the WHEEL part and choose Visible from the pop-up menu. Remember, it is still there and will move when the AXLE gets assembled to the frame. Use an Insert constraint between the stepped face on the end of the AXLE which fits up against the inner face of one of the BUSHING parts. If you are in the classic front-right isometric, use the far end of the AXLE and the BUSHING on the far YOKE support. That should finish all the parts being assembled and constrained. Make the WHEEL visible, render, and use 3D Orbit to view your work. See **Figure 15-18.** Now is a good time to save your work.

Making a Revision

It is normal in engineering for a part or assembly to go through several revisions before it is finalized. Regarding the assembly just created, perhaps the wheel would be better if it had a tire covering it. We happen to have one in our engineering library. Use the **AMCATALOG** command to attach the tire. Find TIRE in the directory listing, right-click on it, choose **Attach**. Now, place it near the assembly and exit the dialog box.

Figure 15-18.
The completed
assembly to this
point.

The constraints are identical to the ones used to assemble the AXLE to the WHEEL. You might find it useful to move the WHEEL out of the assembly for the time being. First, address the coaxial constraint between the centerline of the TIRE and the centerline of the WHEEL. Then, using the same method as before, pick two mirrored points on each part and specify the midpoints with an offset of zero. You will have a lot of trouble with this step because the TIRE has few planar faces. There are the grooves and that is what you should use. Orient your display to a front view by entering 6, and pick on the cylinders' edge lines to effectively select the proper ones. Remember, your first pick will be interpreted as you select the centerline—left-click to cycle past this.

Update the assembly when they are constrained, **3D Orbit** around and you will notice that there is a big problem. This wheel is not rolling very far. We need to increase the length of the side supports to create more clearance. Although they are externally referenced into our assembly, we can still edit them *in-place*. To do this, double-click on the Part YOKE_1 in the **Browser**. Expand the list, if necessary, until you see the feature Support. Double-click on Support and pick **OK** to exit the dialog box that pops up. On screen, find the dimension controlling the length of the support (d18) and change as shown below:

> Select object: *(pick the long dimension between the top of the* YOKE *to its bushing hole— it should be d18)*
> Enter dimension value <WheelDia/2+0.25>: **WheelDia/2+2.0**⏎
> Solved fully constrained sketch.
> Select object:

Update the assembly with the icon under the **Browser**.

> Command: _AMUPDATE
> Enter an option [active Part/aLl parts] <active Part>: _PART
> Computing…
> Active part has been updated.
> Command: _UNDO Enter the number of operations to undo or
> [Auto/Control/BEgin/End/Mark/Back] <1>: _END

The TIRE should have adequate clearance now. See **Figure 15-19.** Although after studying the design, the width of the tire needs to be increased so it lays more flush with the side of the wheel. As before, make the TIRE part active by double-clicking on it in the **Browser**. This is a little more detailed than what was done with the YOKE. It is not necessary, but we need to see how it is done. Expand the plus signs in the **Browser** down to the sketch profile for the Revolution feature in the TIRE part. Double-click on the sketch profile and notice that the tire is replaced with a sketch profile. Switch your display to a plan view of the current sketch plane by entering 9. The current sketch plane is that of the revealed sketch profile. See **Figure 15-20.** Find the

Figure 15-19.
The TIRE constrained
to the wheel.

Figure 15-20.
Change d11 so that
the TIRE is the same
width as the
WHEEL.

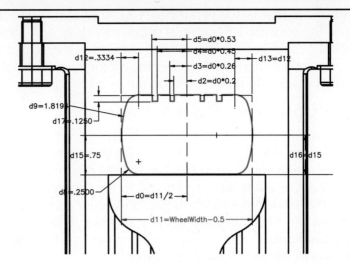

dimension d11 and change it to just WheelWidth. You will have to update the part to regenerate the TIRE part, then you will have to update the assembly and elect to write out the changes to the TIRE part file on the hard disk. It is not necessary to do this for this exercise. Now is a good time to save your work.

You have successfully collected components, built an assembly out of them, added in a new part, and edited the connecting parts to accommodate the revision to the assembly. You have learned to use the **Browser** to work with an assembly and have learned about the intensive interactive process of constraining parts together.

CREATING A TABLE-DRIVEN ASSEMBLY

In this next section, parametrics will be applied to the wheel assembly—changing two key global design variables of Wheelwidth and Wheeldia. Then, we will set up a Microsoft Excel spreadsheet to display the different versions specified in the spreadsheet. Finally, some simple LISP code will be used to make the wheel spin in place.

CAUTION   You must have Microsoft Excel on your computer to complete this section.

Before you can do this, all the parts must be made *local* to this assembly. Using the **AMCATALOG** command, make sure the **All** tab is active. You should see all the parts listed in the **External Assembly Definitions** column. In order, right-click on each part and choose **Localize** from the pop-up menu. Once all parts are done, they should now appear in the right-hand column. Exit the dialog box.

Now examine the design variables that drive these parts. Start **AMVARS** and make the **Global** tab active. There are a few key variables that are common throughout all the parts. Therefore, if we change the value of WheelWidth, for example, we can have all the parts that use that variable to use its new value for their size. Change WheelWidth, which controls how wide the wheel is allowed to be, to some value larger than 1.5 and pick **OK** to exit the dialog box. Any value below 1.5 and the fillets on the cutouts start to fail. Notice the assembly updates and you have a new version of your assembly. This is great for trying out ideas or exploring new possibilities in a product line.

The other key variable is WheelDia and this controls the diameter of the wheel and has similar results on the assembly. The allowable ranges for this wheel are from 4.5 to 17.5. Again, values outside this range will usually result in failed features. More than likely, if this is happening to you, it is the fillets that will fail dimensional changes. Experiment with several different values and observe the results. Notice as the wheel gets larger the number of cutouts in the wheel increase. Notice there are quite a few other variables listed in the dialog box. Most of them will not change the assembly because they are based on the two key variables.

Setting Up a Table-Driven Assembly

You have seen how two key variables can control an entire assembly. If the parts reference them, the design variables will drive them. Setting up the spreadsheet is simply an extension of what you have done manually in the **Design Variables** dialog box. To begin, use **AMVARS** again and in the **Global** tab select the **Setup** button. This accesses the **Table Drive Setup** dialog box. Now click the **Link...** button and go to the location of the files supplied for this chapter. Find and select the spreadsheet file called WheelAssy.xls. Return to the dialog box, pick **Update Link** (just to make sure), and then pick **OK** to exit all dialog boxes.

In the **Browser**, you should have a table-driven assembly listing several types of wheel styles. Double-click on a version other than Standard (which is the default state this assembly was created in) and watch as the assembly again updates. This is merely automation of what you were doing earlier by manually changing the values in **AMVARS**. Try all the versions.

Now you may want to edit the spreadsheet. In the **Browser**, do a double-click on the Global Table feature and Excel is accessed. It may or may not come with its pull-down menus and toolbars—this is widely known as a *bug*. The two key variables (WheelDia and WheelWidth) are in the first two columns. See if you can make a new version yourself. To copy an entire row in Excel, you click on the number next to the row and press [Ctrl]+[C]. Then move down to an empty row, enter [Ctrl]+[V] and you should have a newly populated row. Edit the name of the version first and then change the values for the two variables. Remember the allowable ranges for this wheel assembly. Changing the values for the other variables will have unforeseen results. Notice that all of the equations somehow relate back to the two key variables. If there are no menus visible, save the spreadsheet by pressing [Ctrl]+[S]. Finally, go back to MDT, make sure to update the link, and check out your new version in the **Browser**.

Adding Animation

One final fun example is to get the wheel to actually spin in place on the screen. You will first create a new design variable and use it with an angle constraint. Use **AMVARS** to set up a design variable called **ROTANG** and give it a value of zero. Next, use **AMANGLE** to make the angle between the angled side of one of the YOKE parts and one of the cutout sides of the WHEELS. Refer to **Figure 15-21** for an example of

Figure 15-21.
The two faces to use
for the **Angle**
assembly constraint.

the proper faces. For the angle value, enter in **ROTANG**. You have just used a *global* design variable to drive an assembly constraint. Now, if you drive the value for the design variable, the assembly just might be animated. You have been supplied with a few simple lines of AutoLISP to do this. Use the AutoCAD **APPLOAD** command to find and load the AutoLISP file Spin.lsp. This file is also with the files supplied for this chapter. Once the file is loaded and you have a good isometric view of the assembly, enter SPIN at Command: prompt and watch as the wheel spins. Depending on the speed of your computer and quality of the video card, you should see the wheel spinning smoothly in place. It will stop after a few minutes, but it is important to note that the value for a design variable can be driven by a LISP or VBA application.

Chapter Test

Answer the following questions on a separate sheet of paper.

1. What is the **Catalog** used for when building assemblies?
2. What does the **Catalog** manage and what does the **Browser** manage?
3. Name the four assembly constraints and discuss their uses and applications.
4. How do you edit the offset value of an assembly constraint?
5. How do you delete a constraint?
6. How many degrees of freedom should you try to remove—at a minimum?
7. Is it necessary to constrain the part fully in the order in which they have been listed in the **Browser**? Why or why not?
8. Can you use active part design variables to drive an assembly?
9. How could the number of grooves in the assembly's tire have been parametrically controlled so that they increased or decreased in number as the wheel's width changed size?
10. Describe how AutoLISP or VBA can be used to automatically drive a design variable? Why would this be done?

Chapter Exercises

Exercise 15-1. *All the Assembly Constraints.* Open Exercise-15-01.dwg as shown here. Examine it and note the different combinations available for a given assembly constraint. It has many parts that are constrained to one another using the labeled constraint. See if you can duplicate the effort on some of them. Use the set of blocks that is unconstrained in each example. These are labeled ...PRACTICE in the **Browser**.

FLUSH PLANE TO PLANE		ANGLE LINE TO LINE	ANGLE PLANE TO PLANE		INSERT BOLT TO HOLE
MATE POINT TO POINT	MATE POINT TO LINE/PLANE	MATE LINE TO LINE	MATE LINE TO PLANE	MATE PLANE TO PLANE	MATE CYLINDER TANGENT PLANE

Exercise 15-2. *Sample Assembly Model.* Open Exercise-15-02.dwg and use the assembly model for practice applying assembly constraints. Assemble as shown here. It will be necessary to instance more bolts into the assembly. It will also be necessary to use the sketch profile to remove material so that the INSERT part can fit into the BASE part.

Exercise 15-3. *Intermediate Assembly Model.* Open Exercise-15-03.dwg and use the assembly model for practice applying assembly constraints. The final assembly is shown. Use **Mate Axis-to-Axis** to tie all the pins into their respective holes. Make the outside ends of the pins *flush* to the planes on the shovel's arms. Do not try to remove all six degrees of freedom from the parts. In some cases, one constraint will do. For example, the hydraulic pistons go into the cylinders with a **Mate Axis-to-Axis** constraint.

Exercise 15-4. *Articulated Arm.* Open Exercise-15-04.dwg and examine it. It is a marine instrumentation problem. A bank of instruments has to be mounted into a pedestal with an articulated arm so that they can be brought up out of the pedestal and positioned where a helmsman might be able to see them better. The parts will need to be constrained so that they will work together to bring the instrument panel down into the recess in the pedestal and redeploy when needed. A **Mate Point-to-Point** is used for the spherical connection between the two CONNECTORS. Similarly, a **Mate Point-to-Point** is also used to tie the CONNECTOR_1 to the BASE part and CONNECTOR_2 to the INSTRUMENT part. The two versions of the completed assemblies are shown.

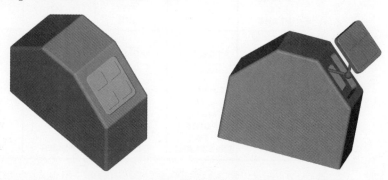

CHAPTER 16

Assembly and Part Analysis

Objectives

After completing this chapter, you will be able to:

* Understand what is meant by part/assembly analysis.
* Analyze weight, volume, centroid, moment of inertia, etc., of a part.
* Analyze weight, volume, centroid, moment of inertia, etc., of an assembly.
* Understand the effect of using different units on the result of the mass properties analysis.
* Determine if there is interference between the parts in an assembly.
* Measure the true 3D distance between two parts.
* Design models that will not fail when they are revised.

Drawings Needed

*The following is a list of files that you will need to work through this chapter. These files can be found on the **User's Drawings** CD included with this text.*

Examples	Exercises
Example-16-01.dwg	Exercise-16-01.dwg
Example-16-02.dwg	Exercise-16-02.dwg
Example-16-03.dwg	Exercise-16-03.dwg

ENGINEERING ANALYSIS OF A PART OR ASSEMBLY

You have a finished part and it looks good. However, will the part be sufficient for the task it was designed for? What will it weigh, how much volume does it take up, what is its surface area? Will the MDT model itself survive revisions or will a critical feature fail during a dimension change? These and other questions are critical to design in the modern engineering world. You may have to revise the design based on the information you receive from these analysis tools.

There is a similar process applied to assemblies, except the analysis applies to all the parts you select in the assembly. In addition, you may want to know if two parts have some volume in common. The interference between parts in an assembly is sometimes intentional, and sometimes not. MDT provides capability to determine interference between selected parts of the assembly. Finally, there will be a small section on model integrity—the actual desktop model. Can the model survive the revision process?

APPLYING MASS PROPERTIES ANALYSIS

There are two different commands for mass properties. The first is used for parts. The other is used for assemblies.

Mass Properties Analysis for Parts

AMPARTPROP

Part
➥ Part
  ➥ Mass
    Properties...

Part Modeling
toolbar

Mass Properties

The **AMPARTPROP** command is used for just a single part. To start, open Example-16-01.dwg. This is the finished wheel assembly from *Chapter 15*. Make sure the Wheel_1 is the *active* part. To access the **AMPARTPROP** command, enter AMPARTPROP at the Command: prompt, select **Mass Properties...** from the **Part** cascading menu in the **Part** pull-down menu, or pick the **Mass Properties** button on the **Part Modeling** toolbar.

This accesses the **Part Mass Properties** dialog box. See **Figure 16-1**. It shows the mass properties of the *active* part—there is no prompting to select a part for analysis. Notice the only edit field with a default value of 1.00. The volume of the part multiplied by this value equals the mass or the weight of the part. You will need to plug in the mass or weight per cubic unit for the part material (per cubic inch, per cubic mm, etc.), and pick **Apply** to calculate the proper value. Remember AutoCAD/MDT is unitless and you need to use the units you had in mind when you designed the part. The volume is simply expressed in drawing units, whatever those may be. Please note the **Surface Area** value. This is unique to part analysis. It will not be found when doing an analysis of an assembly.

Mass Properties Analysis for an Assembly

AMASSMPROP

Assembly
➥ Analysis
  ➥ Mass
    Properties...

Assembly Modeling
toolbar

Mass Properties

Determining the aggregate volume or weight of an assembly is often crucial to the manufacturing process. The **AMASSMPROP** command is used to analyze an assembly. It will deliver much more than just volume or weight. Again, the same model of the wheel assembly is used—it should still be on screen. Make certain the *standard* version of the wheel is current. To access the **AMASSMPROP** command, enter AMASSMPROP at the Command: prompt, select **Mass Properties...** from the **Analysis** cascading menu in the **Assembly** pull-down menu, or pick the **Mass Properties** button on the **Main Assembly Modeling** toolbar.

With the command you can pick one of the parts in the assembly, two of them, or all of the parts for mass properties analysis. When prompted for an option, enter S for Select. Now, window all of the parts in the assembly and press [Enter].

This accesses the **Assembly Mass Properties** dialog box. See **Figure 16-2**. In this dialog box you may change a number of items. The coordinate system can be changed

Figure 16-1.
The mass properties of the active part.

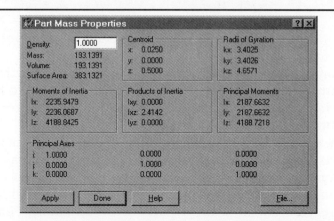

Figure 16-2.
Units, coordinate system, and materials are set in the **Assembly Mass Properties** dialog box.

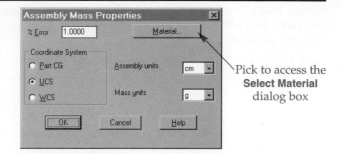

Pick to access the
Select Material
dialog box

in the **Coordinate System** area. For assemblies select the **WCS** radio button. For parts use any of them. This will affect the values displayed in the upcoming report. Above that is the **%Error** text box. This is the allowable error in the reported values. It is never set below its current setting and you would never want to *increase* the error percentage.

Assigning materials

To the right of the **%Error** is the **Material...** button. Pick it to access the **Select Material** dialog box. Pick the arrow to view the drop-down list of the materials that are available. See **Figure 16-3.** The list of available materials in the software is limited.

PROFESSIONAL TIP

It is possible to add your own material. To do this you need to edit MCAD.MAT in the .../desktop/support directory folder using WordPad or a similar word editor. Be careful, if Microsoft Access is loaded on your system, you will not be able to edit the file by double-clicking on it in Explorer.

In the **Part/Subassembly Definition** window is a list of parts. Click on each one and note the material assigned to each one. It is all the same Mild Steel. New materials must be assigned to the parts. Select a part from the list of parts, select a material, and then pick the **Assign** button. Refer to **Figure 16-4** for the proper materials used in our analysis. After all the materials have been assigned, pick **OK** to exit that dialog box.

Figure 16-3.
Each part can be assigned a specific material.

Pick to access
the materials

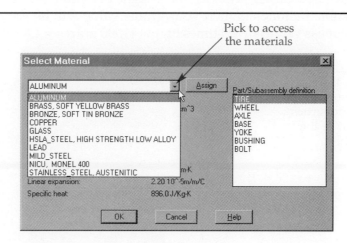

Figure 16-4.
The parts and what
they are made of.

Part	Material
AXLE	HSLA_STEEL
BASE	MILD_STEEL
BOLT	HSLA_STEEL
BUSHING	BRASS
TIRE	ALUMINUM (Closest approximation)
WHEEL	BRONZE
YOKE	MILD STEEL

The reporting of the volume and the mass

The next part is very important because of the often-critical nature of the data and the presentation of it in the dialog box. It deals with specifying the units to be used when the mass properties are reported. Please refer to **Figure 16-5** as you follow the text. It is important that you specify the *same units used when constructing the model.* If the model was made in the English inch system, then that is what should be selected for **Assembly Units** in the **Assembly Mass Properties** dialog box that was shown in **Figure 16-2.** For **Mass Units** you would use Lbs (pounds), also in the English system of measure.

On the other hand, if the model was made in the Metric system of measure, using millimeters, then specify mm (millimeters) for **Assembly Units**. In keeping with the above rule, **Mass Units** would be either g (grams) or Kg (kilograms). If you use In and Lbs for our example you will get realistic results. Please note the mass and the volume of the report. These results will be used to compare all the others and to prove a point. The Mass should be almost 91 pounds (90.7994 lbs) and the volume should be over 400 cubic inches (413.6012 in^3). Remember, this is for all of the parts together in the assembly.

Try selecting m or cm for the units of measure. Notice that the volume always stays the same because AutoCAD/Desktop are unitless. The designers of the software felt that you would know which units you had made the model in. They just append the unit of measure onto the end of the volume data. The mass values might start making sense now. Whatever the **Assembly Units** are set to, the mass is calculated based on the **Mass Unit** you specify times the volume in the current **Assembly Units**. The material you specified earlier provides the weight per cubic unit of measurement. So, if you have cubic millimeters then your grams value will be very small. If you increase to Kg, the mass value is 1000 times bigger (1000 grams = 1 Kg). If you increase to Lbs, the

Figure 16-5.
The results of the
mass properties
analysis of the
assembly.

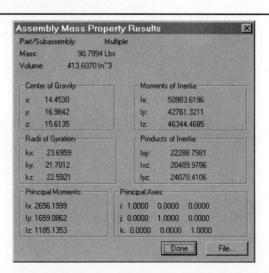

value is 2.2 times bigger (2.2 Kg = 1 Lb). Continue experimenting until it makes sense to you. Throughout all the calculations, the volume never changes. It is the mass that is calculated from that *displayed volume* that changes.

The rest of the dialog box reports calculations expressed in the units based on the coordinate system you specified earlier (WCS for assemblies). The definitions of these calculations are beyond the scope of this text.

PROFESSIONAL TIP

Because the **AMASSMPROP** command can be used to analyze a single part and assign material to it, the only time you need **AMPARTPROP** is when a volume or surface area is needed.

WHEN PARTS COLLIDE

Often in the design process mistakes are made and two or more parts will be sharing some volume. It is an understatement to say it is best to catch this in the design stage before it gets to manufacturing. An *interference* is when two parts share common volume in error. It is rarely required to model intentional interference, as in a press fit between a hole and a shaft. So, most interferences in a solid model are unintentional. MDT provides a way to check for this with the **AMINTERFERE** command. First make sure the wheel assembly is still on screen.

The Wheel and Axle

First you will examine the interference between the wheel and the axle. To make this process easier, right-click on YOKE_2 and then BUSHING_2 in the **Browser** and make them not visible. See **Figure 16-6.**

To access the **AMINTERFERE** command, enter AMINTERFERE at the Command: prompt, select **Check Interference...** from the **Analysis** cascading menu in the **Assembly** pull-down menu, or pick the **Check 3D Interference** button on the **Assembly Modeling** toolbar. Press [Enter] for no, at the Nested part or subassembly selection? prompt. Select the wheel as the first set of data and the axle as the second set of data.

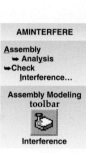

AMINTERFERE

Assembly
  ↳ **Analysis**
↳ Check
    Interference...

**Assembly Modeling
toolbar**

Interference

Figure 16-6.
The wheel assembly
with a yoke and
bushing not visible.

PROFESSIONAL TIP

If you have trouble selecting the axle, remember that if you hold the [Ctrl] key down while selecting you can cycle through several parts until you get the one you want.

You will now receive the following prompts:

```
Select second set of parts or subassemblies: ↵
Comparing 1 parts/subassemblies against 1 parts/subassemblies.
Interference 1:
WHEEL_1
AXLE_1
Create interference solids? [Yes/No] <No>: YES↵
Highlight pairs of interfering parts/subassemblies? [Yes/No] <No>: YES↵
Enter an option [eXit/Next pair] <Next pair>: ↵
Enter an option [eXit/Next pair] <Next pair>: X↵
```

The interference that you found was intentional. There is a press fit between the wheel and the axle. Because you elected to create interference solid, you will have a new piece of geometry in your drawing. It is a very thin hollow tube that represents the very small amount of interference found in the press fit. You can select it with the **Last** option of the **MOVE** command. Move it out to a clear area of the screen and examine it. Notice that it is an AutoCAD solid and not an MDT part. We don't need it so you can erase it.

The Axle and Bushing

Now you will examine the relationship between the axle and the bushing. There should be no interference here because this is a *running/sliding fit*. Before you check it, make BUSHING_2 visible again, and make WHEEL_1 not visible. See **Figure 16-7.** Use the same procedure as before to check the axle and the bushing. You should find an unintentional interference. This will have to be corrected.

Zoom in on the BUSHING_2 and prepare to edit it. Double-click on BUSHING_2 in the **Browser** to make it active. Expand its listing until you see RevolutionAngle1. Double-click on this feature, press **OK**, and look on the screen for the dimension controlling the hole the axle passes through. This should be D7. Refer to **Figure 16-8** for the proper dimensions. Notice that the 0.002 tolerance has been misapplied to the nominal 0.25 value subtracted from AxleDiameter. The value is 0.252, and it should be 0.248—0.002 the other way will provide clearance for a running fit. Change the equation to read AxleDiameter–0.248. It should not be necessary to update the part or the assembly.

Figure 16-7.
The wheel assembly with the wheel and a yoke not visible.

Figure 16-8.
The proper dimensions that will correct the unintentional interference.

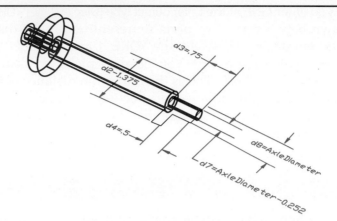

Now enter 88 to orient your view to show the other bushing. Zoom in on the BUSHING_1 and do the interference check between it and the axle. There should not have been any interference. Remember, editing *one instance* of a part affects *all* instances in the assembly.

PRACTICE 16-1

❑ With the wheel assembly on screen, explore the model and check the other parts.
❑ Check interference between one of the bolts and the base.
❑ Check interference between the wheel and the tire.

FINDING THE DISTANCE BETWEEN TWO PARTS

At times it will be important to determine the minimum 3D distance between two parts. You may need to calculate clearance or figure a work envelope for a moving part in an assembly. For example, with the wheel assembly still on screen how can we determine the 3D clearance between the TIRE and the BASE? With MDT you can find out using the **AMDIST** command. To access this command, enter AMDIST at the Command: prompt, select **Minimum 3D Distance...** from the **Analysis** cascading menu in the **Assembly** pull-down menu, or pick the **Minimum 3D Distance** button on the **Assembly Modeling** toolbar. Select the tire as the first set and the base as the second set. Press [Enter] at the Select next instance: prompts. You will now receive the following prompts:

AMDIST

Assembly
➥ **Analysis**
➥ **Minimum 3D Distance...**

Assembly Modeling toolbar

MIN

Minimum 3D Distance

> Enter output type [Display/Line] <Display>: ↵
> Minimum distance:<0.870203> *(your value should be close to this one)*

Figure 16-9.
The base part that will be filleted.

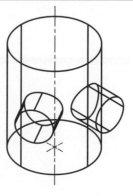

If you enter **LINE** at the Enter output type [Display/Line] <Display>: prompt, a line would be drawn between the two parts representing the 3D distance. Using this method, measure the distance between the YOKE_1 and YOKE_2. Do not be confused by the Select next instance: prompt. You need to pick two distinct sets of data here. The distance should be 3.625. This is the minimum distance between the boss on one yoke and the boss on the other—the effective clearance for the wheel/axle assembly. Knowing and checking this beforehand can save a lot of time and money later on in the manufacturing.

IS THE MODEL OR ASSEMBLY STABLE?

Many times a part will be finished, saved, and then a few weeks later it is loaded back up to do a revision. However, when a change is made to a dimension you get some kind of an error message telling that a feature failed during the command and would you please undo to get back to a normal state. Even though today's solid modeling software is much more robust than previous versions, this type of problem can still occur and the reason is most often due to fillets. *Fillets* are typically the type of feature that fails—they are almost always the culprits in models where they have been applied. However, *extrusions* and *revolutions* are not exempt from failing either.

This behavior during revisions to the model is usually the result of poor parametric logic used in the design of the solid model. This is above and beyond the topic of engineering design inherent in the model. We are referring to the steps and dimensions used to construct the actual MDT model. The problem occurs most often in models that have used fixed numeric values for dimensions—no parametrics. Dimensions control features and features make up the model. Therefore, dimensions and their values are of paramount importance. If the model is to be static with few changes, over time you *should* use numerical values for the dimensions. It is fast and why do more work than you have to. However, there are few projects where a model will not undergo revisions. When numerical values are used for the dimensions and a dimension's value is changed, other related features may fail. This is because there is no *relationship* between a dimension and other dependent dimensions.

Open Example-16-02.dwg. See **Figure 16-9**. This is the "fire-plug" example constructed entirely using simple dimensions with no relationships. Use **AMREPLAY** to review how the part was created. Note that the vertical dimension (d11) locating the small horizontal cylinder is measured from the bottom of the large cylinder. Use **AMFILLET** to apply a 1.0" radius fillet to the top edge of the large cylinder and the two intersections between the large and small cylinders. See **Figure 16-10**.

Figure 16-10.
The base part with three 1.0" fillets that did not fail.

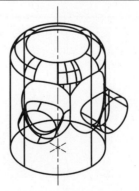

Right-click on ExtrusionBlind1 in the **Browser**, select **Edit** from the pop-up menu, and change the extrusion distance to 7.0 units and update the part. The operation will fail because the fillet on the top of the part intersects the two small cylinders. Fillet1 will be highlighted in red in the **Browser** and you will also receive the following message on the command line:

> Failed to compute Fillet.
> Error: Cannot adjust lateral edge end with mixed convexity.
> Warning - one or more features failed to update. See browser for failed
> features. Undo all steps back to ameditfeat or use amdelfeat to get a stable
> model before continuing or saving. Active part has been updated.

As the message states, please undo back to the **AMEDITFEAT** operation. The solution is to establish a relationship between the two dependent features. To do this, make the height of ExtrusionBlind2 (d11) a function of the height of the large cylinder (d2). Right-click on ExtrusionBlind2 in the **Browser**, select **Edit** from the pop-up menu, click on **OK** in the dialog box, and click on the d11 dimension and enter d2–3.0.

No matter what the height of the large cylinder (d2) is, within reason (d2 is greater than 5.0), the 1.0 radius fillet will not fail and the part will still look like a "fire-plug". If you think ahead and apply parametric relationships, then your models will update more frequently without snags.

PROFESSIONAL TIP

MDT is a *parametric* solid modeling system. It is very important to use the product the way it was intended. One way to check a model as to how it is going to behave during a revision is to make a few "mock" revisions. Save the model under a different name and try to change your design—the length of a face, the angle between two features, the diameter of a cylinder. Change anything that is controlled by a dimension. Change the value of the dimension, update the part and see what happens. If something fails, undo, note there is a problem and continue testing. Try it again, this time replacing the numeric values of the problem spots with equations to establish relationships between features that depend on each other. Do not be afraid to set up design variables and apply them in the equations.

More Problem Spots

Other types of problems may be due to the wrong type of termination used when extruding or revolving a feature. If a slot is extruded into a plate and the *design intent* is for the slot *always* to punch through the plate, then termination should *not* be a numeric value equal to the thickness of the plate. If the thickness of the plate ever changes, the slot will follow true to your design intent. A termination of **Thru** should be used to effect the parametric relationship.

Another example would be centering a slot on the centerline of a part. The trick is to draw a construction circle inside the slot, make it tangent to the sides of the slot, and dimension it to the centerline of the part using an equation like d0/2.

Always think ahead and do not think of your model as a static thing. It is going to have to be flexible enough to undergo revisions. You should have a good idea of which revisions, and thus plan accordingly.

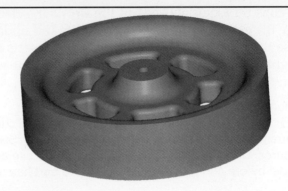

Figure 16-11.
The wheel has been designed so that two key design variables control all else.

In the next example, the engineer specified to the designer that the diameter and width (thickness) of the wheel were the important variables—all else could follow those variables. Open the drawing Example-16-03.dwg. See **Figure 16-11.** This is the wheel used in *Chapter 15* and earlier in this chapter. The model has been designed so that the two key global design variables, WheelWidth and WheelDia, control all else. Change one of the design variables, such as the WheelDia, diameter of the wheel, and all other features change parametrically. It should also be noted in this model that there are limits to the range of allowable values for the diameter and the thickness. Try to change the WheelWidth to something less than the fillets can tolerate—use 1.00 and you will notice the fillets will fail. You may even get dropped right out of MDT. Even using parametrics it is hard getting around this inherent problem in solid modeling. If you make the radius of fillets a function of the thickness of the wheel, many of these problems will be eliminated.

Chapter Test

Answer the following questions on a separate sheet of paper.

1. When would you use the **Name** option for selecting in the **AMASSMPROP** command?
2. If a model is done in English units, what assembly units and mass units would you choose to report in when using the **Assembly Mass Properties** dialog box? Why?
3. When would you want to increase the error percentage (**%Error**) used in the mass properties calculations?
4. What is meant by the part's **Centroid** in the **AMPARTPROP** dialog box?
5. Why would you ever need to use **AMPARTPROP** for a single part when **AMASSMPROP** provides the mass properties for parts as well as assemblies?
6. Why would you find an intentional interference in an assembly? How would you know it was intentional?
7. When you elect to create an interference solid from the common volume of a pair of interfering solids a simple AutoCAD Solid is created. To what use could this nonparametric solid be put?
8. What happens when you choose the **Line** option in the **AMINTERFERE** command?
9. How can you design a MDT solid so that it will update successfully revision after revision?
10. If your part fails during a revision, what can you do to recover?

Chapter Exercises

Exercise 16-1. *Common Shapes and Their Volumes.* Open Exercise-16-01.dwg and notice that there are several primitive geometric shapes. They are all drawn based on one unit (inches). For example, the cube is 1", the cylinder is 1" in diameter with an extrusion height of 1". The sphere is 1" in diameter and the torus has an outside diameter of 1". Using MDT mass properties, check the reported volume versus the calculated value. Do the math.

For the cube: $V = s^3$ [length of a side cubed]

Length of one side = 1.00

For the wedge: $V = s^3/2$ [(length of a side cubed) divided by 2]

Length of one side = 1.00

For the cylinder: $V = pi\ r^2h$ [pi times the circle's radius squared times the height]

Radius = 0.50, Height = 1.00

For the cone: $V = (pi\ r^2h)/3$ [(Pi times the circle's radius squared times the height) divided by 3]

Radius = 0.50, Height = 1.00

For the sphere: $V = 4.1888r^3$ [the radius cubed times 4.1888]

Radius = 0.50

For the torus: $V = 19.739Rr^2$ [(19.739 times the large radius) times the small radius squared]

Large radius = 0.50, small radius = 0.25

Exercise 16-2. *Importing Mass Property Data.* Open Exercise-16-02.dwg. Use **AMPARTPROP** to obtain a mass property report for this part. Before exiting the dialog box, use the **File...** button and save the file as Exercise-16-02.ppr. Create some paperspace views as shown here. On the right-hand side of the sheet, define an **MTEXT** window. Pick **Import Text...** from the right-hand side of the dialog box. Look for the file you exported previously. You will have to change the file search extension in the dialog box because **MTEXT** is looking for *.TXT or *.RTF by default. Exit the **MTEXT** dialog box and you should have the mass property report on the paperspace layout.

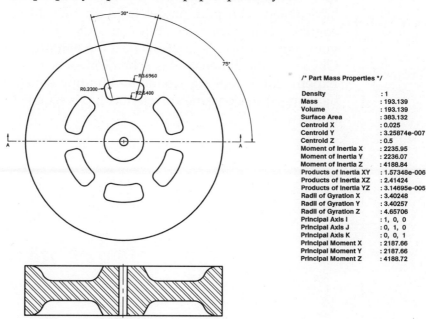

/* Part Mass Properties */

Density	: 1
Mass	: 193.139
Volume	: 193.139
Surface Area	: 383.132
Centroid X	: 0.025
Centroid Y	: 3.25874e-007
Centroid Z	: 0.5
Moment of Inertia X	: 2235.95
Moment of Inertia Y	: 2236.07
Moment of Inertia Z	: 4188.84
Products of Inertia XY	: 1.57348e-006
Products of Inertia XZ	: 2.41424
Products of Inertia YZ	: 3.14695e-005
Radii of Gyration X	: 3.40248
Radii of Gyration Y	: 3.40257
Radii of Gyration Z	: 4.65706
Principal Axis I	: 1, 0, 0
Principal Axis J	: 0, 1, 0
Principal Axis K	: 0, 0, 1
Principal Moment X	: 2187.66
Principal Moment Y	: 2187.66
Principal Moment Z	: 4188.72

SECTION A–A

Exercise 16-3. *Check 3D Distance and Interference.* Open Exercise-16-03.dwg. To provide a better view of the assembly, it is shown in section below. Move the gold INSET_1 part out of the base part to 0,0,4. Use the **AMDIST** command to measure the 3D distance between the two parts. Update the assembly using the button below the **Browser.** Now, repeat the **AMDIST** procedure to measure the distance again. The current value should be 0.00, which means the parts are just touching or they are interfering. **AMDIST** does not provide us with a negative number if interference exists. Use **AMINTERFERE** command to check if there is an interference between the two parts—there should be. Go ahead and choose to create a solid of the interference and finish the command. Use the **AMNEW** command with the **Part** option to select the red interference solid. Give it the name **TOLERANCE**. Make the CUP part active. Use **AMCOMBINE** to cut the TOLERANCE part from the CUP part. Use **AMINTERFERE** again to check for interference between the parts. There should not be any. This is an example of what one could do with that red interference solid.

Exercise 16-4. *Part Integrity.* Open Exercise-16-02.dwg again. We are going to try to get the part to fail in an **AMUPDATE**. Using **AMVARS**, change the value for the design variable WHEELWIDTH to 1.00 and pick **OK**. An automatic **AMUPDATE** issues when you leave **AMVARS**. The part should have failed. The part model is in an unstable state—you cannot continue, you cannot save, but you could quit. To continue working, your only recourse is to **UNDO** past the **AMVARS** command.

Obviously, the dimensioning scheme must be reviewed and changed. Experiment by looking around and finding which dimension should be changed so a parametric relationship is established with the width of the wheel.

CHAPTER 17

Local and External Parts

Objectives

After completing this chapter, you will be able to:

* Understand the difference between a local and external part.
* Understand how to create a subassembly, either local or external.
* Use the catalog to manage the parts and subassemblies in an assembly.
* Use the **REPLACE** function to load in an alternative part.
* Query the assembly to perform an audit on the external parts in the assembly.

Drawings Needed

*The following is a list of files that you will need to work through this chapter. These files can be found on the **User's Drawings** CD included with this text.*

Examples		Exercises
Exercise-17-02.dwg	Powersupply.dwg	Exercise-17-01.dwg
PentiumComputer.dwg	Coolfan.dwg	Exercise-17-02.dwg
Chassis_top.dwg	Drivea.dwg	Exercise-17-02a.dwg
Remove_drive_bay.dwg	Driveb.dwg	Exercise-17-02b.dwg
Hard_drive_bay.dwg	Dvdrom.dwg	12shcs.dwg
Isacard.dwg	Zipdrive.dwg	
Pcicard.dwg	Cdrw.dwg	
Motherboard.dwg	12gbhd.dwg	
Ramdimm.dwg	Pentiumii450.dwg	
24gbhd.dwg	Pentiumiii700.dwg	
Mainboard.dwg		

LOCAL AND EXTERNAL PARTS IN AN ASSEMBLY

An assembly can be made up of local and external parts. A local part has a definition that resides in the assembly drawing—think AutoCAD Blocks. An external part has a parent definition that resides on the hard disk—think AutoCAD XREFs. There are virtually no restrictions to this capability. In a master assembly, local parts may be combined with external subassemblies which are composed of local parts in their drawing (dwg) files. As previous chapters showed, the external parts can be edited *in-place* (within the assembly drawing) without the need to open the external drawing file, make changes, save it, and update the assembly.

In this chapter, you will see that an assembly can contain subassemblies— themselves either being local or external. It can be seen that a fairly simple assembly can become confusing after a while. Please keep in mind that all of the components of the assembly must be constrained to one another, even the subassemblies. This was illustrated in *Chapter 15,* which dealt primarily with constraining parts into an assembly.

Most assemblies done in the real world are based almost entirely on external parts. This is done because the individual parts are usually part of a corporate database and must remain external to the assembly. Maintaining an assembly composed of external parts is easier than one composed of local parts. Not only is an assembly made up of external parts much smaller in file size than one made up of local parts, you are also less likely to lose everything should the assembly become corrupted. Another advantage to an assembly of external parts is that a group of designers can work on the assembly. Each member of the group is responsible for a certain part or subassembly. The project manager is responsible for the assembly and can verify each person's design as it is brought into the assembly. As work progresses, the different parts will change over time and the project manager can update the assembly to incorporate the revisions.

In previous releases of MDT, you had to revise the external models by opening them, editing the geometry, and saving and reopening the assembly model to update the assembly. It is still possible to work this way, but you could keep the master assembly open in MDT multiple design environment (MDE; Windows MDI), and when the parts are edited you could simply reload from the catalog to read in the revised parts. Furthermore, the assembly constraints are not disturbed. The best way, of course, in MDT 5 is simply to "edit-in-place" the external part. When closing the main assembly, you will have the choice of saving the changes back to the original external part file. The system variable **AMXASSEMBLE** has control over this. Look up the system variable in the MDT **Help** and learn about its purpose.

Remember, an assembly is started with the AutoCAD **NEW** command. Local parts are created in the assembly drawing using the methods you have learned previously. External parts are those that have been created previously and are imported, or *instanced*, via MDT's catalog. The external parts are actually AutoCAD XREFs and use the same technology. Please do not use any AutoCAD commands to manage these XREFs—only use MDT's catalog. Once all the parts are either created or imported, you start constraining them together in roughly the same order as they are presented in the **Browser**. Remember, the **Browser** lists the parts in a historical fashion—the first part is listed first, followed by the second, and so on. Keep in mind that you can reorder the list in the **Browser** by simply dragging and dropping a part into a new place in the list. This can even be done after the parts have been constrained into the assembly. This is very useful for organizing your list of parts in a coherent manner. Perhaps the assembly order was not followed when parts were imported. Now is a good time to reorder logically. The new order should be representative of the actual assembly in reality. Pay special attention to the first part in the assembly because it is crucial. This part is special because all others will trace their assembly constraints to it. It is the only part in a constrained assembly that is considered *grounded*. It has no constraints on it and can be moved freely about the screen. All other parts that are constrained to it will follow its movement.

UPGRADING A PENTIUM COMPUTER

In this chapter, you will be upgrading a model of a Pentium computer. You will start out with an assembly that is composed of a few local parts, a local subassembly, external parts, and an external subassembly. These will already be constrained together. You will be updating the components by *replacing* them with other higher performance parts. This upgrading is something we are all familiar with in real life. As usual we need more disk space and power for our computer. So, we will update the hard drives and processor for the computer.

Open the drawing PentiumComputer.dwg. See **Figure 17-1.** Examine what is in the drawing already. Take some time and review the structure of the assembly. Look through the **Browser** and notice the different icons used for local and external parts/subassemblies. Notice that the CASE_1 and COVER_1 have been made not visible.

Use the **AMCATALOG** command and inspect the **All** tab in the dialog box. See **Figure 17-2.** In both panes there are parts listed. These are the parts that make up the assembly. Remember, this is where you manage part definitions, which are the source for the part instances listed in the **Browser**. Notice the subassemblies have pluses next to their names; click on them to expand the listing to include their individual parts. This is where you can *localize* or *externalize* a part. Right-click on any external part and examine the options. Right-click on any local part and examine the options. Compare the differences between the menus presented for external parts as opposed to that for local parts. See **Figure 17-3.**

The Instance option allows you to place another copy of the part into an assembly—it would then have to be constrained in turn. Try this now. Place a few parts into the assembly. Notice the **Browser** being populated by the new instances. The part names are added with sequential numbering. When you are done, press [Enter], and then press the **Cancel** button.

Figure 17-1.
The Pentium computer with covers removed and the **Browser** showing the assembly.

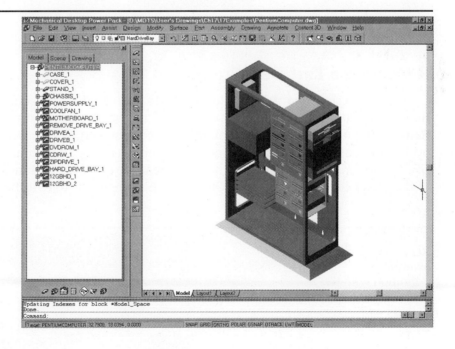

Figure 17-2.
The **Assembly Catalog** dialog box showing the external and local parts.

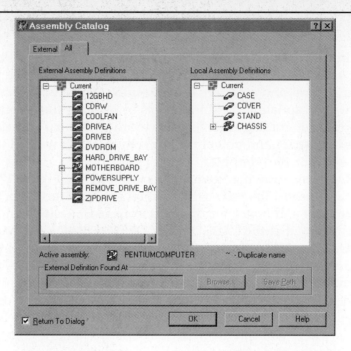

Figure 17-3.
A—The pop-up menu for an external part.
B—The pop-up menu for a local part.

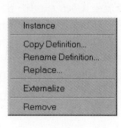

A

B

Renaming a Part Definition

Again access the **AMCATALOG** command. Right-click on DRIVEA and pick **Rename Definition...** in the pop-up menu. This option lets you rename the part. The **Rename Definition** dialog box should be on your screen. See **Figure 17-4.** You have the option of renaming all or none of the instances of a part. Rename the definition of DRIVEA to FLOPPYDRIVEA and choose to rename all of the instances. Now rename all of the instances of DRIVEB to FLOPPYDRIVEB. Press **OK** to exit the dialog box.

Figure 17-4.
Renaming an external part definition with the **Part Definition** dialog box.

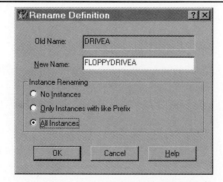

> **NOTE** Renaming the definition does not rename the part file on the hard drive. It merely renames the definition name in MDT. It is still Drivea.dwg on the hard drive.

Replacing the Hard Drive

The hard drive (12GBHD) needs to be upgraded to a larger format (24GBHD). To do this, access the **AMCATALOG** command again, right-click on the 12GBHD, and choose **Replace...** in the pop-up menu. With this option one part in the assembly can be replaced with another. You will notice that 24GBHD is not listed. The part file needs to be loaded first. Press **Cancel** to get back to the **Assembly Catalog** dialog box.

Loading a part file

With the **Assembly Catalog** dialog box still open, access the **External** tab. See **Figure 17-5A.** Now right-click in the top-right pane and select **Add Directory...** from the pop-up menu. This accesses the **Browse for Folder** dialog box. Navigate to where the files for this chapter are located, choose that directory, and pick **OK**. In the left pane is a list of external parts. The parts that are attached to the assembly are banded in white, and those that are not attached to the assembly are banded in dark gray. See **Figure 17-5B.** Find 24GBHD in the listing, right-click on its name, choose **Attach...**, place one instance in the assembly, and press [Enter] to return to the dialog box.

Now that the part is loaded, go back to the **All** tab. Find 12GBHD and right-click on it. Choose **Replace...** to access the **Replace Definition** dialog box. From the drop-down list pick 24GBHD, and then pick **OK**. Pick **OK** to exit the **Assembly Catalog** dialog box. An alert will pop up asking whether you want to delete the part definition for 12GBHD. See **Figure 17-6.** Pick **Yes** because it is not needed.

In the **Browser**, 24GBHD should be listed three times. One is the instanced one and the other two replace the two 12GBHD that were in the assembly. Right-click on 24GBHD_1 and choose **Delete** to remove it. Furthermore, if you expand the listing for the two 24GBHD listings in the **Browser**, you will notice that there are no assembly

---

Figure 17-5.
A—The **Assembly Catalog** dialog box showing the renamed definitions. B—The **Assembly Catalog** dialog box showing the new directory and the external parts.

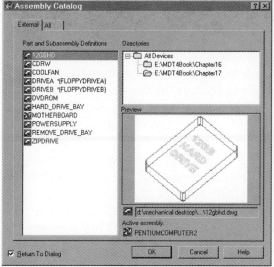

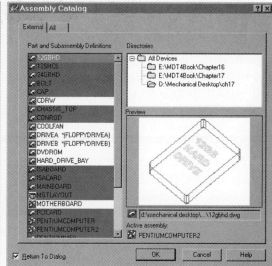

A   B

Figure 17-6.
You are notified that
the old part
definition was
deleted.

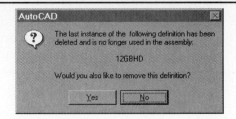

constraints on the hard drives. This is because when you replace a part with another, MDT is unable to reapply the constraints. How could it? A part could have been replaced with one that is a different shape and size—unlike our example where one drive looks very similar to another. When you replace a part in an assembly with another, it will be necessary to reconstrain the new part into the assembly.

Upgrading the Processor

Next we will upgrade the processor for more power for our demanding applications, such as MDT. Because the processor is installed on the external subassembly MOTHERBOARD, we have to open that subassembly and replace the processor there.

Opening a Subassembly

To open MOTHERBOARD subassembly, right-click on it in the **Browser** and pick **Open to Edit** from the pop-up menu. This opens the MOTHERBOARD assembly file. Access the **Assembly Catalog** dialog box and access the **All** tab. Notice that all the parts are local. Go to the **External** tab and navigate to where the files for this chapter are located. Locate the PENTIUMIII700 part file, right-click on it, and pick **Attach...** from the pop-up menu. Place the part anywhere in the assembly and press [Enter] to return to **Assembly Catalog** dialog box. Go to the **All** tab and right-click on the PENTIUMIII700 listing and choose **Localize**. Now right-click on the PENTIUMIII450 listing, choose **Replace...**, and pick the PENTIUMIII700 from the drop-down list in the **Replace Definition** dialog box. Pick **OK** twice and you will be prompted to delete the PENTIUMIII450 part definition from the assembly. Pick **Yes** to do so.

Zoom in on the processor and notice that it has a little more detail than the old one had. Be sure to delete the stray first instance of the processor (PENTIUM700_1). Again notice that you will have to reapply the assembly constraints holding the processor in place. Save the **MOTHERBOARD** assembly file and close it. Go back to the computer assembly and prepare to update the assembly.

Updating the Assembly

AMAUDIT

Assembly
↳ Assembly
Audit...

Assembly Modeling
toolbar

Audit External Refs

To update the assembly, first use the **AMAUDIT** command to determine if there are outstanding files that need to be updated. To access this command, enter AMAUDIT at the Command: prompt, select **Audit...** from the **Assembly** cascading menu in **Assembly** pull-down menu, or pick the **Audit External Refs** button on the **Assembly Modeling** toolbar. This accesses the **Check External References** dialog box. See **Figure 17-7**. Notice that the Motherboard.dwg file was changed and needs to be updated. Press **OK** to exit the dialog box.

Next you will actually update the external subassembly in our main assembly. To update the MOTHERBOARD subassembly in our main assembly, use the **AMREFRESH** command. To access this command, enter AMREFRESH at the Command: prompt, select **Refresh...** from the **Assembly** cascading menu in **Assembly** pull-down menu, or pick the **Update External Refs** on the **Assembly Modeling** toolbar. This accesses the **Refresh External Part/Subassembly** dialog box. See **Figure 17-8**. Make sure the Motherboard.dwg file is highlighted and press **OK**. After MDT chews on this for a while, a new processor should be in the computer assembly.

Figure 17-7.
The **Check External Reference** dialog box is used to see which files needed to be updated.

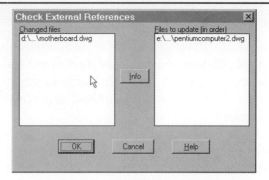

Figure 17-8.
The **Refresh External Part/Subassembly** dialog box is used to update files.

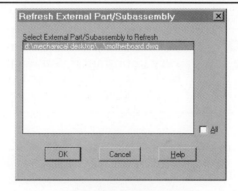

AMREFRESH

Assembly
➥ Assembly
Refresh...

Assembly Modeling
toolbar

Update External Refs

The MOTHERBOARD subassembly could also have been updated from the **Assembly Catalog** dialog box. To do this, right-click on its name in the **All** tab, choose the **Reload** option, and pick **OK** to exit the dialog box. Do this now by undoing the previous **AMREFRESH** command. Then use the **Assembly Catalog** dialog box method. There is no functional difference between the two methods. With both methods, our new computer should now be able to handle Mechanical Desktop Release 5.

Additional Commands for Managing Parts

There are additional commands used to manage your parts in an assembly—whether they are local or external. For external parts, in the **Assembly Catalog** dialog box, there is an option to **Unload** the part file. This is similar to AutoCAD unload of an XREF. It will remove it from the assembly model's display, but not the part definition from the database. In order to remove the part definition of an external file, the **Detach** option must be used.

There are two more commands of note. These commands are **AMCOPYIN** and **AMCOPYOUT**. The **AMCOPYIN** command not only brings in a part definition from the hard drive, it also localizes it for your troubles. This command is accessed by entering AMCOPYIN at the Command: prompt, selecting **Copy In...** from the **Assembly** cascading menu in **Assembly** pull-down menu, or pick the **Input Part Definition** on the **Assembly Modeling** toolbar.

The **AMCOPYOUT** command is used to externalize local parts. This command is accessed by entering AMCOPYOUT at the Command: prompt, selecting **Copy Out...** from the **Assembly** cascading menu in **Assembly** pull-down menu, or pick the **Output Part Definition** on the **Assembly Modeling** toolbar.

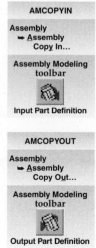

AMCOPYIN

Assembly
➥ Assembly
Copy In...

Assembly Modeling
toolbar

Input Part Definition

AMCOPYOUT

Assembly
➥ Assembly
Copy Out...

Assembly Modeling
toolbar

Output Part Definition

CREATING SUBASSEMBLIES

Creating local subassemblies in a main assembly is very easy. To create an externally referenced subassembly, simply import via the **Assembly Catalog** dialog box. In this example, you initiate a new *local* subassembly, import parts into it, localize the parts, constrain the parts to one another, and constrain the entire subassembly into our main assembly.

Open the drawing Example-17-02.dwg. See **Figure 17-9.** Examine the parts and the assembly constraints already there. In the **Browser**, expand PISTON_1 and SHAFT_1 and notice that they have been constrained together using the midpoint method that was discussed in *Chapter 15*. This assembly constraint method is very important if you want to center parts in other parts.

<table>
<tr>
<td>CAUTION
</td>
<td>The midpoint constraint method, or coincident alignment, is very important if you want to center parts in other parts. It is best that you learn how to do it. If you did not already, please go to Chapter 15 and do so now.</td>
</tr>
</table>

Figure 17-9.
The piston assembly and the **Browser** showing the local parts it is made of.

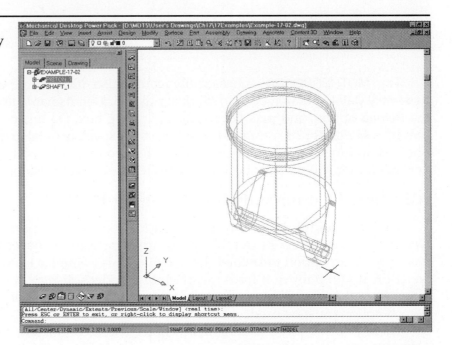

Learning Mechanical Desktop R5

First right-click in the background of the **Browser** and choose **New Subassembly** from the pop-up menu. When prompted for a new subassembly name, enter CONRODASSY. Notice the entry is now shown in the **Browser**. Double-click on CONRODASSY and notice that everything but it is grayed out. Also, notice that the piston and shaft are no longer on screen. This is because you have entered into the subassembly definition. This means that all part instances outside of this definition are not visible until you are done and have backed out. Access the **Assembly Catalog** dialog box. In the **External** tab, add the directory where the parts that will make up the **CONRODASSY** subassembly are located. The parts are the CONROD, CAP, and BOLT. In this order, right-click on each and pick **Attach**. Bring in one instance of the CONROD and CAP. Then bring in two instances of the BOLT.

Now use the **All** tab to make each one of them local. In the order they were instanced, right-click on each and pick **Localize**. Expand CONRODASSY in the top-right pane. Notice that these parts are now part of the subassembly.

Now begin to constrain the pieces together. CONROD_1 is the grounded part for this subassembly. All other parts constrain to it. Constrain the CAP_1 part to the CONROD_1 part using two **Insert** assembly constraints on the bolt holes. Constrain BOLT_1 and BOLT_2 to the CAP part with **Insert** assembly constraints. You have just created a local subassembly. See **Figure 17-10.** The parts of a subassembly do not need to be fully constrained to one another before constraining the subassembly into the main assembly.

PROFESSIONAL TIP
It is often helpful to move one part away from another after the first assembly constraint has been applied. This will facilitate placing subsequent assembly constraints. You may have to do this several times in the course of constraining a part.

Figure 17-10.
The constrained connecting rod subassembly.

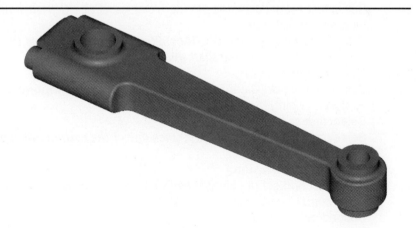

Figure 17-11.
The finished piston
assembly with the
connecting rod
subassembly.

To get back to the main assembly, double-click on the Exercise 17-02 at the top of the **Browser**. The piston assembly should reappear on your screen. Now the connecting rod subassembly needs to be tied to the main piston assembly. First, make the PISTON_1 part invisible by using the **Browser**. Use the **Mate** assembly constraint to tie the axis of the hole in CONROD_1 (there is only one hole) to the axis of the SHAFT_1. Again use the **Mate** assembly constraint to tie the center of the hole in CONROD_1 to the center of the SHAFT_1. The command sequence for this is as follows:

> Command: **AMMATE**
> Select first set of geometry: *(pick one of the cylindrical edges on the shaft)*
> First set = Axis, (arc)
> Select first set or [Clear/fAce/Point/cYcle] <accEpt>: *(left-click until you get the point)*
> First set = Point, (arc)
> Select first set or [Clear/aXis/fAce/cYcle] <accEpt>: *(move the cursor over the other cylindrical edge on the shaft and left-click)*
> First set = Axis, (arc)
> Select first set or [Clear/Midpoint] <accEpt>: **M** *or* **Midpoint.⏎**
> First set = Point, (arc)
> Select first set or [Clear/aXis] <accEpt>: ⏎
> Select second set of geometry: *(pick one of the cylindrical edges on the boss or hole on connecting rod)*
> Second set = Axis, (arc)
> Select second set or [Clear/fAce/Point/cYcle] <accEpt>: *(left-click until you get the point)*
> Second set = Point, (arc)
> Select second set or [Clear/aXis/fAce/cYcle] <accEpt>: *(move the cursor over the other mirrored edge and left-click)*
> Second set = Axis, (arc)
> Select second set or [Clear/Midpoint] <accEpt>: **M** *or* **Midpoint.⏎**
> Second set = Point, (arc)
> Select second set or [Clear/aXis] <accEpt>: ⏎
> Enter offset <0.0000>: ⏎

Make the PISTON_1 visible using the **Browser**. The connecting rod subassembly should be centered and constrained to the main piston assembly. See **Figure 17-11.** That is all there is. You made a subassembly and tied it into a main assembly. Please save your work under a new name.

Chapter Test

Do not write in this text. Answer the following questions on a separate sheet of paper.

1. What are the advantages of using external part files in an assembly? The disadvantages? Why will you encounter assembly files made up almost entirely of external part/subassemblies in industry?

2. When you right-click on an external part file in the **All** tab in the **Assembly Catalog** dialog box, you get a menu for managing parts. What is the difference between **Unload** and **Detach** options?

3. Is the **Detach** option for an external part similar to the **Remove** option for a local part?

4. How does the **AMCOPYOUT** command differ from choosing **Externalize** from the pop-up menu when right-clicking on a local part name in the **All** tab of the **Assembly Catalog** dialog box?

5. What are the steps to beginning a local subassembly? What are the steps for importing an external subassembly?

6. Is it necessary to constrain *all* the parts in a subassembly first before constraining the subassembly into the main assembly?

7. Is there any difference in the assembly constraints used for a local subassembly and those used for an external subassembly?

8. How deep into the nesting can you go when you edit external parts? For example, can you edit the processor on the motherboard in the main modeling example of this chapter? Can you edit its sketch profile? How do you think AutoCAD/MDT achieves this?

9. What happens to a part's paperspace layout when you try to externalize a part?

10. What happens to an external part's paperspace layout when the part is edited in an assembly and the changes are written out to the original external part file?

Chapter Exercises

Exercise 17-1. *Externalizing Local Parts and Paperspace Layouts.* Open Exercise-17-01.dwg as shown below. Notice that it is an assembly file and that it has a paperspace layout in the **Drawing** tab. There is only one local part in the drawing. The paperspace layout is based on this part. Use the **Assembly Catalog** dialog box to externalize the part. What does the error message read? There is not any way of externalizing this part and not losing the paperspace layout. Why do you suppose there is this deficiency in the software? Is it a deficiency?

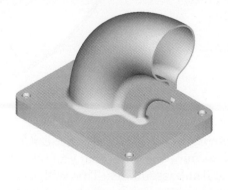

Exercise 17-2.   *Editing, Externalizing, and Paperspace Layouts.* Open Exercise-17-02.dwg as shown below. Notice that it uses the part from the previous exercise in an assembly as an external part. Please open Exercise-17-02a.dwg and notice that it has a paperspace layout in it. Close Exercise-17-02a.dwg at this point. Going back to our assembly, edit the part (Exercise-17-02a) in question and write the changes back out to the original part file. Double-click on the part in the **Browser**—all others should be grayed out in the **Browser**. Double-click on the RevolutionAngle1 feature and bypass the opening dialog box. Find the dimension controlling the diameter of the revolved pipe section. It should be 6.00; change it to 8.00. Afterwards, update the assembly with the **Update Assembly** button below the **Browser**. You will be prompted to see if you would like to save the changes. Go ahead and save the changes. Go back and check to see if Exercise-17-02a.dwg still has a paperspace layout. Open the drawing file and you will notice that there is an alert to let you know that you can only open it in *read-only* mode. Cancel out of this whole operation. In the **Browser**, notice that there is a tiny red padlock on the name EXERCISE-17-02A_1. Right-click on the part name and pick **Purge Lock** from the pop-up menu. Read and answer the alert box. Notice that there is a significant message about your assembly. Now, try to open Exercise-17-02a.dwg. There should be no problem now. Switch over to the **Drawing** mode and notice that there is indeed a paperspace layout intact, but also that it has updated automatically. The pipe section diameter should read 8.00. This answered an important question. Does the paperspace layout survive the writing out of an external part from an assembly?

Why is the paperspace layout preserved in this case and not in Exercise-17-01.dwg? Go ahead and make the same change to the Exercise-17-02b.dwg part and write out the changes. This part file does not have a paperspace layout.

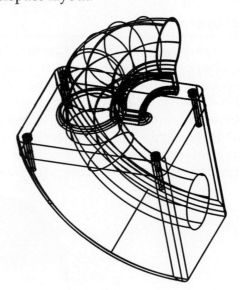

Exercise 17-3.   *Internalizing from Nested Subassemblies.* Open the file PentiumComputer.dwg. Determine the order that you will use to proceed to internalize all parts in all subassemblies. This will, of course, consist of several steps. You will have to start with the subassembly files first.

CHAPTER 18

Assembly Drawings, Balloons, Parts Lists, and Bill of Materials

Objectives

After completing this chapter, you will be able to:

* Create assembly views in the layouts.
* Understand the use of the bill of materials database.
* Edit the bill of materials database.
* Input a parts list into a layout.
* Add and edit part references.
* Edit the format of balloons.
* Place balloons on a layout.

Drawings Needed

*The following is a list of files that you will need to work through this chapter. These files can be found on the **User's Drawings** CD included with this text.*

Examples	Exercises
Example-18-01.dwg	Exercise-18-01.dwg
Example-18-02.dwg	Exercise-18-02.dwg
Example-18-03.dwg	Exercise-18-03.dwg
Example-18-04.dwg	
Example-18-05.dwg	
Example-18-06.dwg	
Example-18-07.dwg	

WHAT WE ARE WORKING TOWARD

The assembly shown in the **Layout** view in **Figure 18-1** has six parts. The first part is the rod and the last is the thumbscrew. There are three views showing all six parts in this adjustable clamp. Each part is labeled with a *balloon*, which is a circle containing a part number and each balloon is attached to its corresponding part with a leader. The part numbers are referenced in a *parts list* in the upper-right corner of the drawing. The list has a controlled number of columns and contains information based on the *bill of materials* database. The database has two kinds of information—*default* *information* taken from the **Assembly Catalog**, such as the part name, and *user-defined data*, such as the vendor name or the part cost.

The parts list is parametric in that changes made to parts, such as changing the name in the **Assembly Catalog** automatically changes the list. Creating the views, adding balloons, and inserting the parts list is easy. Controlling the appearance of them and the information they contain is more complex mostly because of the number of dialog boxes and detail in each one. Because of this complexity, we will start with a very simple example.

CREATING ASSEMBLY VIEWS

Assembly views are created much the same as the part views that you created in *Chapter 4*. For example, open part Example-18-01.dwg and switch to the **Assembly-views** drawing tab that shows a blank ISO A2 sheet (594 × 420mm). Issue the **AMDWGVIEW** command and input the information in the dialog box as shown in **Figure 18-2A.** Pick **Select** in the **Data Set:** drop-down list, pick **Assembly-views** in the **Layout:** drop-down list, set the scale to 1.5, and uncheck **Display Hidden Lines**. Click **OK**, enter ALL when prompted to select objects, and then press [Enter]. Enter Y to select worldYz as the viewing plane. Enter Y again to select worldY as the axis direction, and press [Enter] to select the orientation. Locate the view at the approximate location shown in **Figure 18-1.**

Now create a top orthographic view and a 0.8 relative scale isometric (from the front view). You can hide all the automatically created centerlines with the **AMDT_DRAWING_VISIBILITY** command. To access this command, select **Drawing Visibility** in the **Annotate** pull-down menu or the right-click menu, or pick the **Drawing Visibility** button on the **Drawing Layout** toolbar. Check **Centerlines** as shown in **Figure 18-2B** and click on **OK**.

Figure 18-1.
The assembly used to illustrate how to create assembly drawings.

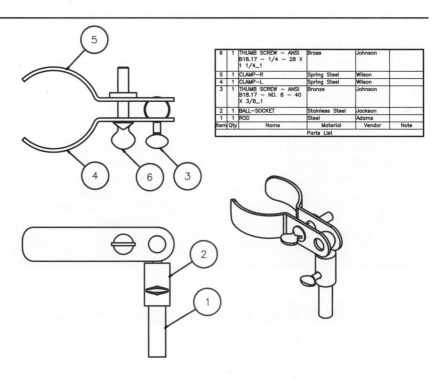

Item	Qty	Name	Material	Vendor	Note
6	1	THUMB SCREW — ANSI B18.17 — 1/4 — 28 X 1 1/4_1	Brass	Johnson	
5	1	CLAMP–R	Spring Steel	Wilson	
4	1	CLAMP–L	Spring Steel	Wilson	
3	1	THUMB SCREW — ANSI B18.17 — NO. 8 — 40 X 3/8_1	Bronze	Johnson	
2	1	BALL–SOCKET	Stainless Steel	Jackson	
1	1	ROD	Steel	Adams	

Parts List

Figure 18-2.
A—The **Create Drawing View** dialog box showing proper values. B—The Desktop Visibility can be used to hide objects.

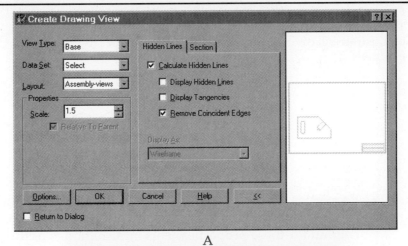

A

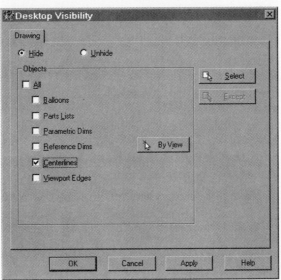

B

THE BILL OF MATERIALS DATABASE

All of the information shown in the *parts list* in **Figure 18-1** is controlled by the bill of materials (BOM) database. This database is made up of columns and rows, the rows being the parts and subassemblies, and the columns the particular piece of information. The information in the database can be output either in BOM Table, which can be displayed on the screen or output to a file, or in the Parts List on the layout. Both of these can contain different sets of some or all of the information in the database. The BOM Table can be exported in a variety of formats including Excel and Access. The Parts List can be inserted into one or more layouts and can also be exported.

The database can contain columns for four types of information about the parts and subassemblies in the assembly. For parts these are (and it is the same for subassemblies):

- **Default.** Information that is automatically included when the database is created. There are three of these and each one is a column in the database:
  - **Part Item Number**—The order number that the part was created on inserted and the number that appears in the default balloon. Moving a part in the **Browser** does not change its **Item** number and changing an **Item** number in the database does not change order in the **Browser**.
  - **Qty (quantity)**—The number of instances of the part.
  - **Part Name**—The name in the catalog. Note that the name in the **Browser** can be edited independent of the catalog name.

Figure 18-3.
The assembly used to illustrate bill of materials.

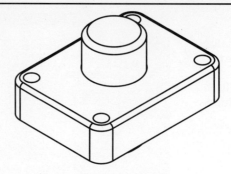

- **Properties.** Information that MDT already knows about the part, such as its volume, but that is not included in the default database. You must add the column for these properties. This information is of three types:
  - **Drawing Properties**—Such as the drawing title.
  - **BOM Attributes**—Such as the part name.
  - **Part Data**—Such as the surface area or material.
- **User Defined.** Added column where information, such as the name of the vendor who supplies the part, is input.
- **Calculations.** Added column for formulas that can include any of the previously defined information and any design variable. For example, the weight of all the instances of a particular bolt would be =QTY*PART:MASS, where the quantity is the number of instances and the PART:MASS is mass or weight of one bolt.

AMBOM

Annotate
➡ Parts List
↳ BOM
   Database

Drawing Layout
toolbar

Edit BOM Database

Open drawing Example-18-02.dwg showing Layout1 as in **Figure 18-3.** The bill of materials (BOM) table is displayed using the **AMBOM** command. To access this command, enter AMBOM at the Command: prompt, select **BOM Database** from the **Parts List** cascading menu in the **Annotate** pull-down menu or the right-click menu, or pick the **Edit BOM Database** button on the **Drawing Layout** toolbar. The first time you access this command in a .dwg file, the database is created. After that the command line will ask if you want to delete or edit the database.

This command accesses the **BOM** dialog box displaying the BOM table. See **Figure 18-4.** Remember that this table is a spreadsheet array of rows and columns representing a selected portion of the information in the database. We will come back to the **BOM** dialog box, but first click on the **Properties** button in the lower-left corner. This displays the **BOM Properties for ANSI** dialog box. See **Figure 18-5.** In this spreadsheet, each *row* represents the format and information for a *column* in the BOM table. There are 14 rows in the default and any of these, except the first three (**Item, Qty**, and **Name**) that are required, and can be deleted. New ones can also be added at the bottom of the list. The words shown in red under the heading "Column" are the "internal names" and cannot be changed for the 14 defaults. Information shown in black can be changed by double-clicking in the cell.

Highlight the row for **Mass** by clicking on the box on the far left side and make the changes as shown in **Figure 18-6** and listed below:
- In the **Format Columns** area, select the **Visible in BOM** check box.
- Select the **Lock Editing** check box so that the format cannot be changed outside this dialog box.
- Set both the **Caption Alignment** and **Data Alignment** to center.
- Set **Data Type** to Numeric and its precision to 2 places.
- Set the **Width** (of the column) to 1.0.
- Then click in the caption cell and change the caption to WEIGHT.
- Do not change the formula for calculating the mass.
- Click on the far right column under the summation sign (Σ), which will calculate a total weight of all the parts in the BOM table.

Figure 18-4.
The bill of materials database is shown in the **BOM** dialog box.

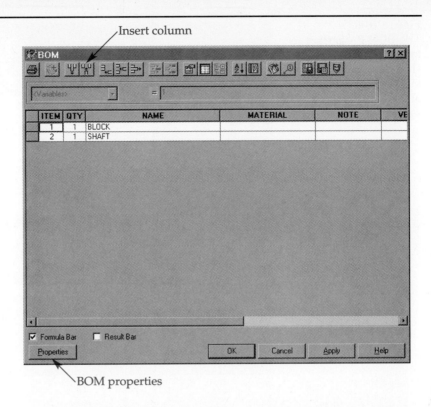

Figure 18-5.
The **BOM Properties for ANSI** dialog box is used to insert columns and define properties of the new column.

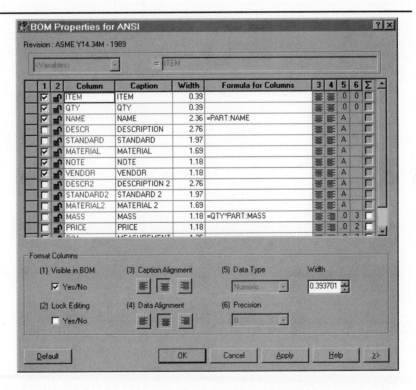

Once all changes have been made, pick **OK** to return to the **BOM** dialog box. Note that the Weight column has been inserted and the total weight of both parts is displayed on the summation row along the bottom. See **Figure 18-7.**

> **NOTE**
> There are 18 buttons across the top of this dialog box. A brief summary of some of their functions is shown in **Figure 18-8.**

Figure 18-6.
The **BOM Properties for ANSI** dialog box showing the new MASS/WEIGHT row and the proper settings.

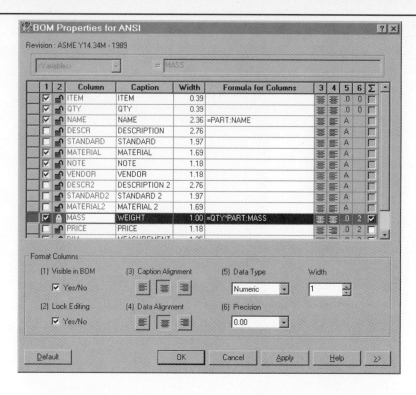

Figure 18-7.
The **BOM** dialog box showing the new WEIGHT column and total weight.

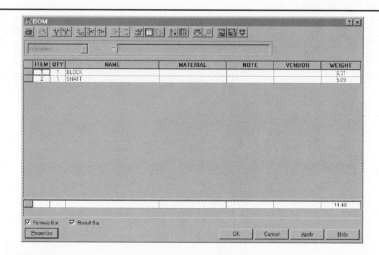

ASSIGNING MATERIALS TO THE PARTS

While the **BOM** dialog box is still open, notice the Material column. This column is related to data in the **AMASSMPROP** command. If you analyze the mass properties of a part or assembly with this command and assign a material to a part, then that material name will display in the **BOM** dialog box. In *Chapter 16* you learned how to assign units to an assembly and material types to the parts and find the geometric properties, such as the mass, with the **AMASSMPROP** command. Pick **OK** in the **BOM** dialog box and then issue the **AMASSMPROP** command. Before issuing this command, make sure you are in **Model Mode**.

Once the **AMASSMPROP** command is issued, press [Enter] for Select and pick both of the parts. This accesses the **Assembly Mass Properties** dialog box. See **Figure 18-9A**. Change the **Assembly units** to In (inches) and the Mass units to Lb (pounds). Click the **Material...** button to access the **Select Material** dialog box. See **Figure 18-9B**. Assign mild steel to the shaft by highlighting SHAFT in the **Part/Subassembly definition** area

Figure 18-8.
The buttons at the top of the **BOM** dialog box.

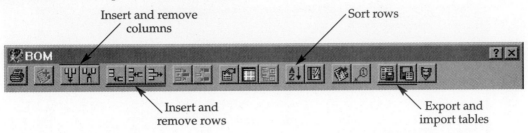

Figure 18-9.
A—The **Assembly Mass Properties** dialog box is used to change mass properties.
B—The **Select Material** dialog box is used to change materials.

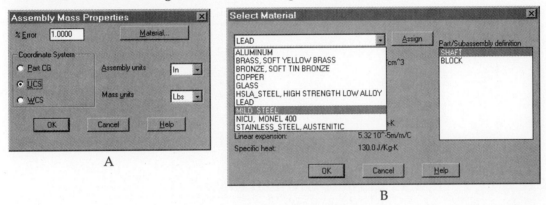

and MILD STEEL in the drop-down list and picking the **Assign** button. Then assign soft yellow brass to the block. Pick **OK** twice and then pick **Done** in the **Assembly Mass Properties Results** dialog box. Now return to the **BOM** dialog box and note that the materials are listed and the mass (weight) has changed.

Additional materials can be added to the **AMASSMPROP** command by using WordPad to open the mcad file in the Program Files\MDT5\Desktop\Support directory. Material names can be added to the end of the list along with their properties. If all you are interested in is the mass of the part, then the only property that needs to be accurate is first one, the density in grams per cubic centimeter, and all the others can be set to 1.0. See **Figure 18-10** for the addition of Nylon. If you are going to apply other analyses to the assembly, such as finite element analysis, then the other properties have to be correct. You can find these values at websites such as WWW.MATWEB.COM, which lists the properties of more than 21,000 materials.

Figure 18-10.
Materials can be added by editing the mcad file in Wordpad.

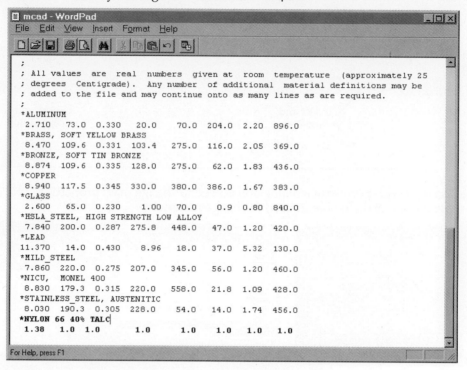

```
; mcad - WordPad                                                    _ □ ×
File  Edit  View  Insert  Format  Help

  D  ☞ 🖫  🖨 🔍  🏛  ✂ 🗐 🗎 ↶  🗏

;
; All values are real numbers given at room temperature (approximately 25
; degrees Centigrade).  Any number of additional material definitions may be
; added to the file and may continue onto as many lines as are required.
;
*ALUMINUM
 2.710   73.0  0.330    20.0     70.0  204.0  2.20  896.0
*BRASS, SOFT YELLOW BRASS
 8.470  109.6  0.331   103.4    275.0  116.0  2.05  369.0
*BRONZE, SOFT TIN BRONZE
 8.874  109.6  0.335   128.0    275.0   62.0  1.83  436.0
*COPPER
 8.940  117.5  0.345   330.0    380.0  386.0  1.67  383.0
*GLASS
 2.600   65.0  0.230     1.00    70.0    0.9  0.80  840.0
*HSLA_STEEL, HIGH STRENGTH LOW ALLOY
 7.840  200.0  0.287   275.8    448.0   47.0  1.20  420.0
*LEAD
11.370   14.0  0.430     8.96    18.0   37.0  5.32  130.0
*MILD_STEEL
 7.860  220.0  0.275   207.0    345.0   56.0  1.20  460.0
*NICU,  MONEL 400
 8.830  179.3  0.315   220.0    558.0   21.8  1.09  428.0
*STAINLESS_STEEL, AUSTENITIC
 8.030  190.3  0.305   228.0     54.0   14.0  1.74  456.0
*NYLON 66 40% TALC
 1.38    1.0   1.0       1.0      1.0    1.0  1.0   1.0

For Help, press F1
```

Creating Formulas in the BOM Database

Formulas can be written in the BOM database using variables from the drawing properties, BOM attributes, and part data lists and any active part or global design variable. Open part Example-18-03.dwg and access the **BOM** dialog box. The **Item** numbers were changed and then sorted. Materials were added with the **AMASSMPROP** command, and some notes and vendor names were added in the database. However, because this drawing was originally created in Mechanical Desktop Release 4, there is no column for the weight or mass of the parts.

Access the **BOM Properties for ANSI** dialog box and click on the blank row at the bottom of the list. Input WEIGHT for both the column name and the caption and set up the format as shown in **Figure 18-11A.** Click in the cell for **Formula for Column** and then in the formula box at the top of dialog box. Type in =QTY*PART:MASS as the formula.

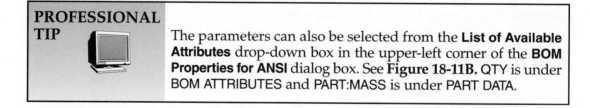

PROFESSIONAL TIP

The parameters can also be selected from the **List of Available Attributes** drop-down box in the upper-left corner of the **BOM Properties for ANSI** dialog box. See **Figure 18-11B.** QTY is under BOM ATTRIBUTES and PART:MASS is under PART DATA.

Click on **OK.** Notice the **WEIGHT** column has been added and the total weight has been calculated.

Figure 18-11.
A—The **BOM Properties for ANSI** dialog box showing the proper settings. B—The parameters can be entered using the **List of Available Attributes** drop-down box.

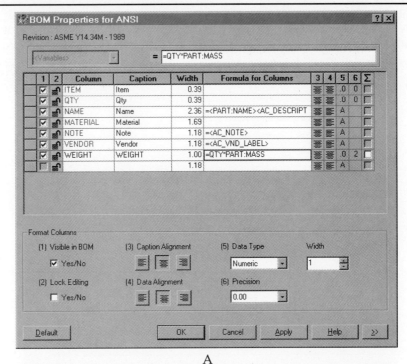

A

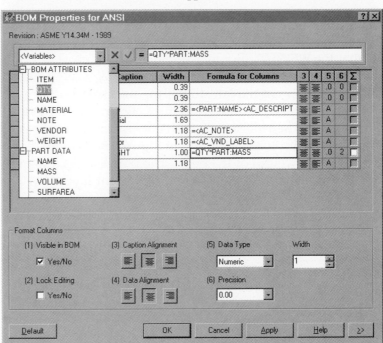

B

Inserting the Parts List and Controlling Its Appearance

Continue with part Example-18-03.dwg and access the **Parts List** dialog box with the **AMPARTLIST** command. Enter AMPARTLIST at the Command: prompt, select **Parts List** from the **Parts List** cascading menu in the **Annotate** pull-down menu, or pick the **Place Parts List** button on the **Drawing Layout** toolbar. To access this command with the **BOM** dialog box still open, simply pick the **Insert parts list** button. When there is a part list inserted in the layout, the **Parts List** dialog box can be displayed by double-clicking on part list or right-clicking Part List in the **Browser**.

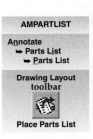

AMPARTLIST

Annotate
➥ Parts List
➥ Parts List

Drawing Layout toolbar

Place Parts List

Figure 18-12.
A— The **Parts List** dialog box showing that the WEIGHT column is not included and must be added manually. B—The **Select Column** dialog box is used to select which column to add.

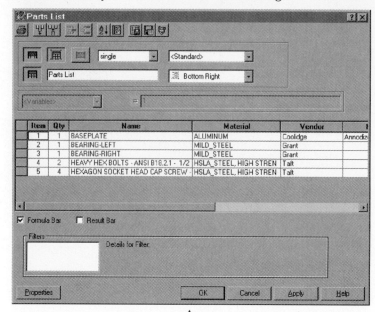

A

B

Issue the **AMPARTLIST** command and the **Parts List** dialog box. See **Figure 18-12A.** Note that the **WEIGHT** column is not included, it must be add manually. Pick the **Insert column** button. In the **Select Columns** dialog box, select WEIGHT. See **Figure 18-12B.** This inserts the column at the left side of the part list.

The column can be moved by first clicking on the **WEIGHT** and highlighting the entire column. Then click on **WEIGHT** again and drag the column to the desired position. The moved column is shown in **Figure 18-13.** Note that the results bar is checked listing the total weight of the parts. The part list has a name and this can be changed so you can place different part lists in different layouts.

Figure 18-13.
The **Parts List** dialog box showing the WEIGHT column in place.

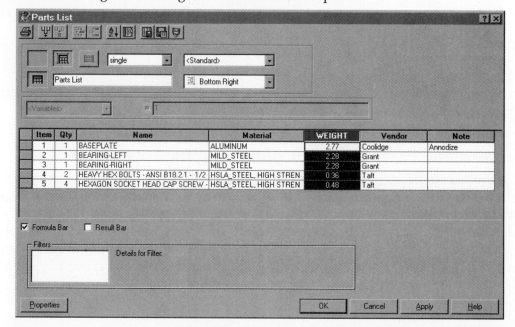

Figure 18-14.
The **List of filters** dialog box is used to add filters, which can control the parts and subassemblies listed in the part list. Notice the filters being used are shown and defined at the bottom of the **Parts List** dialog box.

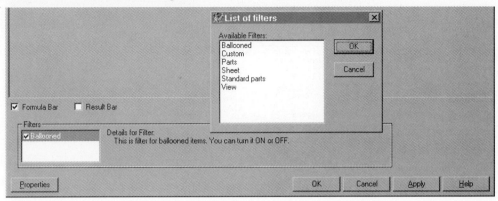

The parts and subassemblies listed in the part list can be controlled by the **Filters**. Right-click in the **Filters** box and pick **Add Filter** from the pop-up menu. Select the parts displayed from one of the six choices. **Figure 18-14** shows the choices and the filter box with the Ballooned items selected. With this choice, only ballooned items in the layout will be included in the part list.

When you pick **OK** in the **Parts List** dialog box, you are then asked to specify a location for the part list. Use your cursor to place the part list in the lower-right corner of the layout. Now Parts List appears in the **Browser**. The part list can be subsequently moved in the layout using grips.

Changing the Format of the Parts List

Access the **Parts List** dialog box and click the **Properties** button. This displays the **Parts List Properties for ANSI** dialog box. See **Figure 18-15**. This dialog box controls the color, line spacing, and text size for both the headings and the data. The **Custom Blocks** tab allows for the use of customized part lists using custom blocks. This process is beyond the scope of this textbook. Pick **Cancel** twice to close these dialog boxes.

Figure 18-15.
The **Parts List Properties for ANSI** dialog box is used to set and change the layout setting of the parts list.

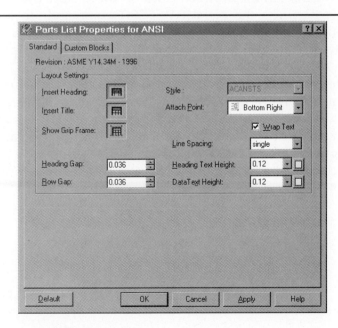

Figure 18-16.
The **Layer/Object**
dialog box is used
to set and change
the color, line type,
line weight, or layer.

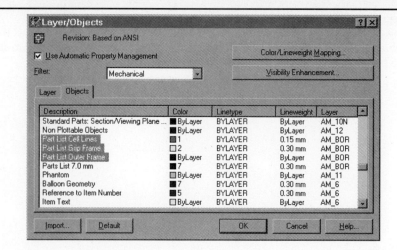

To set the colors and line weights for the borders and lines in the parts, select **Mechanical Options...** from the **Assist** pull-down menu, and then access the **Standards** tab in the dialog box. Double-click on Layer/Objects and access the **Objects** tab in the **Layer/Objects** dialog box. In this list you will find Part List Cell Lines, the Part List Grip Frame, and the Part List Outer Frame. See **Figure 18-16.** Please note that the line color and weight are mapped to one another, so changing one changes the other.

ADDING AND EDITING PART REFERENCES

Every MDT part in the file has a *part reference* that is automatically created and references the BOM database for that part. The database information for a single part can be displayed and edited using the Part Reference symbol. This can be very helpful in an assembly with many parts. In addition, part references can be created for non-MDT objects in the drawing, information added, and displayed in the database.

Open file Example-18-04.dwg. This assembly-in-progress has two MDT parts. One is the *base* and the other is the *insert*. The cap head screw is an AutoCAD solid. Review the BOM database with the **AMBOM** command or by double-clicking on the BOM called ORIGINAL BOM in the **Browser**. Click on **OK** to close the database. To edit the data for a single part, use the **AMPARTREFEDIT** command. To access this command, enter AMPARTREFEDIT at the Command: prompt, select **Part Reference Edit** from the **Parts List** cascading menu in the **Annotate** pull-down menu, or pick the **Part Reference Edit** button on the **Drawing Layout** toolbar.

This command displays the *part references*, which are the cyan circles with an X in them. See **Figure 18-17.** Note that the red cap head screw does not have a reference. Move your cursor to the part reference on the BASE and the name of the part is displayed. Click on the part reference and display the **Part Ref Attributes** dialog box. See **Figure 18-18.** The data can be edited, except for the part name BASE, and additional data can be added. Input ABC Company as the vendor.

AMPARTREFEDIT

Annotate
  ➥ **Parts List**
    ➥ **Part Reference**
       Edit

Drawing Layout
     toolbar

Part Reference Edit

Figure 18-17.
The assembly with
part references
displayed.

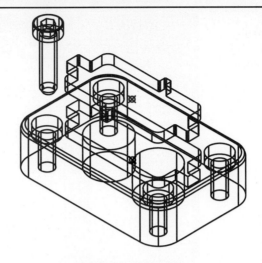

Figure 18-18.
Clicking on a part
reference displays
the **Part Ref
Attributes** dialog
box, which shows
the part's data.

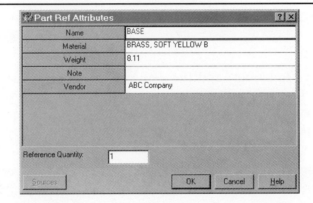

Adding a Reference

The **AMPARTREF** command is used to add a reference to the cap screw so that its
data is included in the BOM and parts list. To access this command, enter AMPARTREF
at the Command: prompt, select **Part Reference** from the **Parts List** cascading menu in the
Annotate pull-down menu, or pick the **Place Reference** button on the **Drawing Layout**
toolbar.

Pick a point on or near the screw and the **Part Ref Attributes** dialog box is displayed.
Enter the Cap Head Screw for the name, Steel as the material, and Jones & Jones as the
vendor. Use **AMBOM** to check the bill of materials and you will see that this information
is now in the database.

ADDING NEW PARTS TO A DRAWING AND UPDATING THE BOM

Open file Example-18-05.dwg. This assembly-in-progress has had the base modified, the screw converted into an MDT part, and assembly constraints have been
applied. To add another part to the assembly, go to the **Assembly Catalog**, right-click
on the Cap Head Screw and pick **Copy Definition...** in the pop-up menu. In the **Copy
Definition** dialog box, enter Long Cap Head as the new definition name and click on
OK. Place one instance of this new part on the drawing as shown in **Figure 18-19.** Its
name will appear in the **Browser** and if you double-click on the ORIGINAL BOM in the
Browser, the part will also appear in the BOM database.

Figure 18-19.
Browser showing
the new part. The
screen also shows
the one instance of
the new part.

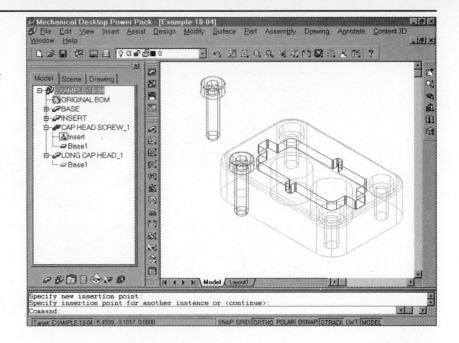

Move to Layout1 and the new part is not included in the drawing view. This is because the view was created with the Select option, which could not include a new part. The view must be deleted and recreated. A better procedure for creating assembly view called Scenes is covered in the next chapter.

ADDING AND FORMATTING BALLOONS IN THE DRAWING

By default, balloons are circles with leaders on the layout that identifies parts and subassemblies with an item number associated with the part list. The default circular balloon shape has a radius 2.5 times the standard text height but the size and the shape can both be changed. See balloons 2 and 3 in **Figure 18-20.** Balloons can contain other BOM database information besides the item number and balloon 1 shows the quantity (**QTY**) of 2 as the lower number. Balloons may have multiple leaders such as number 5, multiple segment leaders such as 3, or no leaders at all—called a *standalone*. Each balloon is related to a part reference and the leader can be attached to any point on the part. Of course the data in balloons is parametric, changes in the BOM database automatically change the balloon data.

Figure 18-20.
The various balloon
types.

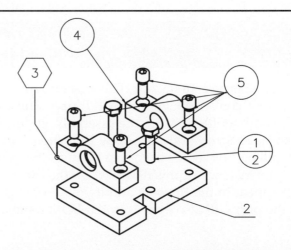

The balloons can be placed in the layout one at a time, which allows a selection of the attachment point, or they can be all placed at once. The automatic selection attaches the leader to the part references and aligns the balloons horizontally, vertically, or on a straight line at an angle.

Open the drawing file Example-18-06.dwg. This layout was created using the **Scene** technique covered in the next chapter. We will place the default balloons in the layout and then modify them to call out both the part number and the quantity. To place balloons on the layout, use the **AMBALLOON** command. To access this command, enter AMBALLOON at the Command: prompt, or select **Balloon** from the **Parts List** cascading menu in the **Annotate** pull-down menu or the right-click menu, pick the **Place Balloon** button on the **Drawing Layout** toolbar.

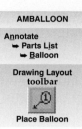

This command gives you four options on the command line for placing balloons. These are auTo, **autoAll**, **Manual**, and **One**. You are also given three options for organizing balloons. These are **Collect**, **Renumber**, and **rEorganize**.

Now input the balloons one at a time so it looks like **Figure 18-21**. Enter the **One** option and all the part references are displayed. Pick one of them and then pick the start point on the part for the leader. You can have as many segments as you want. Press [Enter] to place the balloon and you can then start the next one. Continue until you have placed the five balloons and then press [Enter].

Double-click on the balloon with the 4 in it. This accesses the **Balloon** dialog box. See **Figure 18-22**. This dialog box only controls the characteristics for the *one* balloon you picked. You can change the balloon shape, add multiple leaders, and change the leader arrowhead.

To globally change all the balloons in the layout, click on the **Properties** button. This displays the **Balloon Properties for ANSI** dialog box. See **Figure 18-23A.** Click on the **Override Balloon Type** button and you are given a pop-up menu of balloon types. Select the type that is highlighted in **Figure 18-23B.** Now the **Columns to Display** area shows both the ITEM and the QTY. Click on **OK** twice and all the balloons will change. See **Figure 18-24.**

The balloon properties can be preset by selecting **Mechanical Options** in the **Assist** pull-down menu. Then access the **Standards** tab in the **Mechanical Options** dialog box. Double-click on Balloon to display the **Balloon Properties for ANSI** dialog box similar to the one shown in **Figure 18-23A.**

Open the drawing Example-18-07.dwg. Issue the **AMBALLOON** command but this time select the **autoAll** option. Select one of each of the parts on the part reference symbols and place the balloons horizontally as shown in **Figure 18-25A.** Now double-click on any of the balloons and pick **Properties** in the dialog box. Change the balloon type to the **Polygon** option. Pick **OK** twice. The results are shown in **Figure 18-25.**

Figure 18-21.
The assembly with balloons in place.

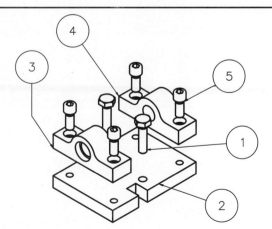

Figure 18-22.
The **Balloon** dialog box is used to make changes to the selected balloon.

Pick to change balloon shape

Pick to add leaders

Pick to change arrowhead

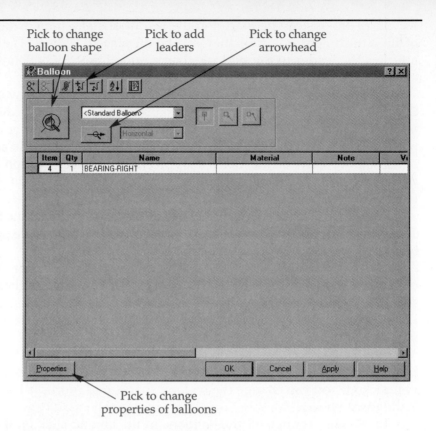

Pick to change properties of balloons

Figure 18-23.
A—The **Balloon Properties for ANSI** dialog box is used to make global changes to all balloons.
B—The override choices of balloon types.

Pick to globally change all balloons

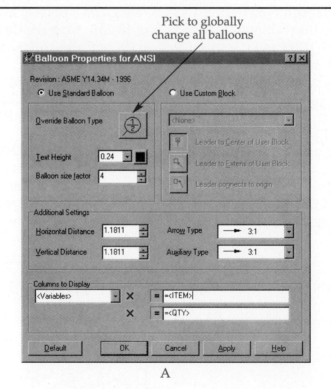

A

B

Figure 18-24.
The assembly after the balloons have been changed.

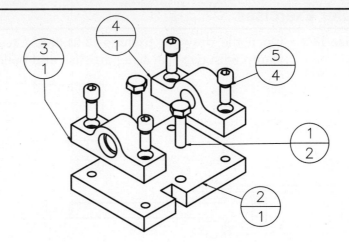

Figure 18-25.
A—The assembly with balloons in place horizontally. B—The assembly after the balloons have been changed.

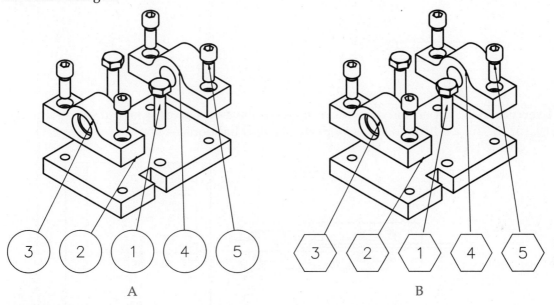

A   B

Chapter Test

Answer the following questions on a separate sheet of paper.

1. What is a *balloon?*
2. How are balloons, part numbers, and parts lists used?
3. What are the four types of information in the bill of material database?
4. How do the four types of information in the bill of material database differ?
5. What command is used to display the bill of material (BOM) database?
6. Describe the differences between *internal columns* and *external columns* in the **BOM** dialog box.
7. The _____ command is used to create a parts list.
8. What command is used to change the colors or values of a parts list?
9. *True or False?* The **AMPARTREF** command is used to edit the data for a single part.
10. The _____ command is used to insert balloons.

Chapter Exercises

Exercise 18-1. Open the drawing Exercise-18-01.dwg. Create a layout of the mold clamp with two views, a parts list, and balloons as shown.

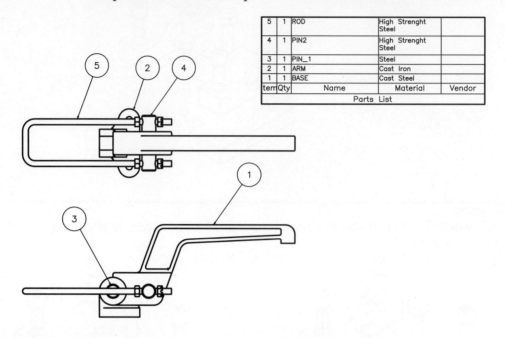

Item	Qty	Name	Material	Vendor
5	1	ROD	High Strenght Steel	
4	1	PIN2	High Strenght Steel	
3	1	PIN_1	Steel	
2	1	ARM	Cast Iron	
1	1	BASE	Cast Steel	
		Parts List		

Exercise 18-2. Open the drawing Exercise-18-02.dwg. Create a layout of the linkage with two views, a parts list, and balloons as shown.

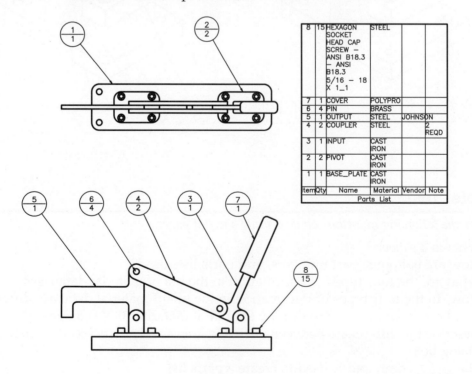

Item	Qty	Name	Material	Vendor	Note
8	15	HEXAGON SOCKET HEAD CAP SCREW — ANSI B18.3 — ANSI B18.3 5/16 — 18 X 1_1	STEEL		
7	1	COVER	POLYPRO		
6	4	PIN	BRASS		
5	1	OUTPUT	STEEL	JOHNSON	
4	2	COUPLER	STEEL		2 REQD
3	1	INPUT	CAST IRON		
2	2	PIVOT	CAST IRON		
1	1	BASE_PLATE	CAST IRON		
		Parts List			

Exercise 18-3. Open the drawing Exercise-18-03.dwg. Add the material Nylon 66 with a density of 1.38 to the material database and assign matterials to the parts. Create a layout of the assembly with one isometric view, a parts list including weights, and balloons as shown.

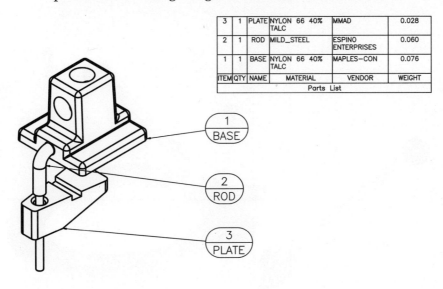

3	1	PLATE	NYLON 66 40% TALC	MMAD	0.028
2	1	ROD	MILD_STEEL	ESPINO ENTERPRISES	0.060
1	1	BASE	NYLON 66 40% TALC	MAPLES−CON	0.076
ITEM	QTY	NAME	MATERIAL	VENDOR	WEIGHT
Parts List					

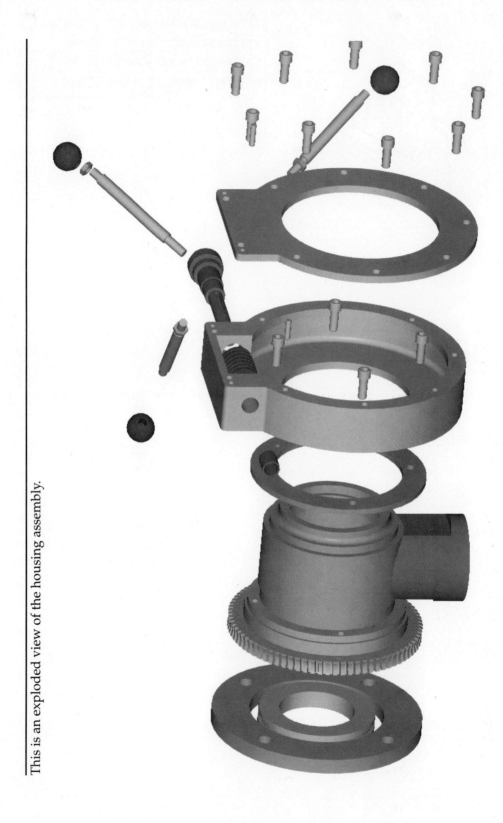

This is an exploded view of the housing assembly.

Scenes and Assembly Drawings

Objectives

After completing this chapter, you will be able to:

* Understand the concept of the MDT scene.
* Create scenes controlling the visibility.
* Understand and use the global and individual explosion factor.
* Add tweaks to exploded views.
* Add and modify trails to exploded views.
* Create layouts and views using scenes.

Drawings Needed

*The following is a list of files that you will need to work through this chapter. These files can be found on the **User's Drawings** CD included with this text.*

Examples	Practice	Exercises
Example-19-01.dwg	Practice-19-01.dwg	Exercise-19-01.dwg
Example-19-02.dwg		Exercise-19-02.dwg
Example-19-03.dwg		Exercise-19-03.dwg
Example-19-04.dwg		Exercise-19-04.dwg
Example-19-05.dwg		Exercise-19-05.dwg

THE MECHANICAL DESKTOP SCENE

There are two scene commands in MDT. The first is the AutoCAD **SCENE** command. This command allows you to add and arrange lights in a 3D view for use with the AutoCAD **RENDER** command. This will not be studied in this chapter. The second scene command, and the one we will be studying, is the **AMNEW** command that creates a MDT scene. **AMNEW** creates a named arrangement of the parts and subassemblies, called a scene. It controls:

* The visibility of each part and subassembly.
* The viewpoint—that is how you are looking at the assembly.
* The zoom scale of the view.
* The suppression of sections in section views in the layout.
* The distance for the automatic separation of parts in exploded views.

- The display of lines showing how parts are assembled.
- The display and location of clipping planes.

It does not control:

- The number or arrangement of viewports.
- The display of the assembly as wireframe, shaded, or hidden.

Viewing Scenes

Open part Example-19-01.dwg. See **Figure 19-1.** This simple assembly has a base plate with two bearing blocks and all three are individual parts with one instance. The white socket head cap screw is a part with four instances and the green hex head bolt is a part with two instances. All the parts are related with 3D constraints and the assembly is shown in the model mode.

Two scenes have already been created. To display one of them, click on the **Scene** tab at the top of the **Browser** or pick the **Scene** icon in the **Desktop Main** toolbar. The **Browser** now shows SCENE1 as the active scene and SCENE2 as the inactive scene. See **Figure 19-2.** The inactive names are shown in gray. Just like you can only have one active part, you can only have one active scene. Expand SCENE1 by clicking the plus sign next to it in the **Browser**. You'll see that the right bearing and the two cap screws have their visibility turned off. The screen shows the assembly in the right front isometric. See **Figure 19-3A.**

Double-click on SCENE2 to make it the active screen. Expand SCENE2 in the **Browser** and note that the left bearing is invisible. The screen shows the assembly in the left front isometric view, zoomed out slightly, and with a different set of parts visible. See **Figure 13-3B.**

Figure 19-1.
Assembly with a base plate with two bearing blocks. All three of these are individual parts.

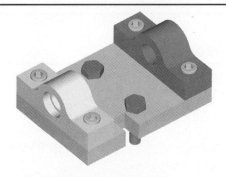

Figure 19-2.
The **Browser** showing SCENE1 as the active scene and SCENE2 as the inactive scene.

Figure 19-3.
A—SCENE1 as the active scene. B—SCENE2 as the active scene.

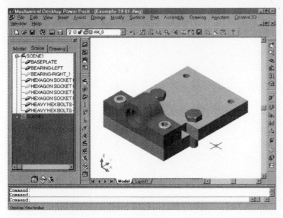

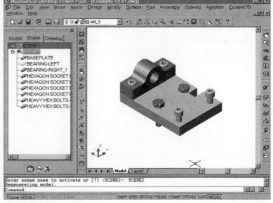

A   B

Creating a New Scene

Now we will create a new scene where all the fasteners are invisible. We will call it NO_BOLTS. The **AMNEW** command is used to create a new scene. To access this command, enter AMNEW at the Command: prompt, select **New Scene** from the **Scene** cascading menu in the **Assembly** pull-down menu, pick the **New Scene** button on the **Scenes** toolbar.

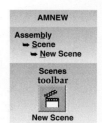

AMNEW

Assembly
➡ Scene
   ➡ New Scene

Scenes
toolbar

New Scene

> **PROFESSIONAL TIP**
>
> You can also access the **AMNEW** command by right-clicking in the **Browser** and selecting **New Scene** in the pop-up menu.

When you're prompted for a scene name, enter No_Bolts. Accept the default explosion factor and activate the new scene. The display shows all the parts at the current viewpoint and zoom. Change the viewpoint to the left back isometric. This will also do an automatic **Zoom Extents**. To set the visibility, select **Scene Visibility** in the **Assembly** pull-down menu, pick the **Scene Visibility** button on the **Scenes** toolbar, or pick the **Visibility** button at the bottom of the **Browser**.

Assembly
➡ Scene Visibility

Scenes
toolbar

Scene Visibility

> **PROFESSIONAL TIP**
>
> The visibility also can be set by picking the **Visibility** button at the bottom of the **Browser**.

This accesses the **Desktop Visibility** dialog box. Pick the **Select** button, select all six fasteners, and then press **OK**. You cannot deselect parts in this command so if you make a mistake remember you can change the visibility of any part by expanding No_Bolts in the **Browser**, right-clicking on the part, and checking **Visible** in the pop-up menu.

If you change a variable that "Scene" controls, such as the zoom factor, the scene automatically updates and remembers the new value. This automatic feature can be turned off in the **Assembly** tab of the **Mechanical Options** dialog box. See **Figure 19-4**. To access this dialog box, select **Scene Options** in the **Assembly** pull-down menu, or pick the **Scene Options** button on the **Scenes** toolbar. On the **Assembly** tab you will find the **Update Scene as Modified** check box in the **Automatic** area. If this check box is unchecked, then changes will not be registered.

Assembly
➡ Scene Options

Scenes
toolbar

Scene Options

Figure 19-4.
Automatic updates of modified scenes can be turned on and off in the **Mechanical Options** dialog box.

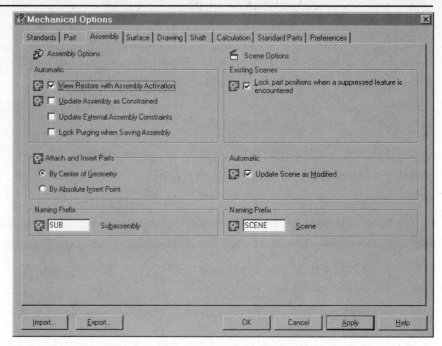

Assembly
↳ Scene Update

Scenes
toolbar

Update Scene

> **PROFESSIONAL TIP**
>
> The **Desktop Options** dialog box can also be accessed by picking the **Options** button at the bottom of the **Browser** and selecting the **Assembly** tab.

When the **Automatic** check box is unchecked, you can update manually with the update scene command. To do this, select **Update Scene** in the **Assembly** pull-down menu or pick the **Update Scene** button on the **Scenes** toolbar.

> **PROFESSIONAL TIP**
>
> The scene update command can also be accessed by picking the **Update Scenes** button at the bottom of the **Browser**.

The Explosion Factor in Scenes

A *scene explosion factor* is used to create exploded views. It sets the distance that all parts are separated and is measured in the current units. The scene explosion factor *only* applies to the **Mate plane-to-plane** 3D constraint between parts. For example, in the drawing you have on the screen, the part BEARING-LEFT (the red bearing) is constrained to the white BASEPLATE with two flush constraints and one **Mate plane-to-plane**. The BEARING-RIGHT_1 (the magenta part) is constrained to the BASEPLATE with two **Mate line-to-line** constraints. When the explosion factor is set, the red bearing will move up separating the **Mate plane-to-plane**, but the magenta bearing will not move. Note that an **Insert** constraint is a combination of a **Mate line-to-line** and a **Mate plane-to-plane**; the explosion factor applies to it as well.

Learning Mechanical Desktop R5

Figure 19-5.
The result of the change in the scene explosion factor.

Set or Modify the Explosion Factor

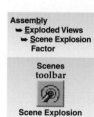

Assembly
→ Exploded Views
→ Scene Explosion Factor

Scenes toolbar

Scene Explosion Factor

First make sure SCENE1 is the active scene. Then to set or modify the explosion factor for this scene, select **Scene Explosion Factor** in the **Exploded Views** cascading menu of the **Assembly** pull-down menu or pick the **Scene Explosion Factor** button on the **Scenes** toolbar.

When you are prompted to select a scene, enter SCENE1. Then enter a value of 2.0 as the new explosion factor; this value is in drawing units. The red part and the bolts will then move apart. See **Figure 19-5.**

PROFESSIONAL TIP

You can also right-click on the scene name in the **Browser** and select **Explode Factor...** in the pop-up menu. This accesses the **Explode Factor** dialog box. Set the explode factor to 3.0, click **OK**, and the red part and the bolts will move further apart.

The factor must be a number and cannot be a design variable. This means that if you are planning to generate exploded views, it helps to use the **Mate plane-to-plane** constraint wisely. The parts move from whichever part is the first part in the **Browser** and moving a part to the top will change the scene. For example, switch to the **Model** view. Hold the [Shift] key down and in the **Browser** click and drag the BASEPLATE below BEARING-RIGHT_1. The red part is now fixed and the white part moves from it.

PRACTICE 19-1

❑ Open the drawing Practice-19-01.dwg. This is a metric part with an ANSI fastener.
❑ Review the 3D constraints on the parts by expanding the **Browser** and clicking on each constraint.
❑ Create a scene with all three parts visible and an explosion factor of 25.

THE PART EXPLOSION FACTOR

AMXFACTOR

Assembly
→ Exploded Views
→ Part Explosion Factor

Scenes toolbar

Part or Subassembly Explosion Factor

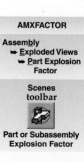

Open part Example-19-02.dwg and click on the **Scene** tab in the **Browser**. This is the same part as the practice example. See **Figure 19-6.** The scene explosion factor is 25mm, but the red ball socket needs to be moved up another 15mm and the green thumbscrew moved in 10mm. One way to do this is by changing the individual *part explosion factor* with the **AMXFACTOR** command. To access this command, enter AMXFACTOR at the Command: prompt, select **Part Explosion Factor** from the **Exploded Views** cascading menu in the **Assembly** pull-down menu, or pick the **Part or Subassembly Explosion Factor** button on the **Scenes** toolbar.

Figure 19-6.
The assembly used to illustrate the use of a part explosion factor.

Pick the ball socket when you are prompted to select a part, subassembly, or scene. Then enter 40 when you are prompted for a new explosion factor. The ball socket will move up 15mm and the thumbscrew will move with it since it is constrained to the ball socket.

 **PROFESSIONAL TIP** You can also right-click on the BALL-SOCKET in the **Browser** and select **Explode Factor...** in the pop-up menu. This accesses the **Explode Factor** dialog box. Set the explode factor to 40, click **OK**, and the ball socket will move.

Now set the explode factor on the thumbscrew to 15. The thumbscrew should move in closer to the red part. The results are shown in **Figure 19-7**.

Figure 19-7.
The result of the change in the part explosion factor for the ball socket and the thumbscrew.

USING TWEAKS TO MOVE PARTS

Open part Example-19-03.dwg as shown in **Figure 19-8**. In this SCENE2, the scene explode factor is 2.0 units and the green bolts have an individual factor of 4.0. The BEARING-RIGHT_1 part (magenta part) did not move, because it has no **Mate plane-to-plane** constraints. To move this part we can use tweaks. A *tweak* is a positional adjustment to a part in the exploded scene with reference to another part. You can add numerous tweaks to a part and they are cumulative.

To add tweaks, use the **AMTWEAK** command. To access this command, enter AMTWEAK at the Command: prompt, select **New Tweak** from the **Exploded Views** cascading menu in the **Assembly** pull-down menu, or pick the **New Tweak** button on the **Scenes** toolbar.

Select the magenta bearing and the **Tweak Part/Subassembly** dialog box is displayed. See **Figure 19-9**. The three choices are **Move** (the default), **Rotate**, and **Transform**. Select **Move** and click on **OK**. If SCENE2 is expanded, you will notice that all the parts below the magenta bearing (BEARING-RIGHT_1) in the **Browser** are now invisible. They are invisible because they cannot be used to reference the tweak.

AMTWEAK

Assembly
➡ Exploded Views
➡ New Tweak

Scenes
toolbar

New Tweak

PROFESSIONAL TIP

You can also right-click on the part name in the **Browser** and select **New Tweak...** in the pop-up menu. This also accesses the **Tweak Part/Subassembly** dialog box.

You are prompted to select the reference geometry. Use the white baseplate. You need to pick an edge of the white part that is collinear with the direction you want the magenta part to move. Pick the upper end of the lower-front, vertical corner of the white baseplate. An easy way of selecting this corner is to use the intersection object snap. Notice the upward pointing arrow. A positive value moves the bearing in the direction of the arrow and a negative value moves it in the opposite direction,

---

Figure 19-8.
The assembly used to illustrate the use of tweaks.

Figure 19-9.
The **Tweak Part/Subassembly** dialog box offers three tweak options.

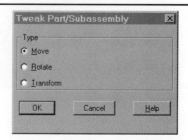

Chapter 19   Scenes and Assembly Drawings

Figure 19-10.
The result of the tweak using the **Move** option.

in this case downward. Input the distance of 2.0 units and the magenta bearing and its associated bolts move up. See **Figure 19-10.**

Using the Transform Option

The **Transform** option is used to accomplish a more complex move with one tweak. You will be able to move the part in all three coordinates (X, Y, and Z) from any reference point. Issue the **AMTWEAK** command, pick the magenta part, and select the **Transform** option. When prompted to select an option, enter M for move. With **Osnap Endpoint**, pick the lower end of the far-right, vertical corner of the white base as the reference point. Then input the relative coordinates @2,2,2 as the end point. The part will move the distance and direction described by the vector (arrow) shown on the screen. Additional movements and rotations are possible before ending the command. When you are done, enter X to exit. See **Figure 19-11.**

> **CAUTION**
>
> Do not try to press [Enter] to end the command, because that will repeat the last transform. Be sure to use the **eXit** option on the command line.

Figure 19-11.
The result of the tweak using the **Transform** option.

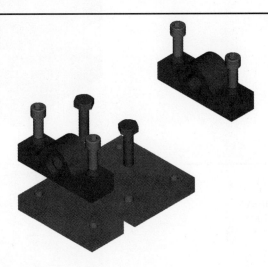

USING TRAILS TO ILLUSTRATE EXPLODED VIEWS

Open part Example-19-04.dwg and activate Scene2. Lines will be added showing how the parts are assembled. See **Figure 19-12.** These lines are called *trails* and they are lines that go from a point on a part in the exploded scene to the same point on the part in the assembled view.

First expand SCENE2 in the **Browser** so that you can see the names of the parts. Trails are constructed one at a time using the **AMTRAIL** command. To access this command, enter AMTRAIL at the Command: prompt, select **New Trail** from the **Exploded Views** cascading menu in the **Assembly** pull-down menu, or pick the **New Trail** button on the **Scenes** toolbar.

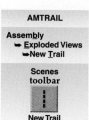

AMTRAIL

Assembly
➥ Exploded Views
➥New Trail

Scenes
toolbar

New Trail

PROFESSIONAL TIP

You can also right-click on the part name in the **Browser** and select **New Trail...** in the pop-up menu.

When prompted to select the reference point on the part or subassembly, pick a point on the lower circular edge of the green bolt nearest the front (HEX BOLT_1). You do not need to use center osnap because this selection automatically selects the center point of the circle. Accept the defaults in the **Trail Offsets** dialog box by clicking **OK**. See **Figure 19-13.** Notice how the trail goes up through the bolt. Trails are automatically placed on the layer AM_TR. Issue the **LAYER** command and change the color of layer AM_TR to cyan and the linetype to PHANTOM2.

Now repeat the AMTRAIL command and select the center of the bottom circle of HEX SOCKET_1. This part has a chamfer on the bottom resulting in two circles—either one will do. If you have difficulty selecting, you may have to zoom in on the part. In the Trail dialog box, select Under Shoot in the Offset at Current Position area. Then set the value of 0.2. This sets the trail so it looks like it just starts at the bottom of the bolt and only goes through the bearing. This is because the bolt is constrained to the bearing and not the baseplate.

Figure 19-12.
The assembly with trails added.

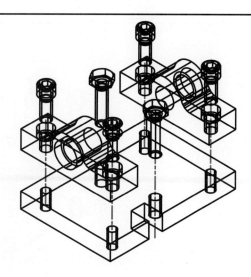

Figure 19-13.
The **Trail Offset**
dialog box is used
to create trails.

Making Changes to the Trail

AMEDITTRAIL

Assembly
↳ **Exploded Views**
  ↳ **Edit Trail**

Scenes
toolbar

Edit Trail

Now make the trail that only goes through the bearing go into the hole in the base plate. To do this, the **AMEDITTRAIL** command is used. To access this command, enter AMEDITTRAIL at the Command: prompt, select **Edit Trail** from the **Exploded Views** cascading menu in the **Assembly** pull-down menu, or pick the **Edit Trail** button on the **Scenes** toolbar.

> **PROFESSIONAL TIP** You can also right-click on the part name in the **Browser** and select **Edit Trail...** in the pop-up menu.

When you are prompted to select the trail to be edited, pick the trail on the hexagon socket head cap screw (HEX SOCKET_1). This accesses the **Trail Offsets** dialog box. In the **Offset at Assembled Position** area, pick the **Over Shoot** radio button and make the distance 2.0. This will extend the trail into the hole in the base plate. See **Figure 19-14.**

Figure 19-14.
Assembly with two
trails added.

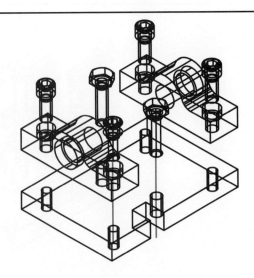

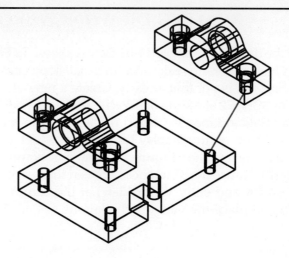

Figure 19-15.
Assembly with tweak and trail added. Notice the improper construction of the trail.

Manipulating Trails and Tweaks

With Example-19-04.dwg still open, double-click on SCENE3 to activate it. The magenta bearing has been moved 2 units in both the Y and Z directions with a **Transform** tweak operation. Put a trail on the bottom circle of the either bolt hole. See **Figure 19-15.** Notice that it goes diagonally to the assembled position rather than in stepped lines, which is the normal convention.

Trails are removed with the **AMDELTRAIL** command (this is singular because it only removes one trail at a time) and tweaks with the **AMDELTWEAKS** command (this is plural because it removes all the tweaks). Remove all of the trails and tweaks from the magenta bearing block.

Add a tweak that moves the magenta part up 1.0 unit and add a trail from the bottom of the bolt hole. Now tweak the part 2.0 units in the Y direction and then 2.0 units in the Z direction. This will generate the stepped trail line shown in **Figure 19-16.** The last two tweaks are done with Transform tweak operations.

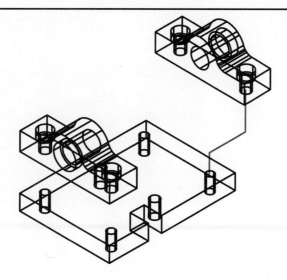

Figure 19-16.
Assembly with tweak and proper constructed trail.

Open the drawing Example-19-05.dwg as shown in **Figure 19-17.** This is SCENE2 and these parts have several tweaks and trails. Click on Layout1 and check the page size. Set it to an A-Size if it is not that already. Create a new view and in the dialog box input a base view of SCENE2 at a scale of 0.5 with the other choices set as shown in **Figure 19-18.** The **Display Hidden Lines** is turned off since exploded views generally do not show hidden lines. Turn on the **Display Tangencies**, which show an edge where a flat and a curved face come together. **Figure 19-19** shows examples where the tangencies are on (A) and off (B). Notice the difference between the bearings and the hex head cap screws.

Click on **OK** and enter V (for View) at the command line. Accept the orientation of the UCS and place the view on the layout.

Figure 19-17.
Assembly with
tweaks and
properly
constructed trails.

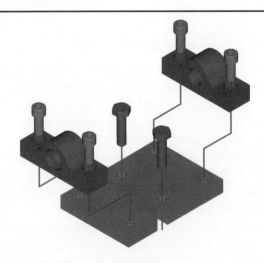

Figure 19-18.
The **Create Drawing View** dialog box is used to create drawings from scenes.

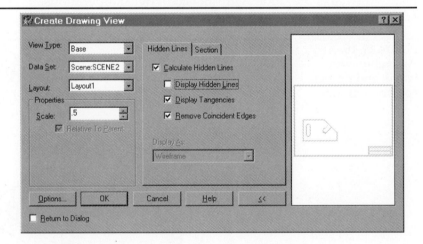

Figure 19-19.
A—The **Display Tangencies** on. B—The **Display Tangencies** off.

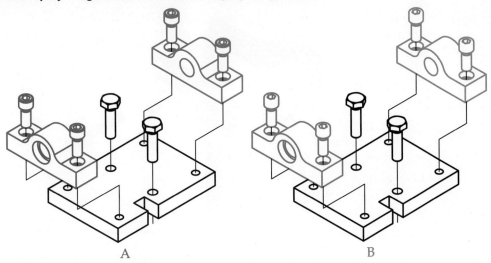

A

B

Chapter Test

Answer the following questions on a separate sheet of paper.

1. *True or False?* There are four scene commands in MDT.
2. Which scene command allows lights to be added and arranged in a 3D view for use with the AutoCAD **RENDER** command?
3. *True or False?* If you change a variable that "Scene" controls, such as the zoom factor, the scene automatically updates and remembers the new value.
4. What is a *scene explosion factor*?
5. The scene explosion factor *only* applies to the _____ 3D constraint between parts.
6. What is the difference between a scene explosion factor and a *part explosion factor*?
7. A(n) _____ is a positional adjustment to a part in the exploded scene with reference to another part.
8. Explain what trails are and their use.
9. What commands are used to edit and delete trails?
10. When creating a scene, what is the effect of the **Display Tangencies** option?

Chapter Exercises

Exercise 19-1. Open the drawing Exercise-19-01.dwg. Expand the base part (called TOP) in the **Browser** and review the three Mate pl/pl constraints by clicking on them. Note the direction of the constraint arrows. Create an exploded scene as shown with an explosion factor of 50.

Exercise 19-2. Open the drawing Exercise-19-02.dwg. Create an exploded scene and add tweaks and trails as shown.

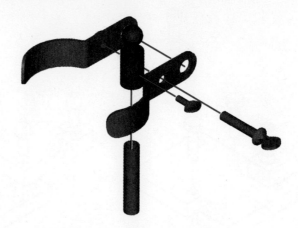

Exercise 19-3. Open the drawing Exercise-19-03.dwg and activate the assembly called FLOOR JACK. Create an exploded scene with a factor of 3.0. Edit the explosion factor on the two nuts to 1.0. Create a drawing view as shown.

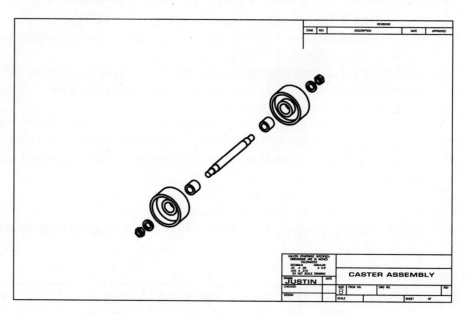

Exercise 19-4. Open the drawing Exercise-19-04.dwg. Create an exploded scene with trails as shown. Use the front view to add tweaks and line up the links.

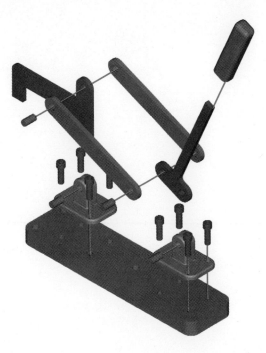

Exercise 19-5. Open the drawing Exercise-19-05.dwg. Create an exploded scene with trails. Create a drawing view as shown.

Appendix

COMMONLY USED COMMANDS AND ALIASES

Aliases for MDT Commands

Alias	Command	Result
CC	**AMUPDATE**	Updates a Part or Assembly
EE	**AMFILLET**	Displays **Fillet** Dialog Box
G	**AMEXTRUDE**	Displays **Extrude** Dialog Box
GG	**AMREVOLVE**	Displays **Revolve** Dialog Box
HH	**AMHOLE**	Displays **Hole** Dialog Box
II	**AMPARDIM**	Create Parametric Dimension
JJ	**AMADDCON**	2D Constraint
K	**AMTWEAK**	Add a Tweak in a Scene
KK	**AMDELTWEAKS**	Delete a Tweak in a Scene
LL	**AMEDITFEAT**	Edit a Feature (command line)
N	**AMNEW**	New Instance or Part (command line)
NN	**AMCATALOG**	Displays **Assembly Catalog** Dialog Box
OO	**3DORBIT**	Displays Orbit Circle
PP	**AMPROFILE**	Profile a Sketch
Q	**AMDWGVIEW**	Displays **Create Drawing View** Dialog Box
QQ	**AMEDITVIEW**	Select Drawing View to Edit
SS	**AMSKPLN**	Create New Sketch Plane
VV	**AMVISIBLE**	**DeskTop Visibility** Dialog Box
WW	**AMVARS**	Displays **Design Variables** Dialog Box
Y	**AMTRAIL**	Creates a Trail in a Scene
YY	**AMDELTRAIL**	Delete a Trail in a Scene

Aliases for AutoCAD Commands

Alias	Command	Result
0 (zero)	HIDE	Hides Hidden Lines
A	ARC	Construct an Arc
B	BREAK	Break a Line, Arc, or Circle
C	CIRCLE	Construct a Circle
E	ERASE	Erase Objects
F	UCSVIEW System Variable	Set system Variable to 1 or 0
H	HATCH	Hatch a Planar Shape
I	INSERT	Insert a Block
L	LINE	Construct a Line
M	MOVE	Move Objects
MM	COPY	Copy Objects
O	OFFSET	Offset a Line, Arc, Circle, or Polyline
P	PAN	Pan the View
R	REDRAW	Redraw the View
RR	REGENALL	Regenerate all Views
S	SPLINE	Construct a Spline
T	MTEXT	Construct Text
TT	TRIM	Trim Lines, Arcs, Circles, Plines
U	UNDO One	Undo Last Command
Z	ZOOM	Zoom View (Command Line)

Aliases for MDT LISP Programs

Alias	Program	Result
AA	AMDTAA	Updates Assembly Constraints
BB	AMDTBB	Activates an Assembly
D	AMDTD	DVIEW Command, Controls for Perspective View (Command Line)
DD	AMDTDD	Sets View Depth (Command Line)
F	AMDTF	Sets UCSVIEW System Variable to 1
FF	AMDTFF	Sets UCSVIEW System Variable to 1
J	AMDTJ	Select Point for Xline
UU	AMDTUU	Sets UCS to the View
W	AMDTW	Switches Between Model and Layout
X	AMDTX	ZOOM 0.5X
XX	AMDTXX	ZOOM 2.0X

Alias	Command	Result
ZZ	AMDTZZ	Real Time Zoom
1	AMDT1	Displays One Viewport on Screen
2	AMDT2	Displays Two Viewports on Screen
3	AMDT3	Displays Three Viewports on Screen
4	AMDT4	Displays Four Viewports on Screen
5	AMDT5	Displays Top View in Current Viewport
55	AMDT55	Displays Bottom View in Current Viewport
6	AMDT6	Displays Front View in Current Viewport
66	AMDT66	Displays Back View in Current Viewport
7	AMDT7	Displays Right Side View in Current Viewport
77	AMDT77	Displays Left Side View in Current Viewport
8	AMDT8	Displays Right Front Isometric View in Current Viewport
88	AMDT88	Displays Left Front Isometric View in Current Viewport
9	AMDT9	Displays Sketch Plane View in Current Viewport
[	AMDT[	Rotates View by 15° About Vertical Axis CCW
]	AMDT]	Rotates View by 15° About Vertical Axis CW
=	AMDT=	Rotates View by 15° About Horizontal Axis CCW
-	AMDT-	Rotates View by 15° About Horizontal Axis CW
5U	AMDT5U	Displays Top View and Sets UCS to the View
6U	AMDT6U	Displays Front View and Sets UCS to the View
7U	AMDT7U	Displays Side View and Sets UCS to the View
5D	AMDT5D	Displays Top View, Sets UCS to View, and Sets Depth (Command Line)
6D	AMDT6D	Displays Front View, Sets UCS to View, and Sets Depth (Command Line)
7D	AMDT7D	Displays Side View, Sets UCS to View, and Sets Depth (Command Line)

GEOMETRIC CONSTRAINTS

Name*	1st Choice – Object to be Reoriented	2nd Choice – Reference Object	Button on Part Modeling Toolbar
Tangent	Line, Circle, or Arc	Line, circle, arc, edge, perpendicular work plane, nonperpendicular work axis	
co**N**centric	Circle or Arc	Circle, arc, or circular edge	
Collinear	Line	Line, linear edges, and objects that appear as lines in the sketch view	
Parallel	Line	Line, linear edges, and objects that appear as lines in the sketch view	
perpendicu**L**ar	Line	Line, linear edges, and objects that appear as lines in the sketch view	
Horizontal	Line	(None)	
Vertical	Line	(None)	
pro**J**ect	Endpoint of a Line or Center of a Circle or Arc	Line, circle, arc, and any edge or work object sketch view that appears as those in the sketch view.	
join (no symbol)	End points of lines and/or arcs		
X value	Line, Circle, or Arc	Line, circle, arc, or edge	
Y value	Line, Circle, or Arc	Line, circle, arc, or edge	
Radius	Circle or Arc	Circle, arc, or circular edge	
E equal length	Line	Line	
Mirror	Line	Line	
	Circle	Circle	
	Arc	Arc	
Fix	Line, Circle, Arc	(None)	

* Boldface letter indicates symbol shown when constraints are visible.

GEOMETRIC DIMENSIONING AND TOLERANCING SYMBOLS

SYMBOL FOR:	ANSI Y14.5M	ASME Y14.5M	ISO
STRAIGHTNESS	—	—	—
FLATNESS	▱	▱	▱
CIRCULARITY	○	○	○
DYLINDRICITY	⌭	⌭	⌭
PROFILE OF A LINE	⌒	⌒	⌒
PROFILE OF A SURFACE	⌓	⌓	⌓
ALL AROUND	↼⊙	↼⊙	↼⊙ (proposed)
ANGULARITY	∠	∠	∠
PERPENDICULARITY	⊥	⊥	⊥
PARALLELISM	//	//	//
POSITION	⊕	⊕	⊕
CONCENTRICITY	◎	◎	◎
SYMMETRY	NONE	≡	≡
CIRCULAR RUNOUT	*↗	*↗	↗
TOTAL RUNOUT	*↗↗	*↗↗	↗↗
AT MAXIMUM MATERIAL CONDITION	Ⓜ	Ⓜ	Ⓜ
AT LEAST MATERIAL CONDITION	Ⓛ	Ⓛ	Ⓛ
REGARDLESS OF FEATURE SIZE	Ⓢ	NONE	NONE
PROJECTED TOLERANCE ZONE	Ⓟ	Ⓟ	Ⓟ
TANGENT PLANE	Ⓣ	Ⓣ	NONE
FREE STATE	Ⓕ	Ⓕ	Ⓕ
DIAMETER	⌀	⌀	⌀
BASIC DIMENSION	50	50	50
REFERENCE DIMENSION	(50)	(50)	(50)
DATUM FEATURE	–A–	*▨Ⓐ	*▨ or *▨Ⓐ
DIMENSION ORIGIN	⊕→	⊕→	⊕→
FEATURE CONTROL FRAME	⊕ ⌀0.5Ⓜ A B C	⊕ ⌀0.5Ⓜ A B C	⊕ ⌀0.5Ⓜ A B C
CONICAL TAPER	▷	▷	▷
SLOPE	◿	◿	◿
COUNTERBORE\SPOTFACE	⌴	⌴	NONE
COUNTERSINK	⌵	⌵	NONE
DEPTH\DEEP	↧	↧	NONE
SQUARE	□	□	□
DIMENSION NOT TO SCALE	15	15	15
NUMBER OF TIMES\PLACES	8X	8X	8X
ARC LENGTH	⏜105	⏜105	⏜105
RADIUS	R	R	R
SPHERICAL RADIUS	SR	SR	SR
SPHERICAL DIAMETER	S⌀	S⌀	S⌀
CONTROLLED RADIUS	NONE	CR	NONE
BETWEEN	NONE	*↔	NONE
STATISTICAL TOLERANCE	NONE	⟨ST⟩	NONE
DATUM TARGET	⌀6/A1	⌀6/A1 or A1–⌀6	⌀6/A1 or A1–⌀6
TARGET POINT	✕	✕	✕

* MAY BE FILLED OR NOT FILLED

DECIMAL AND METRIC EQUIVALENTS

INCHES		MILLI-	INCHES		MILLI-
FRACTIONS	DECIMALS	METERS	FRACTIONS	DECIMALS	METERS
	.00394	.1	15/32	.46875	11.9063
	.00787	.2		.47244	12.00
	.01181	.3	31/64	.484375	12.3031
1/64	.015625	.3969	1/2	.5000	12.70
	.01575	.4		.51181	13.00
	.01969	.5	33/64	.515625	13.0969
	.02362	.6	17/32	.53125	13.4938
	.02756	.7	35/64	.546875	13.8907
1/32	.03125	.7938		.55118	14.00
	.0315	.8	9/16	.5625	14.2875
	.03543	.9	37/64	.578125	14.6844
	.03937	1.00		.59055	15.00
3/64	.046875	1.1906	19/32	.59375	15.0813
1/16	.0625	1.5875	39/64	.609375	15.4782
5/64	.078125	1.9844	5/8	.625	15.875
	.07874	2.00		.62992	16.00
3/32	.09375	2.3813	41/64	.640625	16.2719
7/64	.109375	2.7781	21/32	.65625	16.6688
	.11811	3.00		.66929	17.00
1/8	.125	3.175	43/64	.671875	17.0657
9/64	.140625	3.5719	11/16	.6875	17.4625
5/32	.15625	3.9688	45/64	.703125	17.8594
	.15748	4.00		.70866	18.00
11/64	.171875	4.3656	23/32	.71875	18.2563
3/16	.1875	4.7625	47/64	.734375	18.6532
	.19685	5.00		.74803	19.00
13/64	.203125	5.1594	3/4	.7500	19.05
7/32	.21875	5.5563	49/64	.765625	19.4469
15/64	.234375	5.9531	25/32	.78125	19.8438
	.23622	6.00		.7874	20.00
1/4	.2500	6.35	51/64	.796875	20.2407
17/64	.265625	6.7469	13/16	.8125	20.6375
	.27559	7.00		.82677	21.00
9/32	.28125	7.1438	53/64	.828125	21.0344
19/64	.296875	7.5406	27/32	.84375	21.4313
5/16	.3125	7.9375	55/64	.859375	21.8282
	.31496	8.00		.86614	22.00
21/64	.328125	8.3344	7/8	.875	22.225
11/32	.34375	8.7313	57/64	.890625	22.6219
	.35433	9.00		.90551	23.00
23/64	.359375	9.1281	29/32	.90625	23.0188
3/8	.375	9.525	59/64	.921875	23.4157
25/64	.390625	9.9219	15/16	.9375	23.8125
	.3937	10.00		.94488	24.00
13/32	.40625	10.3188	61/64	.953125	24.2094
27/64	.421875	10.7156	31/32	.96875	24.6063
	.43307	11.00		.98425	25.00
7/16	.4375	11.1125	63/64	.984375	25.0032
29/64	.453125	11.5094	1	1.0000	25.4001

Solutions to Triangles

$A + B + C = 180°$ $S = \dfrac{a+b+c}{2}$		Right	Oblique

Have	Want	**Formulas for Right**	**Formulas for Oblique**
abc	A	$\tan A = a/b$	$1/2A = \sqrt{(s-b)(s-c)/bc}$
	B	$90° - A$ or $\cos B = a/c$	$\sin 1/2B = \sqrt{(s-a)(s-c)/a \times c}$
	C	$90°$	$\sin 1/2C = \sqrt{(s-a)(s-b)/a \times b}$
	Area	$a \times b/2$	$\sqrt{s \times (s-a)(s-b)(s-c)}$
aAC	B	$90° - A$	$180° - (A + C)$
	b	$a \cot A$	$a \sin B/\sin A$
	c	$a/\sin A$	$a \sin C/\sin A$
	Area	$(a^2 \cot A)/2$	$a^2 \sin B \sin C/2 \sin A$
acC	A	$\sin A = a - c$	$\sin A = a \sin C/c$
	B	$90° - A$ or $\cos B = a/c$	$180° - (A + C)$
	b	$\sqrt{c^2 - a^2}$	$c \sin B/\sin C$
	Area	$1/2a \sqrt{c^2 - a^2}$	$1/2 ac \sin B$
abC	A	$\tan A = a/b$	$\tan A = a \sin C/b - a \cos C$
	B	$90° - A$ or $\tan B = b/a$	$180° - (A + C)$
	c	$\sqrt{a^2 - b^2}$	$\sqrt{a^2 + b^2 - 2ab \cos C}$
	Area	$a \times b/2$	$1/2ab \sin C$

Area Equivalents

1	area = radius$^2 \times 3.1416$ or diameter$^2 \times .7854$
1	circumference = diameter $\times 3.1416$ or diameter $\div .3183$
2	when the area of a circle & square are equal, $D = S \times 1.128$
2	when the area of a circle & square are equal, $S = D \times .8862$
3	side of inscribed square – diameter $\times .7071$
3	diameter of circumscribing circle = $S \times 1.1412$
4	surface area of a sphere = diameter $\times$ circumference
4	volume of a sphere = diameter$^3 \times .5236$

Equivalents

Fahrenheit and Celcius

$°F = (1.8 \times °C) + 32$
$°C = (°F - 32) \div 1.8$

Weight

1 gram = .03527 oz (av.)
1 oz = 28.35 grams
1 kilogram = 2.2046 pounds
1 pound = .04536 kilograms
1 metric ton = 2,204.6 pounds
1 ton (2000) lbs in U.S.) = 907.2 kg.

Volume

1 U.S. quart = 0.946 liters
1 U.S. gallon = 3.785 liters
1 liter = 1.0567 U.S. quarts
1 liter = .264 U.S. gallons

Rutland Tool and Supply Co., Inc.

Length Conversions

multiply	by	to obtain
Inches	25.4	Millimeters
Feet	304.8	Millimeters
Inches	2.54	Centimeters
Feet	30.48	Centimeters
Millimeters	.03937008	Inches
Centimeters	.3937008	Inches
Meters	39.37008	Inches
Millimeters	.003280840	Feet
Centimeters	.03280840	Feet
Inches	.0254	Meters

Square Area Conversions

multiply	by	to obtain
Millimeters	.00001076391	Feet
Millimeters	.00155003	Inches
Centimeters	.1550003	Inches
Centimeters	.001076391	Feet
Inches	645.16	Millimeters
Inches	6.4516	Centimeters
Inches	.00064516	Meters
Feet	.09290304	Meters
Feet	929.0304	Centimeters
Feet	92,903.04	Millimeters

Rutland Tool and Supply Co., Inc.

PHYSICAL PROPERTIES OF METALS

Metal	Symbol	Density	Specific Heat	Melting Point*		Lbs. per Cubic Inch
				C°	F°	
Aluminum (cast)	Al	2.56	.2185	658	1217	.0924
Aluminum (rolled)	Al	2.71	–	658	1217	.0978
Antimony	Sb	6.71	.051	630	1166	.2424
Bismuth	Bi	9.80	.031	271	520	.3540
Boron	B	2.30	.3091	2300	4172	.0831
Brass	–	8.51	.094	–	–	.3075
Cadmium	Cd	8.60	.057	321	610	.3107
Calcium	Ca	1.57	.170	810	1490	.0567
Chromium	Cr	6.80	.120	1510	2750	.2457
Cobalt	Co	8.50	.110	1490	2714	.3071
Copper	Cu	8.89	.094	1083	1982	.3212
Columbium	Cb	8.57	–	1950	3542	.3096
Gold	Au	19.32	.032	1063	1945	.6979
Iridium	Ir	22.42	.033	2300	4170	.8099
Iron	Fe	7.86	.110	1520	2768	.2634
Iron (cast)	Fe	7.218	.1298	1375	2507	.2605
Iron (wrought)	Fe	7.70	.1138	1500-1600	2732-2912	.2779
Lead	Pb	11.37	.031	327	621	.4108
Lithium	Li	.057	.941	186	367	.0213
Magnesium	Mg	1.74	.250	651	1204	.0629
Manganese	Mn	8.00	.120	1225	2237	.2890
Mercury	Hg	13.59	.032	−39	−38	.4909
Molybdenum	Mo	10.2	.0647	2620	47.48	.368
Monel metal	–	8.87	.127	1360	2480	.320
Nickel	Ni	8.80	.130	1452	2646	.319
Phosphorus	P	1.82	.177	43	111.4	.0657
Platinum	Pt	21.50	.033	1755	3191	.7767
Potassium	K	0.87	.170	62	144	.0314
Selenium	Se	4.81	.084	220	428	.174
Silicon	Si	2.40	.1762	1427	2600	.087
Silver	Ag	10.53	.056	961	1761	.3805
Sodium	Na	0.97	.290	97	207	.0350
Steel	–	7.858	.1175	1330-1378	2372-2532	.2839
Strontium	Sr	2.54	.074	769	1416	.0918
Tantalum	Ta	10.80	–	2850	5160	.3902
Tin	Sn	7.29	.056	232	450	.2634
Titanium	Ti	5.3	.130	1900	3450	.1915
Tungsten	W	19.10	.033	3000	5432	.6900
Uranium	U	18.70	–	1132	2070	.6755
Vanadium	V	5.50	–	1730	3146	.1987
Zinc	Zn	7.19	.094	419	786	.2598

*Circular of the Bureau of Standards No.35, Department of Commerce and Labor

DENSITY OF SPECIFIC ENGINEERING MATERIALS

THERMOPLASTICS

Material	Density (g/cc)
Polyethylene, high-density	0.935-0.960
Polyethylene, low-density	0.910-0.925
Polystyrene, high-impact	1.05-1.15
ABS	1.05-1.19
Acetal (Delrin)	1.41
Nylon, 66	1.14
Nylon 66, 40% glass	1.44-1.47
Polycarbonate	1.20
Polycarbonate, 30% glass	1.43
PPO	1.04-1.07
PPO, 30% glass	1.28
Polyester teraphthalate, 30% glass	1.70
Polypropylene	0.90
Polypropylene, 40% Talc	1.26

THERMOSET PLASTICS

Carbon fiber, epoxy	1.80
Fiberglass, epoxy	1.90
Fiberglass, polyester	1.90
Phenolics	1.60

STEELS

Stainless, T304	7.90
Low-carbon, 1020	7.87
Low-carbon, cold rolled	7.87
High strength low alloy (high-carbon)	7.90

TITANIUM

Grade 2	4.51
Grade 3	4.50
Grade 4	4.51

ALUMINUM

Material	Density (g/cc)
1100	2.71
6061, T6	2.70
7075, T6	2.81
A360	2.68
A380	2.76
B390	2.71

ZINC

Zamak 2	6.60
Zamak 5	6.70
ZA8	6.30
ZA12	6.04
ZA27	5.02

BRASS

Brass 70/30	8.52
Brass 60/40	8.38
Phosphor bronze	8.92

MAGNESIUM

AZ91D	1.81
AMGOB	3.58

CAST IRON

Gray	7.15-7.87
Nodular	7.30

MONEL

Nickel copper	8.80

COMMON SHAPES OF METALS

Shapes	Length	How Measured	How Purchased
Sheet less than 1/4″ thick	Up to 144″	Thickness × width, widths to 72″	Weight, foot, or piece
Plate more than 1/4″ thick	Up to 20′	Thickness × width	Weight, foot, or piece
Band	Up to 20′	Thickness × width	Weight or piece
Rod	12′ to 20′	Diameter	Weight, foot, or piece
Square	12′ to 20′	Width	Weight, foot, or piece
Flats	Hot rolled 20′-22′ Cold finished	Thickness × width	Weight, foot, or piece
Hexagon	12′ to 20′	Distance across flats	Weight, foot, or piece
Octagon	12′ to 20′	Distance across flats	Weight, foot, or piece
Angle	Up to 40′	Leg length × leg length × thickness of legs	Weight, foot, or piece
Channel	Up to 60′	Depth × web thickness × flange width	Weight, foot, or piece
I-beam	Up to 60′	Height × web thickness × flange width	Weight, foot, or piece

SHEET METAL AND WIRE GAGE DESIGNATION

GAGE NO.	AMERICAN OR BROWN & SHARPE'S A.W.G. OR B. & S.	BIRMING-HAM OR STUBS WIRE B.W.G.	WASHBURN & MOEN OR AMERICAN S.W.G.	UNITED STATES STANDARD	MANU-FACTURERS' STANDARD FOR SHEET STEEL	GAGE NO.
0000000	- - - -	- - - -	.4900	.500	- - - -	0000000
000000	.5800	- - - -	.4615	.469	- - - -	000000
00000	.5165	- - - -	.4305	.438	- - - -	00000
0000	.4600	.454	.3938	.406	- - - -	0000
000	.4096	.425	.3625	.375	- - - -	000
00	.3648	.380	.3310	.344	- - - -	00
0	.3249	.340	.3065	.312	- - - -	0
1	.2893	.300	.2830	.281	- - - -	1
2	.2576	.284	.2625	.266	- - - -	2
3	.2294	.259	.2437	.250	.2391	3
4	.2043	.238	.2253	.234	.2242	4
5	.1819	.220	.2070	.219	.2092	5
6	.1620	.203	.1920	.203	.1943	6
7	.1443	.180	.1770	.188	.1793	7
8	.1285	.165	.1620	.172	.1644	8
9	.1144	.148	.1483	.156	.1495	9
10	.1019	.134	.1350	.141	.1345	10
11	.0907	.120	.1205	.125	.1196	11
12	.0808	.109	.1055	.109	.1046	12
13	.0720	.095	.0915	.0938	.0897	13
14	.0642	.083	.0800	.0781	.0747	14
15	.0571	.072	.0720	.0703	.0673	15
16	.0508	.065	.0625	.0625	.0598	16
17	.0453	.058	.0540	.0562	.0538	17
18	.0403	.049	.0475	.0500	.0478	18
19	.0359	.042	.0410	.0438	.0418	19
20	.0320	.035	.0348	.0375	.0359	20
21	.0285	.032	.0317	.0344	.0329	21
22	.0253	.028	.0286	.0312	.0299	22
23	.0226	.025	.0258	.0281	.0269	23
24	.0201	.022	.0230	.0250	.0239	24
25	.0179	.020	.0204	.0219	.0209	25
26	.0159	.018	.0181	.0188	.0179	26
27	.0142	.016	.0173	.0172	.0164	27
28	.0126	.014	.0162	.0156	.0149	28
29	.0113	.013	.0150	.0141	.0135	29
30	.0100	.012	.0140	.0125	.0120	30
31	.0089	.010	.0132	.0109	.0105	31
32	.0080	.009	.0128	.0102	.0097	32
33	.0071	.008	.0118	.00938	.0090	33
34	.0063	.007	.0104	.00859	.0082	34
35	.0056	.005	.0095	.00781	.0075	35
36	.0050	.004	.0090	.00703	.0067	36
37	.0045	- - - -	.0085	.00624	.0064	37
38	.0040	- - - -	.0080	.00625	.0060	38
39	.0035	- - - -	.0075	- - - -	- - - -	39
40	.0031	- - - -	.0070	- - - -	- - - -	40
41	.0028	- - - -	.0066	- - - -	- - - -	41
42	.0025	- - - -	.0062	- - - -	- - - -	42
43	.0022	- - - -	.0060	- - - -	- - - -	43
44	.0020	- - - -	.0058	- - - -	- - - -	44
45	.0018	- - - -	.0055	- - - -	- - - -	45
46	.0016	- - - -	.0052	- - - -	- - - -	46
47	.0014	- - - -	.0050	- - - -	- - - -	47
48	.0012	- - - -	.0048	- - - -	- - - -	48

METAL SHEET MATERIALS CHART

Material (Sheet less than 1/4″ thick)	How Measured	How Purchased	Characteristics
Copper	Gage number (Brown & Sharpe and Amer. Std.)	24″ × 96″ sheet or 12″ or 18″ by lineal feet on roll	Pure metal
Brass	Gage number (Brown & Sharpe and Amer. Std.)	24″ × 76″ sheet or 12″ or 18″ by lineal feet on roll	Alloy of copper and zinc
Aluminum	Decimal	24″ × 72″ sheet or 12″ or 18″ by lineal feet on roll	Available as commercially pure metal or alloyed for strength, hardness, and ductility
Galvanized steel	Gage number (Amer. Std.)	24″ × 96″ sheet	Mild steel sheet with zinc plating, also available with zinc coating that is part of sheet
Black annealed steel sheet	Gage number (Amer. Std.)	24″ × 96″ sheet	Mild steel with oxide coating, hot-rolled
Cold-rolled steel sheet	Gage number (Amer. Std.)	24″ × 96″ sheet	Oxide removed and cold-rolled to final thickness
Tin plate	Gage number (Amer. Std.)	20″ × 28″ sheet 56 or 112 to pkg.	Mild steel with tin coating
Nickel silver	Gage number (Brown & Sharpe)	6″ or 12″ wide by lineal feet on roll	Copper 50%, zinc 30%, nickel 20%
Expanded	Gage number (Amer. Std.)	36″ × 96″ sheet	Metal is pierced and expanded (stretched) to diamond shape; also available rolled to thickness after it has been expanded
Perforated	Gage number (Amer. Std.)	30″ × 36″ sheet 36″ × 48″ sheet	Design is cut in sheet; many designs available

TAP DRILL SIZES UNIFIED STANDARD SCREW HEADS

Screw Thread		Tap Drill	Screw Thread		Tap Drill
Major Diameter	Threads Per Inch	Size Or Number	Major Diameter	Threads Per Inch	Size Or Number
0	80	3/64	3/8	16	5/16
				24	Q
1	64	53			
	72	53	7/16	14	U
				20	25/64
2	56	50			
	64	50	1/2	13	27/64
				20	29/64
3	48	47			
	56	45	9/16	12	31/64
				18	33/64
4	40	43			
	48	42	5/8	11	17/32
				18	37/64
5	40	38			
	44	37	3/4	10	21/32
				16	11/16
6	32	36			
	40	33	7/8	9	49/64
				14	13/16
8	32	29			
	36	29	1	8	7/8
				12	59/64
10	24	25			
	32	21	1 1/8	7	63/64
				12	1 3/64
12	24	16			
	28	14	1 1/4	7	1 7/64
				12	1 11/64
1/4	20	7			
	28	3	1 3/8	6	1 7/32
				12	1 19/64
5/16	18	F			
	24	I	1 1/2	6	1 11/32
				12	1 27/64

MACHINE SCREW AND CAP SCREW HEADS

	Size	A	B	C	D
Fillister Head	#8	.260	.141	.042	.060
	#10	.302	.164	.048	.072
	1/4	3/8	.205	.064	.087
	5/16	7/16	.242	.077	.102
	3/8	9/16	.300	.086	.125
	1/2	3/4	.394	.102	.168
	5/8	7/8	.500	.128	.215
	3/4	1	.590	.144	.258
	1	1 5/16	.774	.182	.352
Flat Head	#8	.320	.092	.043	.037
	#10	.372	.107	.048	.044
	1/4	1/2	.146	.064	.063
	5/16	5/8	.183	.072	.078
	3/8	3/4	.220	.081	.095
	1/2	7/8	.220	.102	.090
	5/8	1 1/8	.293	.128	.125
	3/4	1 3/8	.366	.144	.153
Round Head	#8	.297	.113	.044	.067
	#10	.346	.130	.048	.073
	1/4	7/16	.183	.064	.107
	5/16	9/16	.236	.072	.150
	3/8	5/8	.262	.081	.160
	1/2	13/16	.340	.102	.200
	5/8	1	.422	.128	.255
	3/4	1 1/4	.526	.144	.320
Hexagon Head	1/4	.494	.170	7/16	
	5/16	.564	.215	1/2	
	3/8	.635	.246	9/16	
	1/2	.846	.333	3/4	
	5/8	1.058	.411	15/16	
	3/4	1.270	.490	1 1/8	
	7/8	1.482	.566	1 5/16	
	1	1.693	.640	1 1/2	
Socket Head	#8	.265	.164	1/8	
	#10	5/16	.190	5/32	
	1/4	3/8	1/4	3/16	
	5/16	7/16	5/16	7/32	
	3/8	9/16	3/8	5/16	
	7/16	5/8	7/16	5/16	
	1/2	3/4	1/2	3/8	
	5/8	7/8	5/8	1/2	
	3/4	1	3/4	9/16	
	7/8	1 1/8	7/8	9/16	
	1	1 5/16	1	5/8	

GRADE MARKING FOR BOLTS

Bolt Head Marking	SAE = Society of Automotive Engineers ASTM = American Society for Testing and Materials	Bolt Material	Minimum Tensile Strength in Pounds per Square Inch (psi)
No Marks	SAE Grade 1 SAE Grade 2 Indeterminate quality	Low-carbon steel Low-carbon steel	65,000 psi
2 Marks	SAE Grade 3	Medium-carbon steel, cold worked	110,000 psi
3 Marks	SAE Grade 5 ASTM – A 325 Common commercial quality	Medium-carbon steel, quenched and tempered	120,000 psi
Letters BB	ASTM – A 354	Low-alloy steel or medium-carbon steel, quenched and tempered	105,000 psi
Letters BC	ASTM – A 354	Low-alloy steel or medium-carbon steel, quenched and tempered	125,000 psi
4 Marks	SAE Grade 6 Better commercial quality	Medium-carbon steel, quenched and tempered	140,000 psi
5 Marks	SAE Grade 7	Medium-carbon alloy steel, quenched and tempered, roll-threaded after heat treatment	133,000 psi
6 Marks	SAE Grade 8 ASTM – A 345 Best commercial quality	Medium-carbon alloy steel, quenched and tempered	150,000 psi

NUMBER AND LETTER DRILLS

Drill No.	Frac	Deci	Drill No.	Frac	Deci	Drill No.	Frac	Deci
80	—	.0135		9/64	.140	S	—	.348
79	—	.0145	28	—	.141	T	—	.358
	1/64	.0156	27	—	.144		23/64	.359
78	—	.0160	26	—	.147	U	—	.368
77	—	.0180	25	—	.150		3/8	.375
76	—	.0200	24	—	.152	V	—	.377
75	—	.0210	23	—	.154	W	—	.386
74	—	.0225		5/32	.156		25/64	.391
73	—	.0240	22	—	.157	X	—	.397
72	—	.0250	21	—	.159	Y	—	.404
71	—	.0260	20	—	.161		13/32	.406
70	—	.0280	19	—	.166	Z	—	.413
69	—	.0292	18	—	.170		27/64	.422
68	—	.0310		11/64	.172		7/16	.438
	1/32	.0313	17	—	.173		29/64	.453
67	—	.0320	16	—	.177		15/32	.469
66	—	.0330	15	—	.180		31/64	.484
65	—	.0350	14	—	.182		1/2	.500
64	—	.0360	13	—	.185		33/64	.516
63	—	.0370		3/16	.188		17/32	.531
62	—	.0380	12	—	.189		35/64	.547
61	—	.0390	11	—	.191		9/16	.562
60	—	.0400	10	—	.194		37/64	.578
59	—	.0410	9	—	.196		19/32	.594
58	—	.0420	8	—	.199		39/64	.609
57	—	.0430	7	—	.201		5/8	.625
56	—	.0465		13/64	.203		41/64	.641
	3/64	.0469	6	—	.204		21/32	.656
55	—	.0520	5	—	.206		43/64	.672
54	—	.0550	4	—	.209		11/16	.688
53	—	.0595	3	—	.213		45/64	.703
	1/16	.0625		7/32	.219		23/32	.719
52	—	.0635	2	—	.221		47/64	.734
51	—	.0670	1	—	.228		3/4	.750
50	—	.0700	A	—	.234		49/64	.766
49	—	.0730		15/64	.234		25/32	.781
48	—	.0760	B	—	.238		51/64	.797
	5/64	.0781	C	—	.242		13/16	.813
47	—	.0785	D	—	.246		53/64	.828
46	—	.0810		1/4	.250		27/32	.844
45	—	.0820	E	—	.250		55/64	.859
44	—	.0860	F	—	.257		7/8	.875
43	—	.0890	G	—	.261		57/64	.891
42	—	.0935		17/64	.266		29/32	.906
	3/32	.0938	H	—	.266		59/64	.922
41	—	.0960	I	—	.272		15/16	.938
40	—	.0980	J	—	.277		61/64	.953
39	—	.0995		9/32	.281		31/32	.969
38	—	.1015	K	—	.281		63/64	.984
37	—	.1040	L	—	.290		1	1.000
36	—	.1065	M	—	.295			
	7/64	.1094		19/64	.297			
35	—	.1100	N	—	.302			
34	—	.1110		5/16	.313			
33	—	.1130	O	—	.316			
32	—	.116	P	—	.323			
31	—	.120		21/64	.328			
	1/8	.125	Q	—	.332			
30	—	.129	R	—	.339			
29	—	.136		11/32	.344			

METRIC DRILLS

MM	DEC.	MM	DEC.	MM	DEC.	MM	DEC.
1.	.0394	3.2	.1260	6.3	.2480	9.5	.3740
1.05	.0413	3.25	.1280	6.4	.2520	9.6	.3780
1.1	.0433	3.3	.1299	6.5	.2559	9.7	.3819
1.15	.0453	3.4	.1339	6.6	.2598	9.75	.3839
1.2	.0472	3.5	.1378	6.7	.2638	9.8	.3858
1.25	.0492	3.6	.1417	6.75	.2657	9.9	.3898
1.3	.0512	3.7	.1457	6.8	.2677	10.	.3937
1.35	.0531	3.75	.1476	6.9	.2717	10.5	.4134
1.4	.0551	3.8	.1496	7.	.2756	11.	.4331
1.45	.0571	3.9	.1535	7.1	.2795	11.5	.4528
1.5	.0591	4.	.1575	7.2	.2835	12.	.4724
1.55	.0610	4.1	.1614	7.25	.2854	12.5	.4921
1.6	.0630	4.2	.1654	7.3	.2874	13.	.5118
1.65	.0650	4.25	.1673	7.4	.2913	13.5	.5315
1.7	.0669	4.3	.1693	7.5	.2953	14.	.5512
1.75	.0689	4.4	.1732	7.6	.2992	14.5	.5709
1.8	.0709	4.5	.1772	7.7	.3031	15.	.5906
1.85	.0728	4.6	.1811	7.75	.3051	15.5	.6102
1.9	.0748	4.7	.1850	7.8	.3071	16.	.6299
1.95	.0768	4.75	.1870	7.9	.3110	16.5	.6496
2.	.0787	4.8	.1890	8.	.3150	17.	.6693
2.05	.0807	4.9	.1929	8.1	.3189	17.5	.6890
2.1	.0827	5.	.1968	8.2	.3228	18.	.7087
2.15	.0846	5.1	.2008	8.25	.3248	18.5	.7283
2.2	.0866	5.2	.2047	8.3	.3268	19.	.7480
2.25	.0886	5.25	.2067	8.4	.3307	19.5	.7677
2.3	.0906	5.3	.2087	8.5	.3346	20.	.7874
2.35	.0925	5.4	.2126	8.6	.3386	20.5	.8071
2.4	.0945	5.5	.2165	8.7	.3425	21.	.8268
2.45	.0965	5.6	.2205	8.75	.3445	21.5	.8465
2.5	.0984	5.7	.2244	8.8	.3465	22.	.8661
2.6	.1024	5.75	.2264	8.9	.3504	22.5	.8858
2.7	.1063	5.8	.2283	9.	.3543	23.	.9055
2.75	.1083	5.9	.2323	9.1	.3583	23.5	.9252
2.8	.1102	6.	.2362	9.2	.3622	24.	.9449
2.9	.1142	6.1	.2402	9.25	.3642	24.5	.9646
3	.1181	6.2	.2441	9.3	.3661	25.	.9843
3.1	.1220	6.25	.2461	9.4	.3701		

UNIFIED STANDARD THREAD SERIES

SIZES Primary	SIZES Secondary	BASIC MAJOR DIAMETER	Coarse UNC	Fine UNF	Extra fine UNEF	4UN	6UN	8UN	12UN	16UN	20UN	28UN	32UN	SIZES
0		0.0600	—	80	—	—	—	—	—	—	—	—	—	0
	1	0.0730	64	72	—	—	—	—	—	—	—	—	—	1
2		0.0860	56	64	—	—	—	—	—	—	—	—	—	2
	3	0.0990	48	56	—	—	—	—	—	—	—	—	—	3
4		0.1120	40	48	—	—	—	—	—	—	—	—	—	4
5		0.1250	40	44	—	—	—	—	—	—	—	—	—	5
6		0.1380	32	40	—	—	—	—	—	—	—	—	UNC	6
8		0.1640	32	36	—	—	—	—	—	—	—	—	UNC	8
10		0.1900	24	32	—	—	—	—	—	—	—	—	UNF	10
	12	0.2160	24	28	32	—	—	—	—	—	—	UNF	UNEF	12
1/4		0.2500	20	28	32	—	—	—	—	—	UNC	UNF	UNEF	1/4
5/16		0.3125	18	24	32	—	—	—	—	—	20	28	UNEF	5/16
3/8		0.3750	16	24	32	—	—	—	—	UNC	20	28	UNEF	3/8
7/16		0.4375	14	20	28	—	—	—	—	16	UNF	UNEF	32	7/16
1/2		0.5000	13	20	28	—	—	—	—	16	UNF	UNEF	32	1/2
9/16		0.5625	12	18	24	—	—	—	UNC	16	20	28	32	9/16
5/8		0.6250	11	18	24	—	—	—	12	16	20	28	32	5/8
	11/16	0.6875	—	—	24	—	—	—	12	16	20	28	32	11/16
3/4		0.7500	10	16	20	—	—	—	12	UNF	UNEF	28	32	3/4
	13/16	0.8125	—	—	20	—	—	—	12	16	UNEF	28	32	13/16
7/8		0.8750	9	14	20	—	—	—	12	16	UNEF	28	32	7/8
	15/16	0.9375	—	—	20	—	—	—	12	16	UNEF	28	32	15/16
1		1.0000	8	12	20	—	—	UNC	UNF	16	UNEF	28	32	1
	1 1/16	1.0625	—	—	18	—	—	8	12	16	20	28	—	1 1/16
1 1/8		1.1250	7	12	18	—	—	8	UNF	16	20	28	—	1 1/8
	1 3/16	1.1875	—	—	18	—	—	8	12	16	20	28	—	1 3/16
1 1/4		1.2500	7	12	18	—	—	8	UNF	16	20	28	—	1 1/4
	1 5/16	1.3125	—	—	18	—	—	8	12	16	20	28	—	1 5/16
1 3/8		1.3750	6	12	18	—	UNC	8	UNF	16	20	28	—	1 3/8
	1 7/16	1.4375	—	—	18	—	6	8	12	16	20	28	—	1 7/16
1 1/2		1.5000	6	12	18	—	UNC	8	UNF	16	20	28	—	1 1/2
	1 9/16	1.5625	—	—	18	—	6	8	12	16	20	—	—	1 9/16
1 5/8		1.6250	—	—	18	—	6	8	12	16	20	—	—	1 5/8
	1 11/16	1.6875	—	—	18	—	6	8	12	16	20	—	—	1 11/16
1 3/4		1.7500	5	—	—	—	6	8	12	16	20	—	—	1 3/4
	1 13/16	1.8125	—	—	—	—	6	8	12	16	20	—	—	1 13/16
1 7/8		1.8750	—	—	—	—	6	8	12	16	20	—	—	1 7/8
	1 15/16	1.9375	—	—	—	—	6	8	12	16	20	—	—	1 15/16
2		2.0000	4½	—	—	—	6	8	12	16	20	—	—	2
	2 1/8	2.1250	—	—	—	—	6	8	12	16	20	—	—	2 1/8
2 1/4		2.2500	4½	—	—	—	6	8	12	16	20	—	—	2 1/4
	2 3/8	2.3750	—	—	—	—	6	8	12	16	20	—	—	2 3/8
2 1/2		2.5000	4	—	—	UNC	6	8	12	16	20	—	—	2 1/2
	2 5/8	2.6250	—	—	—	4	6	8	12	16	20	—	—	2 5/8
2 3/4		2.7500	4	—	—	UNC	6	8	12	16	20	—	—	2 3/4
	2 7/8	2.8750	—	—	—	4	6	8	12	16	20	—	—	2 7/8
3		3.0000	4	—	—	UNC	6	8	12	16	20	—	—	3
	3 1/8	3.1250	—	—	—	4	6	8	12	16	—	—	—	3 1/8
3 1/4		3.2500	4	—	—	UNC	6	8	12	16	—	—	—	3 1/4
	3 3/8	3.3750	—	—	—	4	6	8	12	16	—	—	—	3 3/8
3 1/2		3.5000	4	—	—	UNC	6	8	12	16	—	—	—	3 1/2
	3 5/8	3.6250	—	—	—	4	6	8	12	16	—	—	—	3 5/8
3 3/4		3.7500	4	—	—	UNC	6	8	12	16	—	—	—	3 3/4
	3 7/8	3.8750	—	—	—	4	6	8	12	16	—	—	—	3 7/8
4		4.0000	4	—	—	UNC	6	8	12	16	—	—	—	4
	4 1/8	4.1250	—	—	—	4	6	8	12	16	—	—	—	4 1/8
4 1/4		4.2500	—	—	—	4	6	8	12	16	—	—	—	4 1/4
	4 3/8	4.3750	—	—	—	4	6	8	12	16	—	—	—	4 3/8
4 1/2		4.5000	—	—	—	4	6	8	12	16	—	—	—	4 1/2
	4 5/8	4.6250	—	—	—	4	6	8	12	16	—	—	—	4 5/8
4 3/4		4.7500	—	—	—	4	6	8	12	16	—	—	—	4 3/4
	4 7/8	4.8750	—	—	—	4	6	8	12	16	—	—	—	4 7/8
5		5.0000	—	—	—	4	6	8	12	16	—	—	—	5
	5 1/8	5.1250	—	—	—	4	6	8	12	16	—	—	—	5 1/8
5 1/4		5.2500	—	—	—	4	6	8	12	16	—	—	—	5 1/4
	5 3/8	5.3750	—	—	—	4	6	8	12	16	—	—	—	5 3/8
5 1/2		5.5000	—	—	—	4	6	8	12	16	—	—	—	5 1/2
	5 5/8	5.6250	—	—	—	4	6	8	12	16	—	—	—	5 5/8
5 3/4		5.7500	—	—	—	4	6	8	12	16	—	—	—	5 3/4
	5 7/8	5.8750	—	—	—	4	6	8	12	16	—	—	—	5 7/8
6		6.0000	—	—	—	4	6	8	12	16	—	—	—	6

ISO METRIC SCREW THREAD STANDARD SERIES

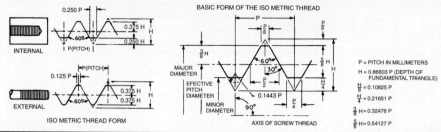

Nominal Size Diam. (mm)			Pitches (mm)														Nominal Size Diam. (mm)
Column[a]			Series With Graded Pitches		Series With Constant Pitches												
1	2	3	Coarse	Fine	6	4	3	2	1.5	1.25	1	0.75	0.5	0.35	0.25	0.2	
0.25			0.075	—	—	—	—	—	—	—	—	—	—	—	—	—	0.25
0.3			0.08	—	—	—	—	—	—	—	—	—	—	—	—	—	0.3
	0.35		0.09	—	—	—	—	—	—	—	—	—	—	—	—	—	0.35
0.4			0.1	—	—	—	—	—	—	—	—	—	—	—	—	—	0.4
	0.45		0.1	—	—	—	—	—	—	—	—	—	—	—	—	—	0.45
0.5			0.125	—	—	—	—	—	—	—	—	—	—	—	—	—	0.5
	.055		0.125	—	—	—	—	—	—	—	—	—	—	—	—	—	0.55
0.6			0.15	—	—	—	—	—	—	—	—	—	—	—	—	—	0.6
	0.7		0.175	—	—	—	—	—	—	—	—	—	—	—	—	—	0.7
0.8			0.2	—	—	—	—	—	—	—	—	—	—	—	—	—	0.8
	0.9		0.225	—	—	—	—	—	—	—	—	—	—	—	—	—	0.9
1			0.25	—	—	—	—	—	—	—	—	—	—	—	—	0.2	1
	1.1		0.25	—	—	—	—	—	—	—	—	—	—	—	—	0.2	1.1
1.2			0.25	—	—	—	—	—	—	—	—	—	—	—	—	0.2	1.2
	1.4		0.3	—	—	—	—	—	—	—	—	—	—	—	—	0.2	1.4
1.6			0.35	—	—	—	—	—	—	—	—	—	—	—	—	0.2	1.6
	1.8		0.35	—	—	—	—	—	—	—	—	—	—	—	—	0.2	1.8
2			0.4	—	—	—	—	—	—	—	—	—	—	—	0.25	—	2
	2.2		0.45	—	—	—	—	—	—	—	—	—	—	—	0.25	—	2.2
2.5			0.45	—	—	—	—	—	—	—	—	—	—	0.35	—	—	2.5
3			0.5	—	—	—	—	—	—	—	—	—	—	0.35	—	—	3
	3.5		0.6	—	—	—	—	—	—	—	—	—	—	0.35	—	—	3.5
4			0.7	—	—	—	—	—	—	—	—	—	0.5	—	—	—	4
	4.5		0.75	—	—	—	—	—	—	—	—	—	0.5	—	—	—	4.5
5			0.8	—	—	—	—	—	—	—	—	—	0.5	—	—	—	5
		5.5	—	—	—	—	—	—	—	—	—	—	0.5	—	—	—	5.5
6			1	—	—	—	—	—	—	—	—	0.75	—	—	—	—	6
		7	1	—	—	—	—	—	—	—	—	0.75	—	—	—	—	7
8			1.25	1	—	—	—	—	—	—	1	0.75	—	—	—	—	8
		9	1.25	—	—	—	—	—	—	—	1	0.75	—	—	—	—	9
10			1.5	1.25	—	—	—	—	—	1.25	1	0.75	—	—	—	—	10
		11	1.5	1.25	—	—	—	—	—	—	1	0.75	—	—	—	—	11
12			1.75	1.25	—	—	—	—	1.5	1.25	1	—	—	—	—	—	12
	14		2	1.5	—	—	—	—	1.5	1.25[b]	1	—	—	—	—	—	14
		15	—	1.5	—	—	—	—	1.5	—	1	—	—	—	—	—	15
16			2	1.5	—	—	—	—	1.5	—	1	—	—	—	—	—	16
		17	—	1.5	—	—	—	—	1.5	—	1	—	—	—	—	—	17
	18		2.5	1.5	—	—	—	2	1.5	—	1	—	—	—	—	—	18
20			2.5	1.5	—	—	—	2	1.5	—	1	—	—	—	—	—	20
	22		2.5	1.5	—	—	—	2	1.5	—	1	—	—	—	—	—	22
24			3	2	—	—	—	2	1.5	—	1	—	—	—	—	—	24
		25	—	—	—	—	—	2	1.5	—	1	—	—	—	—	—	25
		26	—	—	—	—	—	—	1.5	—	—	—	—	—	—	—	26
	27		3	2	—	—	—	2	1.5	—	1	—	—	—	—	—	27
		28	—	—	—	—	—	2	1.5	—	1	—	—	—	—	—	28
30			3.5	2	—	—	(3)	2	1.5	—	1	—	—	—	—	—	30
		32	—	—	—	—	—	2	1.5	—	—	—	—	—	—	—	32
	33		3.5	2	—	—	(3)	2	1.5	—	—	—	—	—	—	—	33
		35[c]	—	—	—	—	—	—	1.5	—	—	—	—	—	—	—	35[c]
36			4	3	—	—	—	2	1.5	—	—	—	—	—	—	—	36
		38	—	—	—	—	—	—	1.5	—	—	—	—	—	—	—	38
	39		4	3	—	—	—	2	1.5	—	—	—	—	—	—	—	39
		40	—	—	—	—	3	2	1.5	—	—	—	—	—	—	—	40
42			4.5	3	—	—	3	2	1.5	—	—	—	—	—	—	—	42
	45		4.5	3	—	4	3	2	1.5	—	—	—	—	—	—	—	45

a Thread diameter should be selected from columns 1, 2, or 3; with preference being given in that order.
b Pitch 1.25 mm in combination with diameter 14 mm has been included for spark plug applications.
c Diameter 35 mm has been included for bearing locknut applications.
The use of pitches shown in parentheses should be avoided wherever possible.
The pitches enclosed in the bold frame, together with the corresponding nominal diameters in Columns 1 and 2, are those combinations which have been established by ISO Recommendations as a selected "coarse" and "fine" series for commercial fasteners. Sizes 0.25 mm through 1.4 mm are covered in ISO Recommendation R 68 and, except for the 0.25 mm size, in **AN Standard ANSI B1.10.**

(ANSI)

DRAFTING STANDARDS AND RELATED DOCUMENTS

The following is a list of ANSI/ASME drafting standards or related documents. They are ANSI/ASME adopted, unless another standard developing organization, such as ANSI/NFPA, is indicated.

Abbreviations

Y1.1-1989, *Abbreviations for Use on Drawings and in Text*

Charts and Graphs (Y15)

Y15.1M-1979 (R1993), *Illustrations for Publication and Projection*

Y15.2M-1979 (R1986), *Time-Series Charts*

Y15.3M-1979 (R1986), *Process Charts*

Dimensions

B4.1-1967 (R1987), *Preferred Limits and Fits for Cylindrical Parts*

B4.2-1978 (R1994), *Preferred Metric Limits and Fits*

B4.3-1978 (R1994), *General Tolerances for Metric Dimensioned Products*

B4.4M-1981 (R1987), *Inspection of Workpieces*

B32.1-1952 (R1994), *Preferred Thickness for Uncoated, Thin, Flat Metals (Under 0.250/in.)*

B32.2-1969 (R1994), *Preferred Diameters for Round Wire—0.500 Inches and Under*

B32.3M-1984 (R1994), *Preferred Metric Sizes for Flat Metal Products*

B32.4M-1980 (R1994), *Preferred Metric Sizes for Round, Square, Rectangle, and Hexagon Metal Products*

B32.5-1977 (R1994), *Preferred Metric Sizes for Tubular Metal Products Other Than Pipe*

B32.6M-1984 (R1994), *Preferred Metric Equivalents of Inch Sizes for Tubular Metal Products Other Than Pipe*

B36.10M-1985, *Welded and Seamless Wrought Steel Pipe*

B36.19M-1985, *Stainless Steel Pipe*

Drafting Standards

Y14.1-1980 (R1987), *Drawing Sheet Size and Format*

Y14.1M-1992, *Metric Drawing Sheet Size and Format*

Y14.2M-1992, *Line Conventions and Lettering*

Y14.3M-1992, *Multi- and Sectional-View Drawings*

Y14.4M-1989 (R1994), *Pictorial Drawings*

Y14.5M-1994, *Dimensioning and Tolerancing*

Y14.5.1-1994, *Mathematical Definition of Y14.5*

Y14.5.2, *Certification of GD&T Professionals*

Y14.6M-1978 (R1993), *Screw Thread Representation*

14.6aM-1981 (R1993), *Engineering Drawing and Related Documentation Practices (Screw Thread Representation) (Metric Supplement)*

Y14.7.1-1971 (R1993), *Gear Drawing Standards—Part 1—Spur, Helical, Double Helical, and Rack*

(Continued)

Drafting Standards, *(Continued)*

Y14.7.2-1978 (R1994), *Gear and Spline Drawing Standards—Part 2—Bevel and Hypoid Gears*

Y14.8M-1989 (R1993), *Castings and Forgings*

Y14.13M-1981 (R1992), *Engineering Drawing and Related Documentation Practices—Mechanical Spring Representation*

Y14.18M-1986 (R1993), *Engineering Drawings and Related Documentation Practices–Optical Parts*

Y14.24M-1989, *Types and Applications of Engineering Drawings*

14.34M-1989 (R1993), *Parts Lists, Data Lists, and Index Lists*

Y14.35M-1992, *Revision of Engineering Drawings and Associated Documents*

Y14.36-1978 (R1993), *Surface Texture Symbols*

Y14 Report 1, *Digital Representation of Physical Object Shapes*

Y14 Report 2, *Guidelines for Documenting of Computer Systems Used in Computer-Aided Preparation of Product Definition Data—User Instructions*

Y14 Report 3, *Guidelines for Documenting of Computer Systems Used in Computer-Aided Preparation of Product Definition Data—Design Requirements*

Y14 Report 4-1989, *A Structural Language Format for Basic Shape Description*

ANSI/US PRO/IPO-100-1993, *Digital Representation for Communication of Product Definition Data (Replaced ANSI Y14.26M-1981)*

Graphic Symbols

Y32.2-1975, *Electrical and Electronic Diagrams*

Y32.2.3-1949 (R1988), *Pipe Fittings, Valves, and Piping*

Y32.2.4-1949 (R1993), *Heating, Ventilating, and Air Conditioning*

Y32.2.6-1950 (R1993), *Heat/Power Apparatus*

Y32.4-1977 (R1987), *Plumbing Fixture Diagrams Used in Architectural and Building Construction*

Y32.7-1972 (R1987), *Railroad Maps and Profiles*

Y32.9-1972 (R1989), *Electrical Wiring and Layout Diagrams Used in Architecture and Building*

Y32.10-1967 (R1987), *Fluid Power Diagrams*

Y32.11-1961 (R1993), *Process Flow Diagrams in the Petroleum and Chemical Industries*

Y32.18-1972 (R1993), *Mechanical and Acoustical Elements as Used in Schematic Diagrams*

ANSI/AWS A2.4-91, *Symbols for Welding, Brazing, and Nondestructive Examination*

ANSI/IEEE 200-1975 (R1989), *Reference Designations for Electrical and Electronics Parts and Equipment*

ANSI/IEEE 315-1975 (R1989), *Electrical and Electronics Diagrams (Including Reference Designation Class Designation Letters)*

ANSI/IEEE 623-1976 (R1989), *Grid and Mapping Used in Cable Television Systems*

ANSI/ISA S5.1-1984 (R1992), *Instrumentation Symbols and Identification*

ANSI/NFPA 170-1991, *Public Fire Safety Symbols*

(Continued)

Letter Symbols

Y10.1-1972 (R1988), *Glossary of Terms Concerning Letter Symbols*
Y10.3M-1984, *Mechanics and Time-Related Phenomena*
Y10.4-1982 (R1988), *Heat and Thermodynamics*
Y10.11-1984, *Acoustics*
Y10.12-1955 (R1988), *Chemical Engineering*
Y10.17-1961 (R1988), *Greek Letters Used as Letter Symbols for Engineering Math*
Y10.18-1967 (R1977), *Illuminating Engineering*
ANSI/IEEE 260-1978 (R1992), *SI Units and Certain Other Units of Measurement*

Metric System

SI-1, *Orientation and Guide for Use of SI (Metric) Units*
SI-2, *SI Units in Strength of Materials*
SI-3, *SI Units in Dynamics*
SI-4, *SI Units in Thermodynamics*
SI-5, *SI Units in Fluid Mechanics*
SI-6, *SI Units in Kinematics*
SI-7, *SI Units in Heat Transfer*
SI-8, *SI Units in Vibration*
SI-9, *Metrification of Codes and Standards SI (Metric) Units*
SI-10, *Steam Charts, SI (Metric) and U.S. Customary Units*

Index

H

hole notes, 83–84, 193–194
holes,
  at a distance from existing holes,
    91–92
  changing color of, 92
  changing size, 84
  constructing, 81–93
  counterbored, 89
  countersunk, 89
  creating multiple holes, 85
  located from two straight edges,
    90–91
  placing hole notes on drawing, 83–84
  threaded, 86
Horizontal constraint, 168
hot keys, 28

I

importing additional parts, 265–267
inner circle, 30
Insert assembly constraint, 260, 265, 293
  and explosion factor, 320
instance, 120
instancing, 257–258
interference, 277–278
ISOLINES, 173, 228
isometric view, 72

K

keyboard aliases, 21
keyway, 30

L

LAYER command, 132, 140, 325
layout,
  creating, 70
  page setup, 69
  renaming, 69
  working in **Drawing** mode, 69
leaders, 195–196
linear fillet, 155, 156
linked spreadsheet,
  driving a part's dimensions, 247–248
  table-driven assembly, 269–271
  table-driven part, 105–106
LIST command, 263

local assembly, 241
local design variables, 99
local part, 285

M

Maintain Orientation option, 124
mass properties analysis, 274–277
Mate assembly constraint, 259, 262–263,
  266, 294
Mate Line to Line constraint, 266
  and explosion factor, 320
Mate Plane to Plane constraint, 264
  and explosion factor, 320–321
middle-out assembly building, 256
midpoint constraint method, 266–267,
  292
Mirror constraint, 177–178
mode selector tabs, 19
Model mode, 19
  moving to and from, 65–66
modes, 19
moldbase, 241–252
  assembling the mold halves, 251–252
  constraining the bottle, 244–246
  hollowing out, 249
  importing in the toolbody, 242–244
  splitting the part, 249–250
  using spreadsheet, 247–248
MOVE command, 169, 263, 278

N

NEW command, 239
new dimensioning, 49

O

offset section view, 76
offset work plane, 134–136
ortho view, 72–78
ORTHO, moving parametric dimensions,
  186–187

P

page setup, 69
Parallel constraint, 170–171
parametric assembly, 16
parametric control, 16

spreadsheet. *See* linked spreadsheet.
springs, 227–229
stability of model or assembly, 280–282
standalone, 310
startup process, 17
static assembly, 16
subassembly, opening, 290
    creating, 292–294
subfeatures, copying, 160
surface finish symbols, 197–198
sweeping,
    features, 223–224
    freeform spline-based, 234–236
    springs, 227–229
    threads, 224–227
    using an edge of a part, 230–232
    using a pipe centerline, 232–234

T

table-driven assembly, 269–271
table-driven parts, 105–106
Tangent constraint, 55–56, 175–176
threaded holes, 86
threads, 224–227
toolbars, 19
toolbody, 217, 240
    importing, 242–243
top-down assembly building, 256
trails, 325–327
transparent zoom (**Z'**), 168
tweaks, 323–324, 327

U

updating,
    assembly, 290–291
    bill of materials (BOM), 309-310
    drawing, 68
    part, 36, 68
upgrading Pentium computer, 287–291

user-defined data, 298
user interface, 18

V

variable radius fillet, 155–156
vertex, 77, 131
Vertical constraint, 168
viewpoint, changing, 21
viewports, changing number of, 21

W

welding symbols, 198
work axis, 241
    creating, 122
    on the sketch plane, 138–139
    through the cylinder, 132, 134
work plane, 31, 129–148
    creating, 32–33, 130–134
    for cylindrical part, 139–140
    editing, 134–136
    extruding to, 31–35
    making invisible, 34
    offset, 134–136
    problems using UCS, 141–142
    terminating extrusions, 136–138
    uses, 130
    using **AMPARTSPLIT** command,
        142–143
work point, 122–123

X

XValue and **YValue** constraints, 176–177

Y

YOKE, attaching, 262–264

FLORIDA TECHNICAL COLLEGE LIBRARY
12689 CHALLENGER PARKWAY
SUITE 130
ORLANDO, FL 32826